Undergraduate Texts in Mathematics

Undergraduate Texts in Mathematics

Abbott: Understanding Analysis.
Anglin: Mathematics: A Concise History and Philosophy.
Readings in Mathematics.
Anglin/Lambek: The Heritage of Thales.
Readings in Mathematics.
Apostol: Introduction to Analytic Number Theory. Second edition.
Armstrong: Basic Topology.
Armstrong: Groups and Symmetry.
Axler: Linear Algebra Done Right. Second edition.
Beardon: Limits: A New Approach to Real Analysis.
Bak/Newman: Complex Analysis. Second edition.
Banchoff/Wermer: Linear Algebra Through Geometry. Second edition.
Berberian: A First Course in Real Analysis.
Bix: Conics and Cubics: A Concrete Introduction to Algebraic Curves.
Brémaud: An Introduction to Probabilistic Modeling.
Bressoud: Factorization and Primality Testing.
Bressoud: Second Year Calculus.
Readings in Mathematics.
Brickman: Mathematical Introduction to Linear Programming and Game Theory.
Browder: Mathematical Analysis: An Introduction.
Buchmann: Introduction to Cryptography.
Buskes/van Rooij: Topological Spaces: From Distance to Neighborhood.
Callahan: The Geometry of Spacetime: An Introduction to Special and General Relativity.
Carter/van Brunt: The Lebesgue–Stieltjes Integral: A Practical Introduction.
Cederberg: A Course in Modern Geometries. Second edition.

Chambert-Loir: A Field Guide to Algebra
Childs: A Concrete Introduction to Higher Algebra. Second edition.
Chung/AitSahlia: Elementary Probability Theory: With Stochastic Processes and an Introduction to Mathematical Finance. Fourth edition.
Cox/Little/O'Shea: Ideals, Varieties, and Algorithms. Second edition.
Croom: Basic Concepts of Algebraic Topology.
Curtis: Linear Algebra: An Introductory Approach. Fourth edition.
Daepp/Gorkin: Reading, Writing, and Proving: A Closer Look at Mathematics.
Devlin: The Joy of Sets: Fundamentals of Contemporary Set Theory. Second edition.
Dixmier: General Topology.
Driver: Why Math?
Ebbinghaus/Flum/Thomas: Mathematical Logic. Second edition.
Edgar: Measure, Topology, and Fractal Geometry.
Elaydi: An Introduction to Difference Equations. Second edition.
Erdős/Surányi: Topics in the Theory of Numbers.
Estep: Practical Analysis in One Variable.
Exner: An Accompaniment to Higher Mathematics.
Exner: Inside Calculus.
Fine/Rosenberger: The Fundamental Theory of Algebra.
Fischer: Intermediate Real Analysis.
Flanigan/Kazdan: Calculus Two: Linear and Nonlinear Functions. Second edition.
Fleming: Functions of Several Variables. Second edition.
Foulds: Combinatorial Optimization for Undergraduates.
Foulds: Optimization Techniques: An Introduction.
Franklin: Methods of Mathematical Economics.

(continued after index)

Kenneth A. Ross

Elementary Analysis
The Theory of Calculus

With 34 Illustrations

 Springer

Kenneth A. Ross
Department of Mathematics
University of Oregon
Eugene, OR 97403
USA

Series Editors

S. Axler
Mathematics Department
San Francisco State
  University
San Francisco, CA 94132
USA

F.W. Gehring
Mathematics Department
East Hall
University of Michigan
Ann Arbor, MI 48109
USA

K.A. Ribet
Mathematics Department
University of California,
  Berkeley
Berkeley, CA 94720-3840
USA

Mathematics Subject Classification (2000): 26-01

Library of Congress Cataloging-in-Publication Data
Ross, Kenneth A.
    Elementary analysis
    Undergraduate texts in mathematics
    Bibliography: p.
    Includes indexes.
    1. Calculus   I. Title
    QA303.R726   515   79-24806

ISBN 0-387-90459-X   Printed on acid-free paper.

Printed in the United States of America.

25 24 23 22 21 20 19 18 17 16

springeronline.com

Preface

Attention: Starting with the 12th printing, this book has been set in LaTeX so that the book will be more readable. In particular, there is less material on each page, so there are more pages. However, these are the only changes from previous printings except that I've updated the bibliography.

Preface to the First Edition

A study of this book, and especially the exercises, should give the reader a thorough understanding of a few basic concepts in analysis such as continuity, convergence of sequences and series of numbers, and convergence of sequences and series of functions. An ability to read and write proofs will be stressed. A precise knowledge of definitions is essential. The beginner should memorize them; such memorization will help lead to understanding.

Chapter 1 sets the scene and, except for the completeness axiom, should be more or less familiar. Accordingly, readers and instructors are urged to move quickly through this chapter and refer back to it when necessary. The most critical sections in the book are Sections 7 through 12 in Chapter 2. If these sections are thoroughly digested and understood, the remainder of the book should be smooth sailing.

The first four chapters form a unit for a short course on analysis. I cover these four chapters (except for the optional sections and Section 20) in about 38 class periods; this includes time for quizzes and examinations. For such a short course, my philosophy is that the students are relatively comfortable with derivatives and integrals but do not really understand sequences and series, much less sequences and series of functions, so Chapters 1–4 focus on these topics. On two or three occasions I draw on the Fundamental Theorem of Calculus or the Mean Value Theorem, which appear later in the book, but of course these important theorems are at least discussed in a standard calculus class.

In the early sections, especially in Chapter 2, the proofs are very detailed with careful references for even the most elementary facts. Most sophisticated readers find excessive details and references a hindrance (they break the flow of the proof and tend to obscure the main ideas) and would prefer to check the items mentally as they proceed. Accordingly, in later chapters the proofs will be somewhat less detailed, and references for the simplest facts will often be omitted. This should help prepare the reader for more advanced books which frequently give very brief arguments.

Mastery of the basic concepts in this book should make the analysis in such areas as complex variables, differential equations, numerical analysis, and statistics more meaningful. The book can also serve as a foundation for an in-depth study of real analysis given in books such as [2], [25], [26], [33], [36], and [38] listed in the bibliography.

Readers planning to teach calculus will also benefit from a careful study of analysis. Even after studying this book (or writing it) it will not be easy to handle questions such as "What is a number?", but at least this book should help give a clearer picture of the subtleties to which such questions lead.

The optional sections contain discussions of some topics that I think are important or interesting. Sometimes the topic is dealt with lightly, and suggestions for further reading are given. Though these sections are not particularly designed for classroom use, I hope that some readers will use them to broaden their horizons and see how this material fits in the general scheme of things.

I have benefitted from numerous helpful suggestions from my colleagues Robert Freeman, William Kantor, Richard Koch, and John Leahy, and from Timothy Hall, Gimli Khazad, and Jorge López. I have also had helpful conversations with my wife Lynn concerning grammar and taste. Of course, remaining errors in grammar and mathematics are the responsibility of the author.

Several users have supplied me with corrections and suggestions that I've incorporated in subsequent printings. I thank them all, including Robert Messer of Albion College who caught a subtle error in the proof of Theorem 12.1.

Kenneth A. Ross
Eugene, Oregon

Contents

Preface   **v**

1 Introduction   **1**

    1   The Set $\mathbb{N}$ of Natural Numbers   1

    2   The Set $\mathbb{Q}$ of Rational Numbers   6

    3   The Set $\mathbb{R}$ of Real Numbers   12

    4   The Completeness Axiom   19

    5   The Symbols $+\infty$ and $-\infty$ 27

    6   * A Development of $\mathbb{R}$ 28

2 Sequences   **31**

    7   Limits of Sequences   31

    8   A Discussion about Proofs   37

    9   Limit Theorems for Sequences   43

    10   Monotone Sequences and Cauchy Sequences . .   54

    11   Subsequences   63

    12   lim sup's and lim inf's   75

    13   * Some Topological Concepts in Metric Spaces .   79

    14   Series   90

    15   Alternating Series and Integral Tests   100

    16   * Decimal Expansions of Real Numbers   105

3 Continuity   **115**
   17   Continuous Functions 115
   18   Properties of Continuous Functions 126
   19   Uniform Continuity 132
   20   Limits of Functions 145
   21   * More on Metric Spaces: Continuity 156
   22   * More on Metric Spaces: Connectedness 164

4 Sequences and Series of Functions   **171**
   23   Power Series 171
   24   Uniform Convergence 177
   25   More on Uniform Convergence 184
   26   Differentiation and Integration of Power Series . . 192
   27   * Weierstrass's Approximation Theorem 200

5 Differentiation   **205**
   28   Basic Properties of the Derivative 205
   29   The Mean Value Theorem 213
   30   * L'Hospital's Rule 222
   31   Taylor's Theorem 230

6 Integration   **243**
   32   The Riemann Integral 243
   33   Properties of the Riemann Integral 253
   34   Fundamental Theorem of Calculus 261
   35   * Riemann-Stieltjes Integrals 268
   36   * Improper Integrals 292
   37   * A Discussion of Exponents and Logarithms . . . 299

Appendix on Set Notation   **309**

Selected Hints and Answers   **311**

References   **341**

Symbols Index   **345**

Index   **347**

Introduction

CHAPTER

The underlying space for all the analysis in this book is the set of real numbers. In this chapter we set down some basic properties of this set. These properties will serve as our axioms in the sense that it is possible to derive all the properties of the real numbers using only these axioms. However, we will avoid getting bogged down in this endeavor. Some readers may wish to refer to the appendix on set notation.

§1 The Set ℕ of Natural Numbers

We denote the set $\{1, 2, 3, \ldots\}$ of all *natural numbers* by ℕ. Elements of ℕ will also be called *positive integers*. Each natural number n has a successor, namely $n + 1$. Thus the successor of 2 is 3, and 37 is the successor of 36. You will probably agree that the following properties of ℕ are obvious; at least the first four are.

N1. 1 belongs to ℕ.
N2. If n belongs to ℕ, then its successor $n + 1$ belongs to ℕ.
N3. 1 is not the successor of any element in ℕ.

1

N4. If n and m in $\mathbb{N}$ have the same successor, then $n = m$.

N5. A subset of $\mathbb{N}$ which contains 1, and which contains $n + 1$ whenever it contains n, must equal $\mathbb{N}$.

Properties N1 through N5 are known as the *Peano Axioms* or *Peano Postulates*. It turns out that most familiar properties of $\mathbb{N}$ can be proved based on these five axioms; see [3] or [28].

Let's focus our attention on axiom N5, the one axiom that may not be obvious. Here is what the axiom is saying. Consider a subset S of $\mathbb{N}$ as described in N5. Then 1 belongs to S. Since S contains $n + 1$ whenever it contains n, it follows that S must contain $2 = 1 + 1$. Again, since S contains $n + 1$ whenever it contains n, it follows that S must contain $3 = 2 + 1$. Once again, since S contains $n + 1$ whenever it contains n, it follows that S must contain $4 = 3 + 1$. We could continue this monotonous line of reasoning to conclude that S contains any number in $\mathbb{N}$. Thus it seems reasonable to conclude that $S = \mathbb{N}$. It is this reasonable conclusion that is asserted by axiom N5.

Here is another way to view axiom N5. Assume axiom N5 is false. Then $\mathbb{N}$ contains a set S such that

(i) $1 \in S$,
(ii) if $n \in S$, then $n + 1 \in S$,

and yet $S \neq \mathbb{N}$. Consider the smallest member of the set $\{n \in \mathbb{N} : n \notin S\}$, call it n_0. Since (i) holds, it is clear that $n_0 \neq 1$. So n_0 must be a successor to some number in $\mathbb{N}$, namely $n_0 - 1$. We must have $n_0 - 1 \in S$ since n_0 is the smallest member of $\{n \in \mathbb{N} : n \notin S\}$. By (ii), the successor of $n_0 - 1$, namely n_0, must also be in S, which is a contradiction. This discussion may be plausible, but we emphasize that we have not *proved* axiom N5 using the successor notion and axioms N1 through N4, because we implicitly used two unproven facts. We assumed that every nonempty subset of $\mathbb{N}$ contains a least element and we assumed that if $n_0 \neq 1$ then n_0 is the successor to some number in $\mathbb{N}$.

Axiom N5 is the basis of *mathematical induction*. Let $P_1, P_2, P_3, \ldots$ be a list of statements or propositions that may or may not be true. The principle of mathematical induction asserts that all the statements $P_1, P_2, P_3, \ldots$ are true provided

(I_1) P_1 is true,

(I$_2$) P_{n+1} is true whenever P_n is true.

We will refer to (I$_1$), i.e., the fact that P_1 is true, as the *basis for induction* and we will refer to (I$_2$) as the *induction step*. For a sound proof based on mathematical induction, properties (I$_1$) and (I$_2$) must both be verified. In practice, (I$_1$) will be easy to check.

Example 1

Prove $1 + 2 + \cdots + n = \frac{1}{2}n(n+1)$ for natural numbers n.

Solution

Our nth proposition is

$$P_n\text{: ``}1 + 2 + \cdots + n = \tfrac{1}{2}n(n+1).\text{''}$$

Thus P_1 asserts that $1 = \frac{1}{2} \cdot 1(1+1)$, P_2 asserts that $1+2 = \frac{1}{2} \cdot 2(2+1)$, P_{37} asserts that $1+2+\cdots+37 = \frac{1}{2} \cdot 37(37+1) = 703$, etc. In particular, P_1 is a true assertion which serves as our basis for induction.

For the induction step, suppose that P_n is true. That is, we suppose

$$1 + 2 + \cdots + n = \tfrac{1}{2}n(n+1)$$

is true. Since we wish to prove P_{n+1} from this, we add $n+1$ to both sides to obtain

$$1 + 2 + \cdots + n + (n+1) = \tfrac{1}{2}n(n+1) + (n+1)$$
$$= \tfrac{1}{2}[n(n+1) + 2(n+1)] = \tfrac{1}{2}(n+1)(n+2)$$
$$= \tfrac{1}{2}(n+1)((n+1) + 1.)$$

Thus P_{n+1} holds if P_n holds. By the principle of mathematical induction, we conclude that P_n is true for all n. $\quad\square$

We emphasize that prior to the last sentence of our solution we did *not* prove "P_{n+1} is true." We merely proved an implication: "if P_n is true, then P_{n+1} is true." In a sense we proved an infinite number of assertions, namely: P_1 is true; if P_1 is true then P_2 is true; if P_2 is true then P_3 is true; if P_3 is true then P_4 is true; etc. Then we applied mathematical induction to conclude P_1 is true, P_2 is true, P_3 is true, P_4 is true, etc. We also confess that formulas like the one just proved are easier to prove than to derive. It can be a tricky matter to guess such a result. Sometimes results such as this are discovered by trial and error.

Example 2
All numbers of the form $7^n - 2^n$ are divisible by 5.

Solution
More precisely, we show that $7^n - 2^n$ is divisible by 5 for each $n \in \mathbb{N}$. Our nth proposition is

$$P_n: \text{``} 7^n - 2^n \quad \text{is divisible by} \quad 5.\text{''}$$

The basis for induction P_1 is clearly true, since $7^1 - 2^1 = 5$. For the induction step, suppose that P_n is true. To verify P_{n+1}, we write

$$7^{n+1} - 2^{n+1} = 7^{n+1} - 7 \cdot 2^n + 7 \cdot 2^n - 2 \cdot 2^n = 7[7^n - 2^n] + 5 \cdot 2^n.$$

Since $7^n - 2^n$ is a multiple of 5 by the induction hypothesis, it follows that $7^{n+1} - 2^{n+1}$ is also a multiple of 5. In fact, if $7^n - 2^n = 5m$, then $7^{n+1} - 2^{n+1} = 5 \cdot [7m + 2^n]$. We have shown that P_n implies P_{n+1}, so the induction step holds. An application of mathematical induction completes the proof. □

Example 3
Show that $|\sin nx| \le n|\sin x|$ for all natural numbers n and all real numbers x.

Solution
Our nth proposition is

$$P_n: \text{``} |\sin nx| \le n|\sin x| \quad \text{for all real numbers} \quad x.\text{''}$$

The basis for induction is again clear. Suppose P_n is true. We apply the addition formula for sine to obtain

$$|\sin(n+1)x| = |\sin(nx + x)| = |\sin nx \cos x + \cos nx \sin x|.$$

Now we apply the Triangle Inequality and properties of the absolute value [see 3.7 and 3.5] to obtain

$$|\sin(n+1)x| \le |\sin nx| \cdot |\cos x| + |\cos nx| \cdot |\sin x|.$$

Since $|\cos y| \le 1$ for all y we see that

$$|\sin(n+1)x| \le |\sin nx| + |\sin x|.$$

Now we apply the induction hypothesis P_n to obtain

$$|\sin(n+1)x| \le n|\sin x| + |\sin x| = (n+1)|\sin x|.$$

Thus P_{n+1} holds. Finally, the result holds for all n by mathematical induction. □

Exercises

1.1. Prove $1^2 + 2^2 + \cdots + n^2 = \frac{1}{6}n(n+1)(2n+1)$ for all natural numbers n.

1.2. Prove $3 + 11 + \cdots + (8n - 5) = 4n^2 - n$ for all natural numbers n.

1.3. Prove $1^3 + 2^3 + \cdots + n^3 = (1 + 2 + \cdots + n)^2$ for all natural numbers n.

1.4. **(a)** Guess a formula for $1 + 3 + \cdots + (2n - 1)$ by evaluating the sum for $n = 1, 2, 3$, and 4. [For $n = 1$, the sum is simply 1.]

(b) Prove your formula using mathematical induction.

1.5. Prove $1 + \frac{1}{2} + \frac{1}{4} + \cdots + \frac{1}{2^n} = 2 - \frac{1}{2^n}$ for all natural numbers n.

1.6. Prove that $(11)^n - 4^n$ is divisible by 7 when n is a natural number.

1.7. Prove that $7^n - 6n - 1$ is divisible by 36 for all positive integers n.

1.8. The principle of mathematical induction can be extended as follows. A list $P_m, P_{m+1}, \ldots$ of propositions is true provided (i) P_m is true, (ii) P_{n+1} is true whenever P_n is true and $n \ge m$.

(a) Prove that $n^2 > n + 1$ for all integers $n \ge 2$.

(b) Prove that $n! > n^2$ for all integers $n \ge 4$. [Recall that $n! = n(n-1) \cdots 2 \cdot 1$; for example, $5! = 5 \cdot 4 \cdot 3 \cdot 2 \cdot 1 = 120$.]

1.9. **(a)** Decide for which integers the inequality $2^n > n^2$ is true.

(b) Prove your claim in (a) by mathematical induction.

1.10. Prove $(2n + 1) + (2n + 3) + (2n + 5) + \cdots + (4n - 1) = 3n^2$ for all positive integers n.

1.11. For each $n \in \mathbb{N}$, let P_n denote the assertion "$n^2 + 5n + 1$ is an even integer."

(a) Prove that P_{n+1} is true whenever P_n is true.

(b) For which n is P_n actually true? What is the moral of this exercise?

1.12. For $n \in \mathbb{N}$, let $n!$ [read "n factorial"] denote the product $1 \cdot 2 \cdot 3 \cdots n$. Also let $0! = 1$ and define

$$\binom{n}{k} = \frac{n!}{k!(n-k)!} \quad \text{for} \quad k = 0, 1, \ldots, n.$$

The *binomial theorem* asserts that

$$(a+b)^n = \binom{n}{0}a^n + \binom{n}{1}a^{n-1}b + \binom{n}{2}a^{n-2}b^2 + \cdots$$
$$+ \binom{n}{n-1}ab^{n-1} + \binom{n}{n}b^n$$
$$= a^n + na^{n-1}b + \tfrac{1}{2}n(n-1)a^{n-2}b^2 + \cdots + nab^{n-1} + b^n.$$

(a) Verify the binomial theorem for $n = 1, 2$, and 3.

(b) Show that $\binom{n}{k} + \binom{n}{k-1} = \binom{n+1}{k}$ for $k = 1, 2, \ldots, n$.

(c) Prove the binomial theorem using mathematical induction and part (b).

§2 The Set $\mathbb{Q}$ of Rational Numbers

Small children first learn to add and to multiply natural numbers. After subtraction is introduced, the need to expand the number system to include 0 and negative numbers becomes apparent. At this point the world of numbers is enlarged to include the set $\mathbb{Z}$ of all *integers*. Thus we have $\mathbb{Z} = \{0, 1, -1, 2, -2, \ldots\}$.

Soon the space $\mathbb{Z}$ also becomes inadequate when division is introduced. The solution is to enlarge the world of numbers to include all fractions. Accordingly, we study the space $\mathbb{Q}$ of all *rational numbers*, i.e., numbers of the form $\frac{m}{n}$ where $m, n \in \mathbb{Z}$ and $n \neq 0$. Note that $\mathbb{Q}$ contains all terminating decimals such as $1.492 = \frac{1492}{1000}$. The connection between decimals and real numbers is discussed in 10.3 and §16. The space $\mathbb{Q}$ is a highly satisfactory algebraic system in which the basic operations addition, multiplication, subtraction and division can be fully studied. No system is perfect, however, and $\mathbb{Q}$

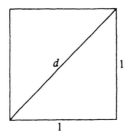

FIGURE 2.1

is inadequate in some ways. In this section we will consider the defects of ℚ. In the next section we will stress the good features of ℚ and then move on to the system of real numbers.

The set ℚ of rational numbers is a very nice algebraic system until one tries to solve equations like $x^2 = 2$. It turns out that no rational number satisfies this equation, and yet there are good reasons to believe that some kind of number satisfies this equation. Consider, for example, a square with sides having length one; see Figure 2.1. If d represents the length of the diagonal, then from geometry we know that $1^2 + 1^2 = d^2$, i.e., $d^2 = 2$. Apparently there is a positive length whose square is 2, which we write as $\sqrt{2}$. But $\sqrt{2}$ cannot be a rational number, as we will show in Example 2. Of course, $\sqrt{2}$ can be approximated by rational numbers. There are rational numbers whose squares are close to 2; for example, $(1.4142)^2 = 1.99996164$ and $(1.4143)^2 = 2.00024449$.

It is evident that there are lots of rational numbers and yet there are "gaps" in ℚ. Here is another way to view this situation. Consider the graph of the polynomial $x^2 - 2$ in Figure 2.2. Does the graph of $x^2 - 2$ cross the x-axis? We are inclined to say it does, because when we draw the x-axis we include "all" the points. We allow no "gaps." But notice that the graph of $x^2 - 2$ slips by all the rational numbers on the x-axis. The x-axis is our picture of the number line, and the set of rational numbers again appears to have significant "gaps."

There are even more exotic numbers such as π and e that are not rational numbers, but which come up naturally in mathematics. The number π is basic to the study of circles and spheres, and e arises in problems of exponential growth.

We return to $\sqrt{2}$. This is an example of what is called an algebraic number because it satisfies the equation $x^2 - 2 = 0$.

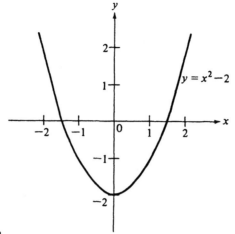

FIGURE 2.2

2.1 Definition.
A number is called an *algebraic number* if it satisfies a polynomial equation

$$a_n x^n + a_{n-1} x^{n-1} + \cdots + a_1 x + a_0 = 0$$

where the coefficients $a_0, a_1, \ldots, a_n$ are integers, $a_n \neq 0$ and $n \geq 1$.

Rational numbers are always algebraic numbers. In fact, if $r = \frac{m}{n}$ is a rational number [$m, n \in \mathbb{Z}$ and $n \neq 0$], then it satisfies the equation $nx - m = 0$. Numbers defined in terms of $\sqrt{}$, $\sqrt[3]{}$, etc. [or fractional exponents, if you prefer] and ordinary algebraic operations on the rational numbers are invariably algebraic numbers.

Example 1
$\frac{4}{17}$, $3^{1/2}$, $(17)^{1/3}$, $(2 + 5^{1/3})^{1/2}$ and $((4 - 2 \cdot 3^{1/2})/7)^{1/2}$ all represent algebraic numbers. In fact, $\frac{4}{17}$ is a solution of $17x - 4 = 0$, $3^{1/2}$ represents a solution of $x^2 - 3 = 0$, and $(17)^{1/3}$ represents a solution of $x^3 - 17 = 0$. The expression $a = (2 + 5^{1/3})^{1/2}$ means $a^2 = 2 + 5^{1/3}$ or $a^2 - 2 = 5^{1/3}$ so that $(a^2 - 2)^3 = 5$. Therefore we have $a^6 - 6a^4 + 12a^2 - 13 = 0$ which shows that $a = (2 + 5^{1/3})^{1/2}$ satisfies the polynomial equation $x^6 - 6x^4 + 12x^2 - 13 = 0$. Similarly, the expression $b = ((4 - 2 \cdot 3^{1/2})/7)^{1/2}$ leads to $7b^2 = 4 - 2 \cdot 3^{1/2}$, hence

$2 \cdot 3^{1/2} = 4 - 7b^2$, hence $12 = (4 - 7b^2)^2$, hence $49b^4 - 56b^2 + 4 = 0$. Thus b satisfies the polynomial equation $49x^4 - 56x^2 + 4 = 0$.

The next theorem may be familiar from elementary algebra. It is the theorem that justifies the following remarks: the only possible rational solutions of $x^3 - 7x^2 + 2x - 12 = 0$ are $\pm 1, \pm 2, \pm 3, \pm 4, \pm 6, \pm 12$, so the only possible (rational) monomial factors of $x^3 - 7x^2 + 2x - 12$ are $x - 1, x + 1, x - 2, x + 2, x - 3, x + 3, x - 4, x + 4, x - 6, x + 6$, $x - 12, x + 12$. We won't pursue these algebraic problems; we merely made these observations in the hope that they would be familiar.

The next theorem also allows one to prove that algebraic numbers that do not look like rational numbers are not rational numbers. Thus $\sqrt{4}$ is obviously a rational number, while $\sqrt{2}, \sqrt{3}, \sqrt{5}$, etc. turn out to be nonrational. See the examples following the theorem. Recall that an integer k is a *factor* of an integer m or *divides* m if $\frac{m}{k}$ is also an integer. An integer $p \geq 2$ is a *prime* provided the only positive factors of p are 1 and p. It can be shown that every positive integer can be written as a product of primes and that this can be done in only one way, except for the order of the factors.

2.2 Rational Zeros Theorem.
Suppose that $a_0, a_1, \ldots, a_n$ are integers and that r is a rational number satisfying the polynomial equation

$$a_n x^n + a_{n-1} x^{n-1} + \cdots + a_1 x + a_0 = 0 \tag{1}$$

where $n \geq 1$, $a_n \neq 0$ and $a_0 \neq 0$. Write $r = \frac{p}{q}$ where p, q are integers having no common factors and $q \neq 0$. Then q divides a_n and p divides a_0.

In other words, the only rational *candidates* for solutions of (1) have the form $\frac{p}{q}$ where p divides a_0 and q divides a_n.

Proof
We are given

$$a_n \left(\frac{p}{q}\right)^n + a_{n-1} \left(\frac{p}{q}\right)^{n-1} + \cdots + a_1 \left(\frac{p}{q}\right) + a_0 = 0.$$

We multiply through by q^n and obtain

$$a_n p^n + a_{n-1} p^{n-1} q + a_{n-2} p^{n-2} q^2 + \cdots + a_2 p^2 q^{n-2} + a_1 p q^{n-1} + a_0 q^n = 0.$$
(2)

If we solve for $a_n p^n$, we obtain

$$a_n p^n = -q[a_{n-1} p^{n-1} + a_{n-2} p^{n-2} q + \cdots + a_2 p^2 q^{n-3} + a_1 p q^{n-2} + a_0 q^{n-1}].$$

It follows that q divides $a_n p^n$. But p and q have no common factors, so q must divide a_n. [Here are more details: p can be written as a product of primes $p_1 p_2 \cdots p_k$ where the p_i's need not be distinct. Likewise q can be written as a product of primes $q_1 q_2 \cdots q_l$. Since q divides $a_n p^n$, the quantity $\frac{a_n p^n}{q} = \frac{a_n p_1^n \cdots p_k^n}{q_1 \cdots q_l}$ must be an integer. Since no p_i can equal any q_j, the unique factorization of a_n as a product of primes must include the product $q_1 q_2 \cdots q_l$. Thus q divides a_n.]

Now we solve (2) for $a_0 q^n$ and obtain

$$a_0 q^n = -p[a_n p^{n-1} + a_{n-1} p^{n-2} q + a_{n-2} p^{n-3} q^2 + \cdots + a_2 p q^{n-2} + a_1 q^{n-1}].$$

Thus p divides $a_0 q^n$. Since p and q have no common factors, p must divide a_0. ∎

Example 2
$\sqrt{2}$ cannot represent a rational number.

Proof
By Theorem 2.2 the only rational numbers that could possibly be solutions of $x^2 - 2 = 0$ are $\pm 1, \pm 2$. [Here $n = 2$, $a_2 = 1$, $a_1 = 0$, $a_0 = -2$. So rational solutions must have the form $\frac{p}{q}$ where p divides $a_0 = -2$ and q divides $a_2 = 1$.] One can substitute each of the four numbers $\pm 1, \pm 2$ into the equation $x^2 - 2 = 0$ to quickly eliminate them as possible solutions of the equation. Since $\sqrt{2}$ represents a solution of $x^2 - 2 = 0$, it cannot represent a rational number. ∎

Example 3
$\sqrt{17}$ cannot represent a rational number.

Proof
The only possible rational solutions of $x^2 - 17 = 0$ are $\pm 1, \pm 17$ and none of these numbers are solutions. ∎

Example 4

$6^{1/3}$ cannot represent a rational number.

Proof

The only possible rational solutions of $x^3 - 6 = 0$ are $\pm 1, \pm 2, \pm 3, \pm 6$. It is easy to verify that none of these eight numbers satisfies the equation $x^3 - 6 = 0$. ∎

Example 5

$a = (2 + 5^{1/3})^{1/2}$ does not represent a rational number.

Proof

In Example 1 we showed that a represents a solution of $x^6 - 6x^4 + 12x^2 - 13 = 0$. By Theorem 2.2, the only possible rational solutions are $\pm 1, \pm 13$. When $x = 1$ or -1, the left hand side of the equation is -6 and when $x = 13$ or -13, the left hand side of the equation turns out to equal 4,657,458. This last computation could be avoided by using a little common sense. Either observe that a is "obviously" bigger than 1 and less than 13, or observe that

$$13^6 - 6 \cdot 13^4 + 12 \cdot 13^2 - 13 = 13(13^5 - 6 \cdot 13^3 + 12 \cdot 13 - 1) \neq 0$$

since the term in parentheses cannot be zero: it is one less than some multiple of 13. ∎

Example 6

$b = ((4 - 2\sqrt{3})/7)^{1/2}$ does not represent a rational number.

Proof

In Example 1 we showed that b is a solution of $49x^4 - 56x^2 + 4 = 0$. The only possible rational solutions are

$$\pm 1, \pm 1/7, \pm 1/49, \pm 2, \pm 2/7, \pm 2/49, \pm 4, \pm 4/7, \pm 4/49.$$

To complete our proof, all we need to do is substitute these eighteen candidates into the equation $49x^4 - 56x^2 + 4 = 0$. This prospect is so discouraging, however, that we choose to find a more clever approach. In Example 1, we also showed that $12 = (4 - 7b^2)^2$. Now if b were rational, then $4 - 7b^2$ would also be rational [Exercise 2.6], so the equation $12 = x^2$ would have a rational solution. But the only possible rational solutions to $x^2 - 12 = 0$ are $\pm 1, \pm 2, \pm 3, \pm 4, \pm 6, \pm 12$, and these all can be eliminated by mentally substituting them into the equation. We conclude that $4 - 7b^2$ cannot be rational, so b cannot be rational. ∎

As a practical matter, many or all of the rational candidates given by the Rational Zeros Theorem can be eliminated by approximating the quantity in question [perhaps with the aid of a calculator]. It is nearly obvious that the values in Examples 2 through 5 are not integers, while all the rational candidates are. My calculator says that b in Example 6 is approximately .2767; the nearest rational candidate is $+2/7$ which is approximately .2857.

Exercises

2.1. Show that $\sqrt{3}$, $\sqrt{5}$, $\sqrt{7}$, $\sqrt{24}$, and $\sqrt{31}$ are not rational numbers.

2.2. Show that $2^{1/3}$, $5^{1/7}$, and $(13)^{1/4}$ do not represent rational numbers.

2.3. Show that $(2 + \sqrt{2})^{1/2}$ does not represent a rational number.

2.4. Show that $(5 - \sqrt{3})^{1/3}$ does not represent a rational number.

2.5. Show that $[3 + \sqrt{2}]^{2/3}$ does not represent a rational number.

2.6. In connection with Example 6, discuss why $4 - 7b^2$ must be rational if b is rational.

§3 The Set $\mathbb{R}$ of Real Numbers

The set $\mathbb{Q}$ is probably the largest system of numbers with which you really feel comfortable. There are some subtleties but you have learned to cope with them. For example, $\mathbb{Q}$ is not simply the *set* $\{\frac{m}{n} : m, n \in \mathbb{Z}, n \neq 0\}$, since we regard some pairs of different looking fractions as equal. For example, $\frac{2}{4}$ and $\frac{3}{6}$ are regarded as the same element of $\mathbb{Q}$. A rigorous development of $\mathbb{Q}$ based on $\mathbb{Z}$, which in turn is based on $\mathbb{N}$, would require us to introduce the notion of equivalence class; see [38]. In this book we assume a familiarity with and understanding of $\mathbb{Q}$ as an algebraic system. However, in order to clarify exactly what we need to know about ℚ, we set down some of its basic axioms and properties.

The basic algebraic operations in ℚ are addition and multiplication. Given a pair a, b of rational numbers, the sum $a + b$ and the product ab also represent rational numbers. Moreover, the following properties hold.

A1. $a + (b + c) = (a + b) + c$ for all a, b, c.
A2. $a + b = b + a$ for all a, b.
A3. $a + 0 = a$ for all a.
A4. For each a, there is an element $-a$ such that $a + (-a) = 0$.
M1. $a(bc) = (ab)c$ for all a, b, c.
M2. $ab = ba$ for all a, b.
M3. $a \cdot 1 = a$ for all a.
M4. For each $a \neq 0$, there is an element a^{-1} such that $aa^{-1} = 1$.
DL $a(b + c) = ab + ac$ for all a, b, c.

Properties A1 and M1 are called the *associative laws*, and properties A2 and M2 are the *commutative laws*. Property DL is the *distributive law*; this is the least obvious law and is the one that justifies "factorization" and "multiplying out" in algebra. A system that has more than one element and satisfies these nine properties is called a *field*. The basic algebraic properties of ℚ can proved solely on the basis of these field properties. We do not want to pursue this topic in any depth, but we illustrate our claim by proving some familiar properties in Theorem 3.1 below.

The set ℚ also has an order structure ≤ satisfying

O1. Given a and b, either $a \leq b$ or $b \leq a$.
O2. If $a \leq b$ and $b \leq a$, then $a = b$.
O3. If $a \leq b$ and $b \leq c$, then $a \leq c$.
O4. If $a \leq b$, then $a + c \leq b + c$.
O5. If $a \leq b$ and $0 \leq c$, then $ac \leq bc$.

Property O3 is called the *transitive law*. This is the characteristic property of an ordering. A field with an ordering satisfying properties O1 through O5 is called an *ordered field*. Most of the algebraic and order properties of ℚ can be established for any ordered field. We will prove a few of them in Theorem 3.2 below.

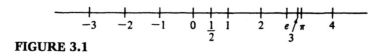

FIGURE 3.1

The mathematical system on which we will do our analysis will be the set $\mathbb{R}$ of all *real numbers*. The set $\mathbb{R}$ will include all rational numbers, all algebraic numbers, π, e, and more. It will be a set that can be drawn as the real number line; see Figure 3.1. That is, every real number will correspond to a point on the number line, and every point on the number line will correspond to a real number. In particular, unlike $\mathbb{Q}$, $\mathbb{R}$ will not have any "gaps." We will also see that real numbers have decimal expansions; see 10.3 and §16. These remarks help describe $\mathbb{R}$, but we certainly have not defined $\mathbb{R}$ as a concise mathematical object. It turns out that $\mathbb{R}$ can be defined entirely in terms of the set $\mathbb{Q}$ of rational numbers; we indicate in the optional §6 one way that this can be done. But then it is a long and tedious task to show how to add and multiply the objects defined in this way and to show that the set $\mathbb{R}$, with these operations, satisfies all the familiar algebraic and order properties that we expect to hold for $\mathbb{R}$. To develop $\mathbb{R}$ properly from $\mathbb{Q}$ in this way and to develop $\mathbb{Q}$ properly from $\mathbb{N}$ would take us several chapters. This would defeat the purpose of this book, which is to accept $\mathbb{R}$ as a mathematical system and to study some important properties of $\mathbb{R}$ and functions on $\mathbb{R}$. Nevertheless, it is desirable to specify exactly what properties of $\mathbb{R}$ we are assuming.

Real numbers, i.e., elements of $\mathbb{R}$, can be added together and multiplied together. That is, given real numbers a and b, the sum $a+b$ and the product ab also represent real numbers. Moreover, these operations satisfy the field properties A1 through A4, M1 through M4, and DL. The set $\mathbb{R}$ also has an order structure $\leq$ that satisfies properties O1 through O5. Thus, like $\mathbb{Q}$, $\mathbb{R}$ is an ordered field.

In the remainder of this section, we will obtain some results for $\mathbb{R}$ that are valid in any ordered field. In particular, these results would be equally valid if we restricted our attention to $\mathbb{Q}$. These remarks emphasize the similarities between $\mathbb{R}$ and $\mathbb{Q}$. We have not yet indicated how $\mathbb{R}$ can be distinguished from $\mathbb{Q}$ as a mathematical object, although we have asserted that $\mathbb{R}$ has no "gaps." We will make this observation much more precise in the next section, and then we will give a "gap filling" axiom that finally will distinguish ℝ from ℚ.

3.1 Theorem.
The following are consequences of the field properties:

(i) $a + c = b + c$ implies $a = b$;

(ii) $a \cdot 0 = 0$ for all a;

(iii) $(-a)b = -ab$ for all a, b;

(iv) $(-a)(-b) = ab$ for all a, b;

(v) $ac = bc$ and $c \neq 0$ imply $a = b$;

(vi) $ab = 0$ implies either $a = 0$ or $b = 0$;

for $a, b, c \in \mathbb{R}$.

Proof

(i) $a + c = b + c$ implies $(a+c)+(-c) = (b+c)+(-c)$, so by A1, we have $a + [c + (-c)] = b + [c + (-c)]$. By A4, this reduces to $a + 0 = b + 0$, so $a = b$ by A3.

(ii) We use A3 and DL to obtain $a \cdot 0 = a \cdot (0 + 0) = a \cdot 0 + a \cdot 0$, so $0 + a \cdot 0 = a \cdot 0 + a \cdot 0$. By (i) we conclude that $0 = a \cdot 0$.

(iii) Since $a + (-a) = 0$, we have $ab + (-a)b = [a + (-a)] \cdot b = 0 \cdot b = 0 = ab + (-(ab))$. From (i) we obtain $(-a)b = -(ab)$.

(iv) **and** (v) are left to Exercise 3.3.

(v) If $ab = 0$ and $b \neq 0$, then $0 = b^{-1} \cdot 0 = 0 \cdot b^{-1} = (ab) \cdot b^{-1} = a(bb^{-1}) = a \cdot 1 = a$. ∎

3.2 Theorem.
The following are consequences of the properties of an ordered field:

(i) *if* $a \leq b$, *then* $-b \leq -a$;

(ii) *if* $a \leq b$ *and* $c \leq 0$, *then* $bc \leq ac$;

(iii) *if* $0 \leq a$ *and* $0 \leq b$, *then* $0 \leq ab$;

(iv) $0 \leq a^2$ *for all* a;

(v) $0 < 1$;

(vi) *if* $0 < a$, *then* $0 < a^{-1}$;

(vii) *if* $0 < a < b$, *then* $0 < b^{-1} < a^{-1}$;

for $a, b, c \in \mathbb{R}$.

Note that $a < b$ means $a \leq b$ and $a \neq b$.

Proof

(i) Suppose that $a \leq b$. By O4 applied to $c = (-a) + (-b)$, we have $a + [(-a) + (-b)] \leq b + [(-a) + (-b)]$. It follows that $-b \leq -a$.

(ii) If $a \leq b$ and $c \leq 0$, then $0 \leq -c$ by (i). Now by O5 we have $a(-c) \leq b(-c)$, i.e., $-ac \leq -bc$. From (i) again, we see that $bc \leq ac$.

(iii) If we put $a = 0$ in property O5, we obtain: $0 \leq b$ and $0 \leq c$ imply $0 \leq bc$. Except for notation, this is exactly assertion (iii).

(iv) For any a, either $a \geq 0$ or $a \leq 0$ by O1. If $a \geq 0$, then $a^2 \geq 0$ by (iii). If $a \leq 0$, then we have $-a \geq 0$ by (i), so $(-a)^2 \geq 0$, i.e., $a^2 \geq 0$.

(v) is left to Exercise 3.4.

(vi) Suppose that $0 < a$ but that $0 < a^{-1}$ fails. Then we must have $a^{-1} \leq 0$ and $0 \leq -a^{-1}$. Now by (iii) $0 \leq a(-a^{-1}) = -1$, so that $1 \leq 0$, contrary to (v).

(vii) is left to Exercise 3.4. ∎

Another important notion that should be familiar is that of absolute value.

3.3 Definition.
We define

$$|a| = a \quad \text{if} \quad a \geq 0 \qquad \text{and} \qquad |a| = -a \quad \text{if} \quad a \leq 0.$$

$|a|$ is called the *absolute value of a*.

Intuitively, the absolute value of a represents the distance between 0 and a, but in fact we will *define* the idea of "distance" in terms of the "absolute value," which in turn was defined in terms of the ordering.

3.4 Definition.
For numbers a and b we define $\text{dist}(a, b) = |a - b|$; $\text{dist}(a, b)$ represents the *distance between a and b*.

The basic properties of the absolute value are given in the next theorem.

3.5 Theorem.
   (i) $|a| \geq 0$ *for all* $a \in \mathbb{R}$.
   (ii) $|ab| = |a| \cdot |b|$ *for all* $a, b \in \mathbb{R}$.
  (iii) $|a + b| \leq |a| + |b|$ *for all* $a, b \in \mathbb{R}$.

Proof
(i) is obvious from the definition. [The word "obvious" as used here signifies that the reader should be able to quickly see why the result is true. Certainly if $a \geq 0$, then $|a| = a \geq 0$, while $a < 0$ implies $|a| = -a > 0$. We will use expressions like "obviously" and "clearly" in place of very simple arguments, but we will not use these terms to obscure subtle points.]

    (ii) There are four easy cases here. If $a \geq 0$ and $b \geq 0$, then $ab \geq 0$, so $|a| \cdot |b| = ab = |ab|$. If $a \leq 0$ and $b \leq 0$, then $-a \geq 0$, $-b \geq 0$ and $(-a)(-b) \geq 0$ so that $|a| \cdot |b| = (-a)(-b) = ab = |ab|$. If $a \geq 0$ and $b \leq 0$, then $-b \geq 0$ and $a(-b) \geq 0$ so that $|a| \cdot |b| = a(-b) = -(ab) = |ab|$. If $a \leq 0$ and $b \geq 0$, then $-a \geq 0$ and $(-a)b \geq 0$ so that $|a| \cdot |b| = (-a)b = -ab = |ab|$.

    (iii) The inequalities $-|a| \leq a \leq |a|$ are obvious, since either $a = |a|$ or else $a = -|a|$. Similarly $-|b| \leq b \leq |b|$. Now four applications of O4 yield

$$-|a| + (-|b|) \leq -|a| + |b| \leq a + b \leq |a| + b \leq |a| + |b|$$

so that

$$-(|a| + |b|) \leq a + b \leq |a| + |b|.$$

This tells us that $a + b \leq |a| + |b|$ and also that $-(a + b) \leq |a| + |b|$. Since $|a + b|$ is equal to either $a + b$ or $-(a + b)$, we conclude that $|a + b| \leq |a| + |b|$. ∎

3.6 Corollary.
$\operatorname{dist}(a, c) \leq \operatorname{dist}(a, b) + \operatorname{dist}(b, c)$ *for all* $a, b, c \in \mathbb{R}$.

Proof
We can apply inequality (iii) of Theorem 3.5 to $a - b$ and $b - c$ to obtain $|(a - b) + (b - c)| \leq |a - b| + |b - c|$ or $\operatorname{dist}(a, c) = |a - c| \leq |a - b| + |b - c| \leq \operatorname{dist}(a, b) + \operatorname{dist}(b, c)$. ∎

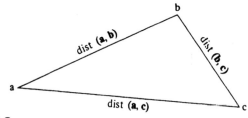

FIGURE 3.2

The inequality in Corollary 3.6 is very closely related to an inequality concerning points **a**, **b**, **c** in the plane, and the latter inequality can be interpreted as a statement about triangles: the length of a side of a triangle is less than or equal to the sum of the lengths of the other two sides. See Figure 3.2. For this reason, the inequality in Corollary 3.6 and its close relative (iii) in 3.5 are often called the *Triangle Inequality*.

3.7 Triangle Inequality.
$|a + b| \leq |a| + |b|$ *for all* a, b.

A useful variant of the triangle inequality is given in Exercise 3.5(b).

Exercises

3.1. (a) Which of the properties A1-A4, M1-M4, DL, O1-O5 fail for $\mathbb{N}$?

   (b) Which of these properties fail for $\mathbb{Z}$?

3.2. (a) The commutative law A2 was used in the proof of (ii) in Theorem 3.1. Where?

   (b) The commutative law A2 was also used in the proof of (iii) in Theorem 3.1. Where?

3.3. Prove (iv) and (v) of Theorem 3.1.

3.4. Prove (v) and (vii) of Theorem 3.2.

3.5. (a) Show that $|b| \leq a$ if and only if $-a \leq b \leq a$.

   (b) Prove that $||a| - |b|| \leq |a - b|$ for all $a, b \in \mathbb{R}$.

3.6. (a) Prove that $|a+b+c| \leq |a|+|b|+|c|$ for all $a, b, c \in \mathbb{R}$. *Hint*: Apply the triangle inequality twice. Do *not* consider eight cases.

(b) Use induction to prove
$$|a_1 + a_2 + \cdots + a_n| \leq |a_1| + |a_2| + \cdots + |a_n|$$
for n numbers $a_1, a_2, \ldots, a_n$.

3.7. (a) Show that $|b| < a$ if and only if $-a < b < a$.

(b) Show that $|a - b| < c$ if and only if $b - c < a < b + c$.

(c) Show that $|a - b| \leq c$ if and only if $b - c \leq a \leq b + c$.

3.8. Let $a, b \in \mathbb{R}$. Show that if $a \leq b_1$ for every $b_1 > b$, then $a \leq b$.

§4 The Completeness Axiom

In this section we give the completeness axiom for $\mathbb{R}$. This is the axiom that will assure us that $\mathbb{R}$ has no "gaps." It has far-reaching consequences and almost every significant result in this book relies on it. Most theorems in this book would be false if we restricted our world of numbers to the set $\mathbb{Q}$ of rational numbers.

4.1 Definition.
Let S be a nonempty subset of $\mathbb{R}$.
 (a) If S contains a largest element s_0 [that is, s_0 belongs to S and $s \leq s_0$ for all $s \in S$], then we call s_0 the *maximum of S* and write $s_0 = \max S$.
 (b) If S contains a smallest element, then we call the smallest element the *minimum of S* and write it as $\min S$.

Example 1
 (a) Every finite nonempty subset of $\mathbb{R}$ has a maximum and a minimum. Thus
$$\max\{1, 2, 3, 4, 5\} = 5 \quad \text{and} \quad \min\{1, 2, 3, 4, 5\} = 1,$$
$$\max\{0, \pi, -7, e, 3, 4/3\} = \pi \quad \text{and} \quad \min\{0, \pi, -7, e, 3, 4/3\} = -7,$$
$$\max\{n \in \mathbb{Z} : -4 < n \leq 100\} = 100 \quad \text{and}$$
$$\min\{n \in \mathbb{Z} : -4 < n \leq 100\} = -3.$$

(b) Consider real numbers a and b where $a < b$. The following notation will be used throughout the book:

$$[a, b] = \{x \in \mathbb{R} : a \le x \le b\}, \qquad (a, b) = \{x \in \mathbb{R} : a < x < b\},$$
$$[a, b) = \{x \in \mathbb{R} : a \le x < b\}, \qquad (a, b] = \{x \in \mathbb{R} : a < x \le b\}.$$

$[a, b]$ is called a *closed interval*, (a, b) is called an *open interval*, while $[a, b)$ and $(a, b]$ are called *half-open* or *semi-open intervals*. Observe that $\max[a, b] = b$ and $\min[a, b] = a$. The set (a, b) has no maximum and no minimum, since the endpoints a and b do not belong to the set. The set $[a, b)$ has no maximum, but a is its minimum.

(c) The sets $\mathbb{Z}$ and $\mathbb{Q}$ have no maximum or minimum. The set $\mathbb{N}$ has no maximum, but $\min \mathbb{N} = 1$.

(d) The set $\{r \in \mathbb{Q} : 0 \le r \le \sqrt{2}\}$ has a minimum, namely 0, but no maximum. This is because $\sqrt{2}$ does not belong to the set, but there are rationals in the set arbitrarily close to $\sqrt{2}$.

(e) Consider the set $\{n^{(-1)^n} : n \in \mathbb{N}\}$. This is shorthand for the set

$$\{1^{-1}, 2, 3^{-1}, 4, 5^{-1}, 6, 7^{-1}, \ldots\} = \{1, 2, \tfrac{1}{3}, 4, \tfrac{1}{5}, 6, \tfrac{1}{7}, \ldots\}.$$

The set has no maximum and no minimum.

4.2 Definition.

Let S be a nonempty subset of $\mathbb{R}$.

(a) If a real number M satisfies $s \le M$ for all $s \in S$, then M is called an *upper bound of S* and the set S is said to be *bounded above*.

(b) If a real number m satisfies $m \le s$ for all $s \in S$, then m is called a *lower bound of S* and the set S is said to be *bounded below*.

(c) The set S is said to be *bounded* if it is bounded above and bounded below. Thus S is bounded if there exist real numbers m and M such that $S \subseteq [m, M]$.

Example 2

(a) The maximum of a set is always an upper bound for the set. Likewise, the minimum of a set is always a lower bound for the set.

(b) Consider a, b in $\mathbb{R}$, $a < b$. The number b is an upper bound for each of the sets $[a, b]$, (a, b), $[a, b)$, $(a, b]$. Every number larger than b is also an upper bound for each of these sets, but b is the smallest or least upper bound.

(c) None of the sets $\mathbb{Z}$, $\mathbb{Q}$ and $\mathbb{N}$ is bounded above. The set $\mathbb{N}$ is bounded below; 1 is a lower bound for $\mathbb{N}$ and so is any number less than 1. In fact, 1 is the largest or greatest lower bound.

(d) Any nonpositive real number is a lower bound for $\{r \in \mathbb{Q} : 0 \leq r \leq \sqrt{2}\}$ and 0 is the set's greatest lower bound. The least upper bound is $\sqrt{2}$.

(e) The set $\{n^{(-1)^n} : n \in \mathbb{N}\}$ is not bounded above. Among its many lower bounds, 0 is the greatest lower bound.

We now formalize two notions that have already appeared in Example 2.

4.3 Definition.
Let S be a nonempty subset of $\mathbb{R}$.
(a) If S is bounded above and S has a least upper bound, then we will call it the *supremum of S* and denote it by sup S.
(b) If S is bounded below and S has a greatest lower bound, then we will call it the *infimum of S* and denote it by inf S.

Note that, unlike max S and min S, sup S and inf S need not belong to S. Note also that a set can have at most one maximum, minimum, supremum and infimum. Sometimes the expressions "least upper bound" and "greatest lower bound" are used instead of the Latin "supremum" and "infimum" and sometimes sup S is written lub S and inf S is written glb S. We have chosen the Latin terminology for a good reason: We will be studying the notions "lim sup" and "lim inf" and this notation is completely standard; no one writes "lim lub" for instance.

Observe that if S is bounded above, then $M = $ sup S if and only if (i) $s \leq M$ for all $s \in S$, and (ii) whenever $M_1 < M$, there exists $s_1 \in S$ such that $s_1 > M_1$.

Example 3
(a) If a set S has a maximum, then max $S = $ sup S. A similar remark applies to sets that have minimums.

(b) If $a, b \in \mathbb{R}$ and $a < b$, then

$$\sup[a, b] = \sup(a, b) = \sup[a, b) = \sup(a, b] = b.$$

(c) As noted in Example 2, we have $\inf \mathbb{N} = 1$.

(d) If $A = \{r \in \mathbb{Q} : 0 \leq r \leq \sqrt{2}\}$, then $\sup A = \sqrt{2}$ and $\inf A = 0$.

(e) We have $\inf\{n^{(-1)^n} : n \in \mathbb{N}\} = 0$.

Notice that, in Examples 2 and 3, every set S that is bounded above possesses a least upper bound, i.e., $\sup S$ exists. This is not an accident. Otherwise there would be a "gap" between the set S and the set of its upper bounds.

4.4 Completeness Axiom.
Every nonempty subset S of $\mathbb{R}$ that is bounded above has a least upper bound. In other words, $\sup S$ exists and is a real number.

The completeness axiom for $\mathbb{Q}$ would assert that every nonempty subset of $\mathbb{Q}$, that is bounded above by some rational number, has a least upper bound that is a rational number. The set $A = \{r \in \mathbb{Q} : 0 \leq r \leq \sqrt{2}\}$ is a set of rational numbers and it is bounded above by some rational numbers [3/2 for example], but A has no least upper bound that is a rational number. Thus the completeness axiom does not hold for $\mathbb{Q}$! Incidentally, the set A can be described entirely in terms of rationals: $A = \{r \in \mathbb{Q} : 0 \leq r \text{ and } r^2 \leq 2\}$.

The completeness axiom for sets bounded below comes free.

4.5 Corollary.
Every nonempty subset S of $\mathbb{R}$ that is bounded below has a greatest lower bound $\inf S$.

EXAM!

Proof
Let $-S$ be the set $\{-s : s \in S\}$; $-S$ consists of the negatives of the numbers in S. Since S is bounded below there is an m in $\mathbb{R}$ such that $m \leq s$ for all $s \in S$. This implies that $-m \geq -s$ for all $s \in S$, so $-m \geq u$ for all u in the set $-S$. Thus $-S$ is bounded above by $-m$. The Completeness Axiom 4.4 applies to $-S$, so $\sup(-S)$ exists. Figure 4.1 suggests that we prove $\inf S = -\sup(-S)$.

Let $s_0 = \sup(-S)$; we need to prove

$$-s_0 \leq s \quad \text{for all} \quad s \in S, \tag{1}$$

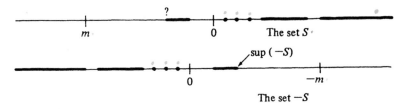

The set S

sup $(-S)$

The set $-S$

FIGURE 4.1

and

$$\text{if } t \le s \quad \text{for all} \quad s \in S, \quad \text{then} \quad t \le -s_0. \qquad (2)$$

The inequality (1) will show that $-s_0$ is a lower bound for S, while (2) will show that $-s_0$ is the *greatest* lower bound, that is, $-s_0 = \inf S$. We leave the proofs of (1) and (2) to Exercise 4.9. ■

It is useful to know:

$$\text{if } a > 0, \quad \text{then} \quad \frac{1}{n} < a \quad \text{for some positive integer} \quad n, \qquad (*)$$

and

$$\text{if } b > 0, \quad \text{then} \quad b < n \quad \text{for some positive integer} \quad n. \qquad (**)$$

These assertions are not as obvious as they may appear. If fact, there exist ordered fields that do not have these properties. In other words, there exists a mathematical system that satisfies all the properties A1-A4, M1-M4, DL and O1-O5 in §3 and yet possesses elements $a > 0$ and $b > 0$ such that $a < 1/n$ and $n < b$ *for all* n. On the other hand, such strange elements cannot exist in $\mathbb{R}$ or $\mathbb{Q}$. We next prove this; in view of the previous remarks we *must* expect to use the Completeness Axiom.

4.6 Archimedean Property.
If $a > 0$ and $b > 0$, then for some positive integer n, we have $na > b$.

This tells us that, even if a is quite small and b is quite large, some integer multiple of a will exceed b. Or, to quote [2], given enough time, one can empty a large bathtub with a small spoon. [Note that if we set $b = 1$, we obtain assertion (*), and if we set $a = 1$, we obtain assertion (**).]

Proof
Assume the Archimedean property fails. Then there exist $a > 0$ and $b > 0$ such that $na \le b$ for all $n \in \mathbb{N}$. In particular, b is an upper bound for the set $S = \{na : n \in \mathbb{N}\}$. Let $s_0 = \sup S$; this is where we are using the completeness axiom. Since $a > 0$, we have $s_0 < s_0 + a$, so $s_0 - a < s_0$. [To be precise, we obtain $s_0 \le s_0 + a$ and $s_0 - a \le s_0$ by property O4 and the fact that $a + (-a) = 0$. Then we conclude $s_0 - a < s_0$ since $s_0 - a = s_0$ implies $a = 0$ by Theorem 3.1(i).] Since s_0 is the least upper bound for S, $s_0 - a$ cannot be an upper bound for S. It follows that $s_0 - a < n_0 a$ for some $n_0 \in \mathbb{N}$. This implies that $s_0 < (n_0 + 1)a$. Since $(n_0 + 1)a$ is in S, s_0 is not an upper bound for S and we have reached a contradiction. Our assumption that the Archimedean property fails must be in error. ∎

We give one more result that seems obvious from our experience with the real number line, but which cannot be proved for an arbitrary ordered field.

4.7 Denseness of $\mathbb{Q}$.
If $a, b \in \mathbb{R}$ and $a < b$, then there is a rational $r \in \mathbb{Q}$ such that $a < r < b$.

Proof
We need to show that $a < \frac{m}{n} < b$ for some integers m and n where $n > 0$, and thus we need

$$an < m < bn. \tag{1}$$

Since $b - a > 0$, the Archimedean property shows that there exists an $n \in \mathbb{N}$ such that $n(b - a) > 1$. Since $bn - an > 1$, it is fairly evident that there is an integer m between an and bn, so that (1) holds. However, the proof that such an m exists is a little delicate. We argue as follows. By the Archimedean property again, there exists an integer $k > \max\{|an|, |bn|\}$, so that

$$-k < an < bn < k.$$

Then the set $\{j \in \mathbb{Z} : -k < j \le k \text{ and } an < j\}$ is finite and nonempty and we can set

$$m = \min\{j \in \mathbb{Z} : -k < j \le k \text{ and } an < j\}.$$

Then $an < m$ but $m - 1 \leq an$. Also, we have

$$m = (m - 1) + 1 \leq an + 1 < an + (bn - an) = bn,$$

so (1) holds. ∎

Exercises

4.1. For each set below that is bounded above, list three upper bounds for the set. Otherwise write "NOT BOUNDED ABOVE" or "NBA."

(a) $[0, 1]$
(b) $(0, 1)$
(c) $\{2, 7\}$
(d) $\{\pi, e\}$
(e) $\{\frac{1}{n} : n \in \mathbb{N}\}$
(f) $\{0\}$
(g) $[0, 1] \cup [2, 3]$
(h) $\cup_{n=1}^{\infty} [2n, 2n + 1]$
(i) $\cap_{n=1}^{\infty} [-\frac{1}{n}, 1 + \frac{1}{n}]$
(j) $\{1 - \frac{1}{3^n} : n \in \mathbb{N}\}$
(k) $\{n + \frac{(-1)^n}{n} : n \in \mathbb{N}\}$
(l) $\{r \in \mathbb{Q} : r < 2\}$
(m) $\{r \in \mathbb{Q} : r^2 < 4\}$
(n) $\{r \in \mathbb{Q} : r^2 < 2\}$
(o) $\{x \in \mathbb{R} : x < 0\}$
(p) $\{1, \frac{\pi}{3}, \pi^2, 10\}$
(q) $\{0, 1, 2, 4, 8, 16\}$
(r) $\cap_{n=1}^{\infty} (1 - \frac{1}{n}, 1 + \frac{1}{n})$
(s) $\{\frac{1}{n} : n \in \mathbb{N}$ and n is prime$\}$
(t) $\{x \in \mathbb{R} : x^3 < 8\}$
(u) $\{x^2 : x \in \mathbb{R}\}$
(v) $\{\cos(\frac{n\pi}{3}) : n \in \mathbb{N}\}$
(w) $\{\sin(\frac{n\pi}{3}) : n \in \mathbb{N}\}$

4.2. Repeat Exercise 4.1 for lower bounds.

4.3. For each set in Exercise 4.1, give its supremum if it has one. Otherwise write "NO sup."

4.4. Repeat Exercise 4.3 for infima [plural of infimum].

4.5. Let S be a nonempty subset of $\mathbb{R}$ that is bounded above. Prove that if $\sup S$ belongs to S, then $\sup S = \max S$. *Hint*: Your proof should be very short.

4.6. Let S be a nonempty bounded subset of $\mathbb{R}$.

    (a) Prove that $\inf S \leq \sup S$. *Hint*: This is almost obvious; your proof should be short.

    (b) What can you say about S if $\inf S = \sup S$?

4.7. Let S and T be nonempty bounded subsets of $\mathbb{R}$.

    (a) Prove that if $S \subseteq T$, then $\inf T \leq \inf S \leq \sup S \leq \sup T$.

(b) Prove that $\sup(S \cup T) = \max\{\sup S, \sup T\}$. *Note:* In part (b), do *not* assume $S \subseteq T$.

4.8. Let S and T be nonempty subsets of $\mathbb{R}$ with the following property: $s \leq t$ for all $s \in S$ and $t \in T$.

   (a) Observe that S is bounded above and that T is bounded below.

   (b) Prove that $\sup S \leq \inf T$.

   (c) Give an example of such sets S and T where $S \cap T$ is nonempty.

   (d) Give an example of sets S and T where $\sup S = \inf T$ and $S \cap T$ is the empty set.

4.9. Complete the proof that $\inf S = -\sup(-S)$ in Corollary 4.5 by proving (1) and (2).

4.10. Prove that if $a > 0$, then there exists $n \in \mathbb{N}$ such that $\frac{1}{n} < a < n$.

4.11. Consider $a, b \in \mathbb{R}$ where $a < b$. Use Denseness of $\mathbb{Q}$ 4.7 to show that there are infinitely many rationals between a and b.

4.12. Let $\mathbb{I}$ be the set of real numbers that are not rational; elements of $\mathbb{I}$ are called *irrational numbers*. Prove that if $a < b$, then there exists $x \in \mathbb{I}$ such that $a < x < b$. *Hint:* First show $\{r + \sqrt{2} : r \in \mathbb{Q}\} \subseteq \mathbb{I}$.

4.13. Prove that the following are equivalent for real numbers a, b, c. [*Equivalent* means that either all the properties hold or none of the properties hold.]

   (a) $|a - b| < c$,

   (b) $b - c < a < b + c$,

   (c) $a \in (b - c, b + c)$.

   Hint: Use Exercise 3.7(b).

4.14. Let A and B be nonempty bounded subsets of $\mathbb{R}$, and let S be the set of all sums $a + b$ where $a \in A$ and $b \in B$.

   (a) Prove that $\sup S = \sup A + \sup B$.

   (b) Prove that $\inf S = \inf A + \inf B$.

4.15. Let $a, b \in \mathbb{R}$. Show that if $a \leq b + \frac{1}{n}$ for all $n \in \mathbb{N}$, then $a \leq b$. Compare Exercise 3.8.

4.16. Show that $\sup\{r \in \mathbb{Q} : r < a\} = a$ for each $a \in \mathbb{R}$.

§5   The Symbols $+\infty$ and $-\infty$

The symbols $+\infty$ and $-\infty$ are extremely useful even though they are *not* real numbers. We will often write $+\infty$ as simply ∞. We will adjoin $+\infty$ and $-\infty$ to the set $\mathbb{R}$ and extend our ordering to the set $\mathbb{R}\cup\{-\infty, +\infty\}$. Explicitly, we will agree that $-\infty \leq a \leq +\infty$ for all a in $\mathbb{R}\cup\{-\infty, \infty\}$. This provides the set $\mathbb{R}\cup\{-\infty, +\infty\}$ with an ordering that satisfies properties O1, O2 and O3 of §3. We emphasize that we will *not* provide the set $\mathbb{R} \cup \{-\infty, +\infty\}$ with any algebraic structure. We may use the symbols $+\infty$ and $-\infty$, but we must continue to remember that they do not represent real numbers. Do *not* apply a theorem or exercise that is stated for real numbers to the symbols $+\infty$ or $-\infty$.

It is convenient to use the symbols $+\infty$ and $-\infty$ to extend the notation established in Example 1(b) of §4 to unbounded intervals. For real numbers $a, b \in \mathbb{R}$, we adopt the following notation:

$$[a, \infty) = \{x \in \mathbb{R} : a \leq x\}, \quad (a, \infty) = \{x \in \mathbb{R} : a < x\},$$
$$(-\infty, b] = \{x \in \mathbb{R} : x \leq b\}, \quad (-\infty, b) = \{x \in \mathbb{R} : x < b\}.$$

We occasionally also write $(-\infty, \infty)$ for $\mathbb{R}$. $[a, \infty)$ and $(-\infty, b]$ are called *closed intervals* or *unbounded closed intervals*, while (a, ∞) and $(-\infty, b)$ are called *open intervals* or *unbounded open intervals*.

Consider a nonempty subset S of $\mathbb{R}$. Recall that if S is bounded above, then $\sup S$ exists and represents a real number by the completeness axiom 4.4. We define

$$\sup S = +\infty \quad \text{if } S \text{ is not bounded above.}$$

Likewise, if S is bounded below, then $\inf S$ exists and represents a real number [Corollary 4.5]. And we define

$$\inf S = -\infty \quad \text{if } S \text{ is not bounded below.}$$

For emphasis, we recapitulate:

Let S be any nonempty subset of $\mathbb{R}$. The symbols $\sup S$ and $\inf S$ always make sense. If S is bounded above, then $\sup S$ is a real number; otherwise $\sup S = +\infty$. If S is bounded below, then $\inf S$ is a real number; otherwise $\inf S = -\infty$. Moreover, we have $\inf S \leq \sup S$.

The exercises for this section clear up some loose ends. Most of them extend results in §4 to sets that are not necessarily bounded.

Exercises

5.1. Write the following sets in interval notation:

    (a) $\{x \in \mathbb{R} : x < 0\}$ **(b)** $\{x \in \mathbb{R} : x^3 \leq 8\}$

    (c) $\{x^2 : x \in \mathbb{R}\}$ **(d)** $\{x \in \mathbb{R} : x^2 < 8\}$

5.2. Give the infimum and supremum of each set listed in Exercise 5.1.

5.3. Give the infimum and supremum of each unbounded set listed in Exercise 4.1.

5.4. Let S be a nonempty subset of $\mathbb{R}$, and let $-S = \{-s : s \in S\}$. Prove that $\inf S = -\sup(-S)$. *Hint*: For the case $-\infty < \inf S$, simply state that this was proved in Exercise 4.9.

5.5. Prove that $\inf S \leq \sup S$ for every nonempty subset of $\mathbb{R}$. Compare Exercise 4.6(a).

5.6. Let S and T be nonempty subsets of $\mathbb{R}$ such that $S \subseteq T$. Prove that $\inf T \leq \inf S \leq \sup S \leq \sup T$. Compare Exercise 4.7(a).

§6   * A Development of $\mathbb{R}$

There are several ways to give a careful development of $\mathbb{R}$ based on $\mathbb{Q}$. We will briefly discuss one of them and give suggestions for further reading on this topic. [See the remarks about optional sections in the preface.]

To motivate our development we begin by observing that

$$a = \sup\{r \in \mathbb{Q} : r < a\} \quad \text{for each} \quad a \in \mathbb{R};$$

see Exercise 4.16. Note the intimate relationship: $a \leq b$ if and only if $\{r \in \mathbb{Q} : r < a\} \subseteq \{r \in \mathbb{Q} : r < b\}$ and, moreover, $a = b$ if and only if $\{r \in \mathbb{Q} : r < a\} = \{r \in \mathbb{Q} : r < b\}$. Subsets α of $\mathbb{Q}$ having the form $\{r \in \mathbb{Q} : r < a\}$ satisfy these properties:

    (i) $\alpha \neq \mathbb{Q}$ and α is not empty,

    (ii) if $r \in \alpha$, $s \in \mathbb{Q}$ and $s < r$, then $s \in \alpha$,

   (iii) α contains no largest rational.

Moreover, every subset α of $\mathbb{Q}$ that satisfies (i)–(iii) has the form $\{r \in \mathbb{Q} : r < a\}$ for some $a \in \mathbb{R}$; in fact, $a = \sup \alpha$. Subsets α of $\mathbb{Q}$ satisfying (i)–(iii) are called *Dedekind cuts*.

The remarks in the last paragraph relating real numbers and Dedekind cuts are based on our knowledge of $\mathbb{R}$, including the completeness axiom. But they can also motivate a development of $\mathbb{R}$ based solely on $\mathbb{Q}$. In such a development we make no *a priori* assumptions about $\mathbb{R}$. We assume only that we have the ordered field $\mathbb{Q}$ and that $\mathbb{Q}$ satisfies the Archimedean property 4.6. A Dedekind cut is a subset α of $\mathbb{Q}$ satisfying (i)–(iii). The set $\mathbb{R}$ of real numbers is *defined* as the space of all Dedekind cuts. Thus elements of $\mathbb{R}$ are *defined* as certain subsets of $\mathbb{Q}$. The rational numbers are identified with certain Dedekind cuts in the natural way: each rational s corresponds to the Dedekind cut $s^* = \{r \in \mathbb{Q} : r < s\}$. In this way $\mathbb{Q}$ is regarded as a subset of $\mathbb{R}$, that is, $\mathbb{Q}$ is identified with the set $\mathbb{Q}^* = \{s^* : s \in \mathbb{Q}\}$.

The set $\mathbb{R}$ defined in the last paragraph is given an order structure as follows: if α and β are Dedekind cuts, then we define $\alpha \leq \beta$ to signify that $\alpha \subseteq \beta$. Properties O1, O2 and O3 in §3 hold for this ordering. Addition is defined in $\mathbb{R}$ as follows: if α and β are Dedekind cuts, then

$$\alpha + \beta = \{r_1 + r_2 : r_1 \in \alpha \text{ and } r_2 \in \beta\}.$$

It turns out that $\alpha + \beta$ is a Dedekind cut [hence in $\mathbb{R}$] and that this definition of addition satisfies properties A1–A4 in §3. Multiplication of Dedekind cuts is a tedious business and has to be defined first for Dedekind cuts that are $\geq 0^*$. For a naive attempt, see Exercise 6.4. After the product of Dedekind cuts has been defined, the remaining properties of an ordered field can be verified for $\mathbb{R}$. The ordered field $\mathbb{R}$ constructed in this manner from $\mathbb{Q}$ is complete: the completeness property in 4.4 can be *proved* rather than taken as an axiom.

The development of $\mathbb{R}$ outlined above is given in [34] and [36]. The real numbers are developed from Cauchy sequences in $\mathbb{Q}$ in [23], §5. A thorough development of $\mathbb{R}$ based on Peano's axioms is given in [28].

Exercises

6.1. Consider $s, t \in \mathbb{Q}$. Show that

    (a) $s \leq t$ if and only if $s^* \subseteq t^*$;

(b) $s = t$ if and only if $s^* = t^*$;

(c) $(s + t)^* = s^* + t^*$. Note that $s^* + t^*$ is a sum of Dedekind cuts.

6.2. Show that if α and β are Dedekind cuts, then so is $\alpha + \beta = \{r_1 + r_2 : r_1 \in \alpha \text{ and } r_2 \in \beta\}$.

6.3. **(a)** Show that $\alpha + 0^* = \alpha$ for all Dedekind cuts α.

   (b) We claimed, without proof, that addition of Dedekind cuts satisfies property A4. Thus if α is a Dedekind cut, there must exist a Dedekind cut $-\alpha$ such that $\alpha + (-\alpha) = 0^*$. How would you define $-\alpha$?

6.4. Let α and β be Dedekind cuts and define the "product": $\alpha \cdot \beta = \{r_1 r_2 : r_1 \in \alpha \text{ and } r_2 \in \beta\}$.

   (a) Calculate some "products" of Dedekind cuts using the Dedekind cuts 0^*, 1^* and $(-1)^*$.

   (b) Discuss why this definition of "product" is totally unsatisfactory for defining multiplication in $\mathbb{R}$.

6.5. **(a)** Show that $\{r \in \mathbb{Q} : r^3 < 2\}$ is a Dedekind cut, but that $\{r \in \mathbb{Q} : r^2 < 2\}$ is not a Dedekind cut.

   (b) Does the Dedekind cut $\{r \in \mathbb{Q} : r^3 < 2\}$ correspond to a rational number in $\mathbb{R}$?

   (c) Show that $0^* \cup \{r \in \mathbb{Q} : r \geq 0 \text{ and } r^2 < 2\}$ is a Dedekind cut. Does it correspond to a rational number in $\mathbb{R}$?

6.6. Let $\alpha = 0^* \cup \{p \in \mathbb{Q} : p \geq 0 \text{ and } p^2 < 2\}$. Prove that α is a Dedekind cut and also that it has the property $\alpha \cdot \alpha = 2^*$; that is, the square of α is 2^*. *Note*: This seems to be surprisingly tricky, as pointed out by Linda Hill and Robert J. Fisher at Idaho State University. Their solution is available from them or from the author.

2

CHAPTER

Sequences

§7 Limits of Sequences

A *sequence* is a function whose domain is a set that has the form $\{n \in \mathbb{Z} : n \geq m\}$; m is usually 1 or 0. Thus a sequence is a function that has a specified value for each integer $n \geq m$. It is customary to denote a sequence by a letter such as s and to denote its value at n as s_n rather than $s(n)$. It is often convenient to write the sequence as $(s_n)_{n=m}^{\infty}$ or $(s_m, s_{m+1}, s_{m+2}, \ldots)$. If $m = 1$ we may write $(s_n)_{n \in \mathbb{N}}$ or of course $(s_1, s_2, s_3, \ldots)$. Sometimes we will write (s_n) when the domain is understood or when the results under discussion do not depend on the specific value of m. In this chapter, we will be interested in sequences whose range values are real numbers, i.e., each s_n represents a real number.

Example 1

(a) Consider the sequence $(s_n)_{n \in \mathbb{N}}$ where $s_n = \frac{1}{n^2}$. This is the sequence $(1, \frac{1}{4}, \frac{1}{9}, \frac{1}{16}, \frac{1}{25}, \ldots)$. Formally, of course, this is the function with domain $\mathbb{N}$ whose value at each n is $\frac{1}{n^2}$. The *set* of values is $\{1, \frac{1}{4}, \frac{1}{9}, \frac{1}{16}, \frac{1}{25}, \ldots\}$.

(b) Consider the sequence given by $a_n = (-1)^n$ for $n \geq 0$, i.e., $(a_n)_{n=0}^{\infty}$ where $a_n = (-1)^n$. Note that the first term of the sequence is $a_0 = 1$ and the sequence is $(1, -1, 1, -1, 1, -1, 1, \ldots)$. Formally, this is a function whose domain is $\{0, 1, 2, \ldots\}$ and whose *set* of values is $\{-1, 1\}$.

It is important to distinguish between a sequence and its set of values, since the validity of many results in this book depends on whether we are working with a sequence or a set. We will always use parentheses ( ) to signify a sequence and braces { } to signify a set. The sequence given by $a_n = (-1)^n$ has an infinite number of terms even though their values are repeated over and over. On the other hand, the *set* $\{(-1)^n : n = 0, 1, 2, \ldots\}$ is exactly the set $\{-1, 1\}$ consisting of two numbers.

(c) Consider the sequence $\cos(\frac{n\pi}{3})$, $n \in \mathbb{N}$. The first term of this sequence is $\cos(\frac{\pi}{3}) = \cos 60° = \frac{1}{2}$ and the sequence looks like

$$(\tfrac{1}{2}, -\tfrac{1}{2}, -1, -\tfrac{1}{2}, \tfrac{1}{2}, 1, \tfrac{1}{2}, -\tfrac{1}{2}, -1, -\tfrac{1}{2}, \tfrac{1}{2}, 1, \tfrac{1}{2}, -\tfrac{1}{2}, -1, \ldots).$$

The set of values is $\{\cos(\frac{n\pi}{3}) : n \in \mathbb{N}\} = \{\tfrac{1}{2}, -\tfrac{1}{2}, -1, 1\}$.

(d) If $a_n = n^{1/n}$, $n \in \mathbb{N}$, the sequence is $(1, \sqrt{2}, 3^{1/3}, 4^{1/4}, \ldots)$. If we approximate values to four decimal places, the sequence looks like

$$(1, 1.4142, 1.4422, 1.4142, 1.3797, 1.3480, 1.3205, 1.2968, \ldots).$$

It turns out that a_{100} is approximately 1.0471 and that a_{1000} is approximately 1.0069.

(e) Consider the sequence $b_n = (1 + \frac{1}{n})^n$, $n \in \mathbb{N}$. This is the sequence $(2, (\frac{3}{2})^2, (\frac{4}{3})^3, (\frac{5}{4})^4, \ldots)$. If we approximate the values to four decimal places, we obtain

$$(2, 2.25, 2.3704, 2.4414, 2.4883, 2.5216, 2.5465, 2.5658, \ldots).$$

Also b_{100} is approximately 2.7048 and b_{1000} is approximately 2.7169.

The "limit" of a sequence (s_n) is a real number that the values s_n are close to for large values of n. For instance, the values of the sequence in Example 1(a) are close to 0 for large n and the values of the sequence in Example 1(d) appear to be close to 1 for large n. The sequence (a_n) given by $a_n = (-1)^n$ requires some thought. We might say that 1 is a limit because in fact $a_n = 1$ for the large values of n that are even. On the other hand, $a_n = -1$ [which is quite a distance from 1] for other large values of n. We need a concise definition in order to decide whether 1 is a limit of $a_n = (-1)^n$. It turns out that our definition will require the values to be close to the limit value for *all* large n, so 1 will *not* be a limit of the sequence $a_n = (-1)^n$.

7.1 Definition.
A sequence (s_n) of real numbers is said to *converge* to the real number s provided that

> for each $\epsilon > 0$ there exists a number N such that $n > N$ implies $|s_n - s| < \epsilon$. (1)

If (s_n) converges to s, we will write $\lim_{n\to\infty} s_n = s$, or $s_n \to s$. The number s is called the *limit* of the sequence (s_n). A sequence that does not converge to some real number is said to *diverge*.

Several comments are in order. First, in view of the Archimedean property, the number N in Definition 7.1 can be taken to be a natural number if we wish. Second, the symbol ϵ [lower case Greek epsilon] in this definition represents a positive number, not some new exotic number. However, it is traditional in mathematics to use ϵ and δ [lower case Greek delta] in situations where the interesting or challenging values are the small positive values. Third, condition (1) is an infinite number of statements, one for each positive value of ϵ. The condition states that to each $\epsilon > 0$ there corresponds a number N with a certain property, namely $n > N$ implies $|s_n - s| < \epsilon$. The value N depends on the value ϵ, and normally N must be large if ϵ is small. We illustrate these remarks in the next example.

Example 2
Consider the sequence $s_n = \frac{3n+1}{7n-4}$. If we write s_n as $\frac{3+\frac{1}{n}}{7-\frac{4}{n}}$ and note that $\frac{1}{n}$ and $\frac{4}{n}$ are very small for large n, it seems reasonable to conclude that $\lim s_n = \frac{3}{7}$. In fact, this reasoning will be completely valid after we have the limit theorems in §9:

$$\lim s_n = \lim \left[\frac{3+\frac{1}{n}}{7-\frac{4}{n}}\right] = \frac{\lim 3 + \lim(\frac{1}{n})}{\lim 7 - 4\lim(\frac{1}{n})} = \frac{3+0}{7-4\cdot 0} = \frac{3}{7}.$$

However, for now we are interested in analyzing exactly what we mean by $\lim s_n = \frac{3}{7}$. By Definition 7.1, $\lim s_n = \frac{3}{7}$ means that

for each $\epsilon > 0$ there exists a number N such that
$n > N$ implies $\left| \frac{3n+1}{7n-4} - \frac{3}{7} \right| < \epsilon.$
(1)

As ϵ varies, N varies. In Example 2 of the next section we will show that, for this particular sequence, N can be taken to be $\frac{19}{49\epsilon} + \frac{4}{7}$. Using this observation and a calculator, we find that for ϵ equal to 1, 0.1, 0.01, 0.001 and 0.000001, respectively, N can be taken to be approximately 0.96, 4.45, 39.35, 388.33 and 387,755.67, respectively. Since we are interested only in integer values of n, we may as well drop the fractional part of N. Then we see that five of the infinitely many statements given by (1) are:

$$n > 0 \quad \text{implies} \quad \left| \frac{3n+1}{7n-4} - \frac{3}{7} \right| < 1; \qquad (2)$$

$$n > 4 \quad \text{implies} \quad \left| \frac{3n+1}{7n-4} - \frac{3}{7} \right| < 0.1; \qquad (3)$$

$$n > 39 \quad \text{implies} \quad \left| \frac{3n+1}{7n-4} - \frac{3}{7} \right| < 0.01; \qquad (4)$$

$$n > 388 \quad \text{implies} \quad \left| \frac{3n+1}{7n-4} - \frac{3}{7} \right| < 0.001; \qquad (5)$$

$$n > 387,755 \quad \text{implies} \quad \left| \frac{3n+1}{7n-4} - \frac{3}{7} \right| < 0.000001. \qquad (6)$$

Table 7.1 partially confirms assertions (2) through (6). We could go on and on with these numerical illustrations, but it should be clear that we need a more theoretical approach if we are going to *prove* results about limits.

Table 7.1

n	$s_n = \frac{3n+1}{7n-4}$ approximately	$\|s_n - \frac{3}{7}\|$ approximately
1	1.3333	.9047
2	0.7000	.2714
3	0.5882	.1597
4	0.5417	.1131
5	0.5161	.0876
6	0.5000	.0714
40	0.4384	.0098
400	0.4295	.0010

Example 3
We return to the examples in Example 1.
  (a) $\lim \frac{1}{n^2} = 0$. This will be proved in Example 1 of the next section.
  (b) The sequence (a_n) where $a_n = (-1)^n$ does not converge. Thus the expression "$\lim a_n$" is meaningless in this case. We will discuss this example again in Example 4 of the next section.
  (c) The sequence $\cos(\frac{n\pi}{3})$ does not converge. See Exercise 8.7.
  (d) The sequence $n^{1/n}$ appears to converge to 1. We will prove $\lim n^{1/n} = 1$ in 9.7(c).
  (e) The sequence (b_n) where $b_n = (1 + \frac{1}{n})^n$ converges to the number e that should be familiar from calculus. The limit $\lim b_n$ and the number e will be discussed further in the optional §37. Recall that e is approximately 2.7182818.

We conclude this section by showing that limits are unique. That is, if $\lim s_n = s$ and $\lim s_n = t$, then we must have $s = t$. In short, the values s_n cannot be getting arbitrarily close to different values for large n. To prove this, consider $\epsilon > 0$. By the definition of limit there must exist N_1 so that

$$n > N_1 \quad \text{implies} \quad |s_n - s| < \frac{\epsilon}{2}$$

and there must exist N_2 so that

$$n > N_2 \quad \text{implies} \quad |s_n - t| < \frac{\epsilon}{2}.$$

For $n > \max\{N_1, N_2\}$, the Triangle Inequality 3.7 shows that

$$|s - t| = |(s - s_n) + (s_n - t)| \leq |s - s_n| + |s_n - t| \leq \frac{\epsilon}{2} + \frac{\epsilon}{2} = \epsilon.$$

This shows that $|s - t| < \epsilon$ for all $\epsilon > 0$. It follows that $|s - t| = 0$, hence $s = t$.

Exercises

7.1. Write out the first five terms of the following sequences.
    (a) $s_n = \frac{1}{3n+1}$
    (b) $b_n = \frac{3n+1}{4n-1}$
    (c) $c_n = \frac{n}{3^n}$
    (d) $\sin(\frac{n\pi}{4})$

7.2. For each sequence in Exercise 7.1, determine whether it converges. If it converges, give its limit. No proofs are required.

7.3. For each sequence below, determine whether it converges and, if it converges, give its limit. No proofs are required.
    (a) $a_n = \frac{n}{n+1}$
    (b) $b_n = \frac{n^2+3}{n^2-3}$
    (c) $c_n = 2^{-n}$
    (d) $t_n = 1 + \frac{2}{n}$
    (e) $x_n = 73 + (-1)^n$
    (f) $s_n = (2)^{1/n}$
    (g) $y_n = n!$
    (h) $d_n = (-1)^n n$
    (i) $\frac{(-1)^n}{n}$
    (j) $\frac{7n^3+8n}{2n^3-31}$
    (k) $\frac{9n^2-18}{6n+18}$
    (l) $\sin(\frac{n\pi}{2})$
    (m) $\sin(n\pi)$
    (n) $\sin(\frac{2n\pi}{3})$
    (o) $\frac{1}{n}\sin n$
    (p) $\frac{2^{n+1}+5}{2^n-7}$
    (q) $\frac{3^n}{n!}$
    (r) $(1 + \frac{1}{n})^2$
    (s) $\frac{4n^2+3}{3n^2-2}$
    (t) $\frac{6n+4}{9n^2+7}$

7.4. Give examples of

    (a) a sequence (x_n) of irrational numbers having a limit $\lim x_n$ that is a rational number.

    (b) a sequence (r_n) of rational numbers having a limit $\lim r_n$ that is an irrational number.

7.5. Determine the following limits. No proofs are required, but show any relevant algebra.

    (a) $\lim s_n$ where $s_n = \sqrt{n^2+1} - n$,

    (b) $\lim(\sqrt{n^2+n} - n)$,

(c) $\lim(\sqrt{4n^2 + n} - 2n)$.
Hint for (a): First show that $s_n = \frac{1}{\sqrt{n^2+1}+n}$.

§8 A Discussion about Proofs

In this section we give several examples of proofs using the definition of the limit of a sequence. With a little study and practice, students should be able to do proofs of this sort themselves. We will sometimes refer to a proof as a *formal proof* to emphasize that it is a rigorous mathematical proof.

Example 1
Prove that $\lim \frac{1}{n^2} = 0$.

Discussion. Our task is to consider an arbitrary $\epsilon > 0$ and show that there exists a number N [which will depend on ϵ] such that $n > N$ implies $|\frac{1}{n^2} - 0| < \epsilon$. So we expect our formal proof to begin with "Let $\epsilon > 0$" and to end with something like "Hence $n > N$ implies $|\frac{1}{n^2} - 0| < \epsilon$." In between the proof should specify an N and then verify that N has the desired property, namely that $n > N$ does indeed imply $|\frac{1}{n^2} - 0| < \epsilon$.

As is often the case with trigonometric identities, we will initially work backward from our desired conclusion, but in the formal proof we will have to be sure that our steps are reversible. In the present example, we want $|\frac{1}{n^2} - 0| < \epsilon$ and we want to know how big n must be. So we will operate on this inequality algebraically and try to "solve" for n. Thus we want $\frac{1}{n^2} < \epsilon$. By multiplying both sides by n^2 and dividing both sides by ϵ, we find that we want $\frac{1}{\epsilon} < n^2$ or $\frac{1}{\sqrt{\epsilon}} < n$. If our steps are reversible, we see that $n > \frac{1}{\sqrt{\epsilon}}$ implies $|\frac{1}{n^2} - 0| < \epsilon$. This suggests that we put $N = \frac{1}{\sqrt{\epsilon}}$.

Formal Proof
Let $\epsilon > 0$. Let $N = \frac{1}{\sqrt{\epsilon}}$. Then $n > N$ implies $n > \frac{1}{\sqrt{\epsilon}}$ which implies $n^2 > \frac{1}{\epsilon}$ and hence $\epsilon > \frac{1}{n^2}$. Thus $n > N$ implies $|\frac{1}{n^2} - 0| < \epsilon$. This proves that $\lim \frac{1}{n^2} = 0$. ∎

Example 2

Prove that $\lim \frac{3n+1}{7n-4} = \frac{3}{7}$.

Discussion. For each $\epsilon > 0$, we need to decide how big n must be to guarantee that $|\frac{3n+1}{7n-4} - \frac{3}{7}| < \epsilon$. Thus we want

$$\left| \frac{21n + 7 - 21n + 12}{7(7n - 4)} \right| < \epsilon \quad \text{or} \quad \left| \frac{19}{7(7n - 4)} \right| < \epsilon.$$

Since $7n - 4 > 0$, we can drop the absolute value and manipulate the inequality further to "solve" for n:

$$\frac{19}{7\epsilon} < 7n - 4 \quad \text{or} \quad \frac{19}{7\epsilon} + 4 < 7n \quad \text{or} \quad \frac{19}{49\epsilon} + \frac{4}{7} < n.$$

Our steps are reversible, so we will put $N = \frac{19}{49\epsilon} + \frac{4}{7}$. Incidentally, we could have chosen N to be any number larger than $\frac{19}{49\epsilon} + \frac{4}{7}$.

Formal Proof

Let $\epsilon > 0$ and let $N = \frac{19}{49\epsilon} + \frac{4}{7}$. Then $n > N$ implies $n > \frac{19}{49\epsilon} + \frac{4}{7}$, hence $7n > \frac{19}{7\epsilon} + 4$, hence $7n - 4 > \frac{19}{7\epsilon}$, hence $\frac{19}{7(7n-4)} < \epsilon$, and hence $|\frac{3n+1}{7n-4} - \frac{3}{7}| < \epsilon$. This proves $\lim \frac{3n+1}{7n-4} = \frac{3}{7}$. ∎

Example 3

Prove that $\lim \frac{4n^3+3n}{n^3-6} = 4$.

Discussion. For each $\epsilon > 0$, we need to determine how large n must be to imply

$$\left| \frac{4n^3 + 3n}{n^3 - 6} - 4 \right| < \epsilon \quad \text{or} \quad \left| \frac{3n + 24}{n^3 - 6} \right| < \epsilon.$$

By considering $n > 1$, we may drop the absolute values; thus we need to find how big n must be to give $\frac{3n+24}{n^3-6} < \epsilon$. This time it would be very difficult to "solve" for or isolate n. Recall that we need to find some N such that $n > N$ implies $\frac{3n+24}{n^3-6} < \epsilon$, but we do not need to find the least such N. So we will simplify matters by making estimates. The idea is that $\frac{3n+24}{n^3-6}$ is bounded by some constant times $\frac{n}{n^3} = \frac{1}{n^2}$ for sufficiently large n. To find such a bound we will find an upper bound for the numerator and a lower bound for the denominator. For example, since $3n + 24 \leq 27n$, it suffices for us to get $\frac{27n}{n^3-6} < \epsilon$. To make the denominator smaller and yet a constant multiple of n^3, we note that $n^3 - 6 \geq \frac{n^3}{2}$ provided n is sufficiently large; in fact, all we need is $\frac{n^3}{2} \geq 6$ or $n^3 \geq 12$ or $n > 2$. So it suffices to get $\frac{27n}{n^3/2} < \epsilon$ or $\frac{54}{n^2} < \epsilon$ or $n > \sqrt{\frac{54}{\epsilon}}$, provided that $n > 2$.

Formal Proof
Let $\epsilon > 0$ and let $N = \max\{2, \sqrt{\frac{54}{\epsilon}}\}$. Then $n > N$ implies $n > \sqrt{\frac{54}{\epsilon}}$, hence $\frac{54}{n^2} < \epsilon$, hence $\frac{27n}{n^3/2} < \epsilon$. Since $n > 2$, we have $\frac{n^3}{2} \leq n^3 - 6$ and also $27n \geq 3n + 24$. Thus $n > N$ implies

$$\frac{3n + 24}{n^3 - 6} \leq \frac{27n}{\frac{1}{2}n^3} = \frac{54}{n^2} < \epsilon,$$

and hence

$$\left| \frac{4n^3 + 3n}{n^3 - 6} - 4 \right| < \epsilon,$$

as desired. ∎

Example 3 illustrates that direct proofs of even rather simple limits can get complicated. With the limit theorems of §9 we would just write

$$\lim \left[\frac{4n^3 + 3n}{n^3 - 6} \right] = \lim \left[\frac{4 + \frac{3}{n^2}}{1 - \frac{6}{n^3}} \right] = \frac{\lim 4 + 3 \cdot \lim(\frac{1}{n^2})}{\lim 1 - 6 \cdot \lim(\frac{1}{n^3})} = 4.$$

Example 4
Show that the sequence $a_n = (-1)^n$ does not converge.
Discussion. We will assume that $\lim(-1)^n = a$ and obtain a contradiction. No matter what a is, either 1 or -1 will have distance at least 1 from a. Thus the inequality $|(-1)^n - a| < 1$ will not hold for all large n.

Formal Proof
Assume that $\lim(-1)^n = a$ for some $a \in \mathbb{R}$. Letting $\epsilon = 1$ in the definition of the limit, we see that there exists N such that

$$n > N \quad \text{implies} \quad |(-1)^n - a| < 1.$$

By considering both an even and an odd $n > N$, we see that

$$|1 - a| < 1 \quad \text{and} \quad |-1 - a| < 1.$$

Now by the Triangle Inequality 3.7

$$2 = |1 - (-1)| = |1 - a + a - (-1)| \leq |1 - a| + |a - (-1)| < 1 + 1 = 2.$$

This absurdity shows that our assumption that $\lim(-1)^n = a$ must be wrong, so the sequence $(-1)^n$ does not converge. ∎

Example 5

Let (s_n) be a sequence of nonnegative real numbers and suppose that $s = \lim s_n$. Note that $s \geq 0$; see Exercise 8.9(a). Prove that $\lim \sqrt{s_n} = \sqrt{s}$.

Discussion. We must consider $\epsilon > 0$ and show that there exists N such that

$$n > N \quad \text{implies} \quad |\sqrt{s_n} - \sqrt{s}| < \epsilon.$$

This time we cannot expect to obtain N explicitly in terms of ϵ because of the general nature of the problem. But we can hope to show such N exists. The trick here is to violate our training in algebra and "irrationalize the denominator":

$$\sqrt{s_n} - \sqrt{s} = \frac{(\sqrt{s_n} - \sqrt{s})(\sqrt{s_n} + \sqrt{s})}{\sqrt{s_n} + \sqrt{s}} = \frac{s_n - s}{\sqrt{s_n} + \sqrt{s}}.$$

Since $s_n \to s$ we will be able to make the numerator small [for large n]. Unfortunately, if $s = 0$ the denominator will also be small. So we consider two cases. If $s > 0$, the denominator is bounded below by $\sqrt{s}$ and our trick will work:

$$|\sqrt{s_n} - \sqrt{s}| \leq \frac{|s_n - s|}{\sqrt{s}},$$

so we will select N so that $|s_n - s| < \sqrt{s}\epsilon$ for $n > N$. Note that N exists, since we can apply the definition of limit to $\sqrt{s}\epsilon$ just as well as to ϵ. For $s = 0$, it can be shown directly that $\lim s_n = 0$ implies $\lim \sqrt{s_n} = 0$; the trick of "irrationalizing the denominator" is not needed in this case.

Formal Proof

Case I: $s > 0$. Let $\epsilon > 0$. Since $\lim s_n = s$, there exists N such that

$$n > N \quad \text{implies} \quad |s_n - s| < \sqrt{s}\epsilon.$$

Now $n > N$ implies

$$|\sqrt{s_n} - \sqrt{s}| = \frac{|s_n - s|}{\sqrt{s_n} + \sqrt{s}} \le \frac{|s_n - s|}{\sqrt{s}} < \frac{\sqrt{s}\epsilon}{\sqrt{s}} = \epsilon.$$

Case II: $s = 0$. This case is left to Exercise 8.3. ∎

Example 6
Let (s_n) be a convergent sequence of real numbers such that $s_n \ne 0$ for all $n \in \mathbb{N}$ and $\lim s_n = s \ne 0$. Prove that $\inf\{|s_n| : n \in \mathbb{N}\} > 0$.

Discussion. The idea is that "most" of the terms s_n are close to s and hence not close to 0. More explicitly, "most" of the terms s_n are within $\frac{1}{2}|s|$ of s, hence most s_n satisfy $|s_n| \ge \frac{1}{2}|s|$. This seems clear from Figure 8.1, but a formal proof will use the triangle inequality.

Formal Proof
Let $\epsilon = \frac{1}{2}|s| > 0$. Since $\lim s_n = s$, there exists N in $\mathbb{N}$ so that

$$n > N \quad \text{implies} \quad |s_n - s| < \frac{|s|}{2}.$$

Now

$$n > N \quad \text{implies} \quad |s_n| \ge \frac{|s|}{2}, \tag{1}$$

since otherwise the triangle inequality would imply

$$|s| = |s - s_n + s_n| \le |s - s_n| + |s_n| < \frac{|s|}{2} + \frac{|s|}{2} = |s|$$

which is absurd. If we set

$$m = \min\left\{\frac{|s|}{2}, |s_1|, |s_2|, \ldots, |s_N|\right\},$$

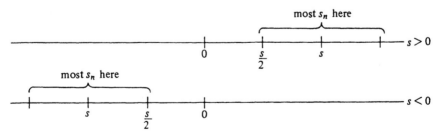

FIGURE 8.1

then we clearly have $m > 0$ and $|s_n| \geq m$ for *all* $n \in \mathbb{N}$ in view of (1). Thus $\inf\{|s_n| : n \in \mathbb{N}\} \geq m > 0$, as desired. ∎

Formal proofs are required in the following exercises.

Exercises

8.1. Prove the following:

(a) $\lim \frac{(-1)^n}{n} = 0$

(b) $\lim \frac{1}{n^{1/3}} = 0$

(c) $\lim \frac{2n-1}{3n+2} = \frac{2}{3}$

(d) $\lim \frac{n+6}{n^2-6} = 0$

8.2. Determine the limits of the following sequences, and then prove your claims.

(a) $a_n = \frac{n}{n^2+1}$

(b) $b_n = \frac{7n-19}{3n+7}$

(c) $c_n = \frac{4n+3}{7n-5}$

(d) $d_n = \frac{2n+4}{5n+2}$

(e) $s_n = \frac{1}{n} \sin n$

8.3. Let (s_n) be a sequence of nonnegative real numbers, and suppose that $\lim s_n = 0$. Prove that $\lim \sqrt{s_n} = 0$. This will complete the proof for Example 5.

8.4. Let (t_n) be a bounded sequence, i.e., there exists M such that $|t_n| \leq M$ for all n, and let (s_n) be a sequence such that $\lim s_n = 0$. Prove that $\lim(s_n t_n) = 0$.

8.5. (a) Consider three sequences (a_n), (b_n) and (s_n) such that $a_n \leq s_n \leq b_n$ for all n and $\lim a_n = \lim b_n = s$. Prove that $\lim s_n = s$.

(b) Suppose that (s_n) and (t_n) are sequences such that $|s_n| \leq t_n$ for all n and $\lim t_n = 0$. Prove that $\lim s_n = 0$.

8.6. Let (s_n) be a sequence in $\mathbb{R}$.

(a) Prove that $\lim s_n = 0$ if and only if $\lim |s_n| = 0$.

(b) Observe that if $s_n = (-1)^n$, then $\lim |s_n|$ exists, but $\lim s_n$ does not exist.

8.7. Show that the following sequences do not converge.

(a) $\cos(\frac{n\pi}{3})$

(b) $s_n = (-1)^n n$

(c) $\sin(\frac{n\pi}{3})$

8.8. Prove the following [see Exercise 7.5]:

(a) $\lim[\sqrt{n^2+1} - n] = 0$

(b) $\lim[\sqrt{n^2+n} - n] = \frac{1}{2}$

(c) $\lim[\sqrt{4n^2+n} - 2n] = \frac{1}{4}$

8.9. Let (s_n) be a sequence that converges.

    (a) Show that if $s_n \geq a$ for all but finitely many n, then $\lim s_n \geq a$.

    (b) Show that if $s_n \leq b$ for all but finitely many n, then $\lim s_n \leq b$.

    (c) Conclude that if all but finitely many s_n belong to $[a, b]$, then $\lim s_n$ belongs to $[a, b]$.

8.10. Let (s_n) be a convergent sequence, and suppose that $\lim s_n > a$. Prove that there exists a number N such that $n > N$ implies $s_n > a$.

§9   Limit Theorems for Sequences

In this section we prove some basic results that are probably already familiar to the reader. First we prove that convergent sequences are bounded. A sequence (s_n) of real numbers is said to be *bounded* if the set $\{s_n : n \in \mathbb{N}\}$ is a bounded set, i.e., if there exists a constant M such that $|s_n| \leq M$ for all n.

9.1 Theorem.
Convergent sequences are bounded.

Proof
Let (s_n) be a convergent sequence, and let $s = \lim s_n$. Applying Definition 7.1 with $\epsilon = 1$ we obtain N in $\mathbb{N}$ so that

$$n > N \quad \text{implies} \quad |s_n - s| < 1.$$

From the triangle inequality we see that $n > N$ implies $|s_n| < |s| + 1$. Define $M = \max\{|s| + 1, |s_1|, |s_2|, \ldots, |s_N|\}$. Then we have $|s_n| \leq M$ for all $n \in \mathbb{N}$, so (s_n) is a bounded sequence.   ■

    In the proof of Theorem 9.1 we only needed to use property 7.1(1) for a single value of ϵ. Our choice of $\epsilon = 1$ was quite arbitrary.

9.2 Theorem.
If the sequence (s_n) converges to s and $k \in \mathbb{R}$, then the sequence (ks_n) converges to ks. That is, $\lim(ks_n) = k \lim s_n$.

Proof

We assume $k \neq 0$, since this result is trivial for $k = 0$. Let $\epsilon > 0$ and note that we need to show that $|ks_n - ks| < \epsilon$ for large n. Since $\lim s_n = s$, there exists N such that

$$n > N \quad \text{implies} \quad |s_n - s| < \frac{\epsilon}{|k|}.$$

Then

$$n > N \quad \text{implies} \quad |ks_n - ks| < \epsilon. \quad\blacksquare$$

9.3 Theorem.

If (s_n) converges to s and (t_n) converges to t, then $(s_n + t_n)$ converges to $s + t$. That is,

$$\lim(s_n + t_n) = \lim s_n + \lim t_n.$$

Proof

Let $\epsilon > 0$; we need to show that

$$|s_n + t_n - (s + t)| < \epsilon \quad \text{for large} \quad n.$$

We note that $|s_n + t_n - (s + t)| \leq |s_n - s| + |t_n - t|$. Since $\lim s_n = s$, there exists N_1 such that

$$n > N_1 \quad \text{implies} \quad |s_n - s| < \frac{\epsilon}{2}.$$

Likewise, there exists N_2 such that

$$n > N_2 \quad \text{implies} \quad |t_n - t| < \frac{\epsilon}{2}.$$

Let $N = \max\{N_1, N_2\}$. Then clearly

$$n > N \quad \text{implies} \quad |s_n + t_n - (s + t)| \leq |s_n - s| + |t_n - t| < \frac{\epsilon}{2} + \frac{\epsilon}{2} = \epsilon. \quad\blacksquare$$

9.4 Theorem.

If (s_n) converges to s and (t_n) converges to t, then $(s_n t_n)$ converges to st. That is,

$$\lim(s_n t_n) = (\lim s_n)(\lim t_n).$$

Discussion. The trick here is to look at the inequality

$$|s_n t_n - st| = |s_n t_n - s_n t + s_n t - st|$$
$$\leq |s_n t_n - s_n t| + |s_n t - st| = |s_n| \cdot |t_n - t| + |t| \cdot |s_n - s|.$$

For large n, $|t_n - t|$ and $|s_n - s|$ are small and $|t|$ is, of course, constant. Fortunately, Theorem 9.1 shows that $|s_n|$ is bounded, so we will be able to show that $|s_n t_n - st|$ is small.

Proof

Let $\epsilon > 0$. By Theorem 9.1 there is a constant $M > 0$ such that $|s_n| \leq M$ for all n. Since $\lim t_n = t$ there exists N_1 such that

$$n > N_1 \quad \text{implies} \quad |t_n - t| < \frac{\epsilon}{2M}.$$

Also, since $\lim s_n = s$ there exists N_2 such that

$$n > N_2 \quad \text{implies} \quad |s_n - s| < \frac{\epsilon}{2(|t| + 1)}.$$

[We used $\frac{\epsilon}{2(|t|+1)}$ instead of $\frac{\epsilon}{2|t|}$, since t could be 0.] Now if $N = \max\{N_1, N_2\}$, then $n > N$ implies

$$|s_n t_n - st| \leq |s_n| \cdot |t_n - t| + |t| \cdot |s_n - s|$$
$$\leq M \cdot \frac{\epsilon}{2M} + |t| \cdot \frac{\epsilon}{2(|t| + 1)} < \frac{\epsilon}{2} + \frac{\epsilon}{2} = \epsilon. \qquad \blacksquare$$

To handle quotients of sequences, we first deal with reciprocals.

9.5 Lemma.

If (s_n) converges to s, if $s_n \neq 0$ for all n, and if $s \neq 0$, then $(1/s_n)$ converges to $1/s$.

Discussion. We begin by considering the equality

$$\left| \frac{1}{s_n} - \frac{1}{s} \right| = \left| \frac{s - s_n}{s_n s} \right|.$$

For large n, the numerator is small. The only possible difficulty would be if the denominator were also small for large n. This difficulty is solved in Example 6 of §8 where it is proved that $m =$

$\inf\{|s_n| : n \in \mathbb{N}\} > 0$. Thus

$$\left| \frac{1}{s_n} - \frac{1}{s} \right| \leq \frac{|s - s_n|}{m|s|},$$

and it is clear how our proof should proceed.

Proof

Let $\epsilon > 0$. By Example 6 of §8, there exists $m > 0$ such that $|s_n| \geq m$ for all n. Since $\lim s_n = s$ there exists N such that

$$n > N \quad \text{implies} \quad |s - s_n| < \epsilon \cdot m|s|.$$

Then $n > N$ implies

$$\left| \frac{1}{s_n} - \frac{1}{s} \right| = \frac{|s - s_n|}{|s_n s|} \leq \frac{|s - s_n|}{m|s|} < \epsilon. \qquad \blacksquare$$

9.6 Theorem.
Suppose that (s_n) converges to s and (t_n) converges to t. If $s \neq 0$ and $s_n \neq 0$ for all n, then (t_n/s_n) converges to t/s.

Proof
By Lemma 9.5 $(1/s_n)$ converges to $1/s$, so

$$\lim \frac{t_n}{s_n} = \lim \frac{1}{s_n} \cdot t_n = \frac{1}{s} \cdot t = \frac{t}{s}$$

by Theorem 9.4. $\qquad \blacksquare$

The preceding limit theorems and a few standard examples allow one to easily calculate many limits.

9.7 Basic Examples.
 (a) $\lim_{n \to \infty}(\frac{1}{n^p}) = 0$ for $p > 0$.
 (b) $\lim_{n \to \infty} a^n = 0$ if $|a| < 1$.
 (c) $\lim(n^{1/n}) = 1$.
 (d) $\lim_{n \to \infty}(a^{1/n}) = 1$ for $a > 0$.

Proof
 (a) Let $\epsilon > 0$ and let $N = (\frac{1}{\epsilon})^{1/p}$. Then $n > N$ implies $n^p > \frac{1}{\epsilon}$ and hence $\epsilon > \frac{1}{n^p}$. Since $\frac{1}{n^p} > 0$, this shows that $n > N$ implies

$|\frac{1}{n^p} - 0| < \epsilon$. [The meaning of n^p when p is not an integer will be discussed in §37.]

(b) We may suppose that $a \neq 0$, because $\lim_{n\to\infty} a^n = 0$ is obvious for $a = 0$. Since $|a| < 1$, we can write $|a| = \frac{1}{1+b}$ where $b > 0$. By the binomial theorem [Exercise 1.12], $(1 + b^n) \geq 1 + nb > nb$, so

$$|a^n - 0| = |a^n| = \frac{1}{(1+b)^n} < \frac{1}{nb}.$$

Now consider $\epsilon > 0$ and let $N = \frac{1}{\epsilon b}$. Then $n > N$ implies $n > \frac{1}{\epsilon b}$ and hence $|a^n - 0| < \frac{1}{nb} < \epsilon$.

(c) Let $s_n = (n^{1/n}) - 1$ and note that $s_n \geq 0$ for all n. By Theorem 9.3 it suffices to show that $\lim s_n = 0$. Since $1 + s_n = (n^{1/n})$, we have $n = (1 + s_n)^n$. For $n \geq 2$ we use the binomial expansion of $(1 + s_n)^n$ to conclude

$$n = (1 + s_n)^n \geq 1 + ns_n + \frac{1}{2}n(n-1)s_n^2 > \frac{1}{2}n(n-1)s_n^2.$$

Thus $n > \frac{1}{2}n(n-1)s_n^2$, so $s_n^2 < \frac{2}{n-1}$. Consequently, we have $s_n < \sqrt{\frac{2}{n-1}}$ for $n \geq 2$. A standard argument now shows that $\lim s_n = 0$; see Exercise 9.7.

(d) First suppose $a \geq 1$. Then for $n \geq a$ we have $1 \leq a^{1/n} \leq n^{1/n}$. Since $\lim n^{1/n} = 1$, it follows easily that $\lim(a^{1/n}) = 1$; compare Exercise 8.5(a). Suppose that $0 < a < 1$. Then $\frac{1}{a} > 1$, so $\lim(\frac{1}{a})^{1/n} = 1$ from above. Lemma 9.5 now shows that $\lim(a^{1/n}) = 1$. ∎

Example 1

Prove that $\lim s_n = \frac{1}{4}$, where

$$s_n = \frac{n^3 + 6n^2 + 7}{4n^3 + 3n - 4}.$$

Solution

We have

$$s_n = \frac{1 + \frac{6}{n} + \frac{7}{n^3}}{4 + \frac{3}{n^2} - \frac{4}{n^3}}.$$

By 9.7(a) we have $\lim \frac{1}{n} = 0$ and $\lim \frac{1}{n^3} = 0$. Hence by Theorems 9.3 and 9.2 we have

$$\lim \left(1 + \frac{6}{n} + \frac{7}{n^3} \right) = \lim(1) + 6 \cdot \lim \left(\frac{1}{n} \right) + 7 \cdot \lim \left(\frac{1}{n^3} \right) = 1.$$

Similarly, we have

$$\lim \left(4 + \frac{3}{n^2} - \frac{4}{n^3} \right) = 4.$$

Hence Theorem 9.6 implies that $\lim s_n = \frac{1}{4}$. □

Example 2
Find $\lim \frac{n-5}{n^2+7}$.

Solution
Let $s_n = \frac{n-5}{n^2+7}$. We can write s_n as $\frac{1 - \frac{5}{n}}{n + \frac{7}{n}}$, but then the denominator does not converge. So we write

$$s_n = \frac{\frac{1}{n} - \frac{5}{n^2}}{1 + \frac{7}{n^2}}.$$

Now $\lim(\frac{1}{n} - \frac{5}{n^2}) = 0$ by 9.7(a) and Theorems 9.3 and 9.2. Likewise $\lim(1 + \frac{7}{n^2}) = 1$, so Theorem 9.6 tells us that $\lim s_n = \frac{0}{1} = 0$. □

Example 3
Find $\lim \frac{n^2+3}{n+1}$.

Solution
We can write $\frac{n^2+3}{n+1}$ as

$$\frac{n + \frac{3}{n}}{1 + \frac{1}{n}} \quad \text{or} \quad \frac{1 + \frac{3}{n^2}}{\frac{1}{n} + \frac{1}{n^2}}.$$

Both fractions lead to problems: either the numerator does not converge or else the denominator converges to 0. It turns out that $\frac{n^2+3}{n+1}$ does not converge and the symbol $\lim \frac{n^2+3}{n+1}$ is undefined, at least for the present; see Example 6. The reader may have the urge to use the symbol $+\infty$ here. Our next task is to make such use of the symbol $+\infty$ legitimate. For a sequence (s_n), $\lim s_n = +\infty$ will signify that the terms s_n are eventually all large. Here is the concise definition. □

9.8 Definition.
For a sequence (s_n), we write $\lim s_n = +\infty$ provided

> for each $M > 0$ there is a number N such that
> $n > N$ implies $s_n > M$.

In this case we say that the sequence *diverges to* $+\infty$.
Similarly, we write $\lim s_n = -\infty$ provided

> for each $M < 0$ there is a number N such that
> $n > N$ implies $s_n < M$.

Henceforth we will say that (s_n) has a *limit* or that the *limit exists* provided (s_n) converges or diverges to $+\infty$ or diverges to $-\infty$. In the definition of $\lim s_n = +\infty$ the challenging values of M are large positive numbers: the larger M is the larger N will need to be. In the definition of $\lim s_n = -\infty$ the challenging values of M are "large" negative numbers like $-10,000,000,000$.

Example 4
We have $\lim n^2 = +\infty$, $\lim(-n) = -\infty$, $\lim 2^n = +\infty$ and $\lim(\sqrt{n} + 7) = +\infty$. Of course, many sequences do not have limits $+\infty$ or $-\infty$ even if they are unbounded. For example, the sequences defined by $s_n = (-1)^n n$ and $t_n = n\cos^2(\frac{n\pi}{2})$ are unbounded, but they do not diverge to $+\infty$ or $-\infty$, so the expressions $\lim[(-1)^n n]$ and $\lim[n\cos^2(\frac{n\pi}{2})]$ are meaningless. Note that $t_n = n$ when n is even and $t_n = 0$ when n is odd.

The strategy for proofs involving infinite limits is very much the same as for finite limits. We give some examples.

Example 5
Give a formal proof that $\lim(\sqrt{n} + 7) = +\infty$.

Discussion. We must consider an arbitrary $M > 0$ and show that there exists N [which will depend on M] such that

$$n > N \quad \text{implies} \quad \sqrt{n} + 7 > M.$$

To see how big N must be we "solve" for n in the inequality $\sqrt{n}+7 > M$. This inequality holds provided $\sqrt{n} > M-7$ or $n > (M-7)^2$. Thus we will take $N = (M-7)^2$.

Formal Proof
Let $M > 0$ and let $N = (M-7)^2$. Then $n > N$ implies $n > (M-7)^2$, hence $\sqrt{n} > M-7$, hence $\sqrt{n}+7 > M$. This shows that $\lim(\sqrt{n}+7) = +\infty$. ∎

Example 6
Give a formal proof that $\lim \frac{n^2+3}{n+1} = +\infty$; see Example 3.

Discussion. Consider $M > 0$. We need to determine how large n must be to guarantee that $\frac{n^2+3}{n+1} > M$. The idea is to bound the fraction $\frac{n^2+3}{n+1}$ below by some multiple of $\frac{n^2}{n} = n$; compare Example 3 of §8. Since $n^2 + 3 > n^2$ and $n+1 \le 2n$, we have $\frac{n^2+3}{n+1} > \frac{n^2}{2n} = \frac{1}{2}n$, and it suffices to arrange for $\frac{1}{2}n > M$.

Formal Proof
Let $M > 0$ and let $N = 2M$. Then $n > N$ implies $\frac{1}{2}n > M$, which implies

$$\frac{n^2 + 3}{n + 1} > \frac{n^2}{2n} = \frac{1}{2}n > M.$$

Hence $\lim \frac{n^2+3}{n+1} = +\infty$. ∎

The limit in Example 6 would be easier to handle if we could apply a limit theorem. But the limit theorems 9.2–9.6 do not apply.

WARNING. Do not attempt to apply the limit theorems 9.2–9.6 to infinite limits. Use Theorem 9.9 or 9.10 below or Exercises 9.9–9.12.

9.9 Theorem.
Let (s_n) and (t_n) be sequences such that $\lim s_n = +\infty$ and $\lim t_n > 0$ [$\lim t_n$ can be finite or $+\infty$]. Then $\lim s_n t_n = +\infty$.

Discussion. Let $M > 0$. We need to show that $s_n t_n > M$ for large n. We have $\lim s_n = +\infty$, and we need to be sure that the t_n's are bounded away from 0 for large n. We will choose a real number m

so that $0 < m < \lim t_n$ and observe that $t_n > m$ for large n. Then all we need is $s_n > \frac{M}{m}$ for large n.

Proof
Let $M > 0$. Select a real number m so that $0 < m < \lim t_n$. Whether $\lim t_n = +\infty$ or not, it is clear that there exists N_1 such that

$$n > N_1 \quad \text{implies} \quad t_n > m;$$

see Exercise 8.10. Since $\lim s_n = +\infty$, there exists N_2 so that

$$n > N_2 \quad \text{implies} \quad s_n > \frac{M}{m}.$$

Put $N = \max\{N_1, N_2\}$. Then $n > N$ implies $s_n t_n > \frac{M}{m} \cdot m = M.$ ∎

Example 7
Use Theorem 9.9 to prove that $\lim \frac{n^2+3}{n+1} = +\infty$; see Example 6.

Solution
We observe that $\frac{n^2+3}{n+1} = \frac{n+\frac{3}{n}}{1+\frac{1}{n}} = s_n t_n$ where $s_n = n + \frac{3}{n}$ and $t_n = \frac{1}{1+\frac{1}{n}}$. It is easy to show that $\lim s_n = +\infty$ and $\lim t_n = 1$. So by Theorem 9.9, we have $\lim s_n t_n = +\infty$. □

Here is another useful theorem.

9.10 Theorem.
For a sequence (s_n) of positive real numbers, we have $\lim s_n = +\infty$ if and only if $\lim(\frac{1}{s_n}) = 0$.

Proof
Let (s_n) be a sequence of positive real numbers. We have to show

$$\lim s_n = +\infty \quad \text{implies} \quad \lim\left(\frac{1}{s_n}\right) = 0 \tag{1}$$

and

$$\lim\left(\frac{1}{s_n}\right) = 0 \quad \text{implies} \quad \lim s_n = +\infty. \tag{2}$$

In this case the proofs will appear very similar, but the thought processes will be quite different.

To prove (1), suppose that $\lim s_n = +\infty$. Let $\epsilon > 0$ and let $M = \frac{1}{\epsilon}$. Since $\lim s_n = +\infty$, there exists N such that $n > N$ implies $s_n > M = \frac{1}{\epsilon}$. Therefore $n > N$ implies $\epsilon > \frac{1}{s_n} > 0$, so

$$n > N \quad \text{implies} \quad \left| \frac{1}{s_n} - 0 \right| < \epsilon.$$

That is, $\lim(\frac{1}{s_n}) = 0$. This proves (1).

To prove (2), we abandon the notation of the last paragraph and begin anew. Suppose that $\lim(\frac{1}{s_n}) = 0$. Let $M > 0$ and let $\epsilon = \frac{1}{M}$. Then $\epsilon > 0$, so there exists N such that $n > N$ implies $|\frac{1}{s_n} - 0| < \epsilon = \frac{1}{M}$. Since $s_n > 0$, we can write

$$n > N \quad \text{implies} \quad 0 < \frac{1}{s_n} < \frac{1}{M}$$

and hence

$$n > N \quad \text{implies} \quad M < s_n.$$

That is, $\lim s_n = +\infty$ and (2) holds. ∎

Exercises

9.1. Using the limit theorems 9.2–9.6 and 9.7, prove the following. Justify all steps.

(a) $\lim \frac{n+1}{n} = 1$

(b) $\lim \frac{3n+7}{6n-5} = \frac{1}{2}$

(c) $\lim \frac{17n^5+73n^4-18n^2+3}{23n^5+13n^3} = \frac{17}{23}$

9.2. Suppose that $\lim x_n = 3$, $\lim y_n = 7$ and that all y_n are nonzero. Determine the following limits:

(a) $\lim(x_n + y_n)$

(b) $\lim \frac{3y_n - x_n}{y_n^2}$

9.3. Suppose that $\lim a_n = a$, $\lim b_n = b$, and that $s_n = \frac{a_n^3+4a_n}{b_n^2+1}$. Prove $\lim s_n = \frac{a^3+4a}{b^2+1}$ carefully, using the limit theorems.

9.4. Let $s_1 = 1$ and for $n \geq 1$ let $s_{n+1} = \sqrt{s_n + 1}$.

(a) List the first four terms of (s_n).

(b) It turns out that (s_n) converges. Assume this fact and prove that the limit is $\frac{1}{2}(1 + \sqrt{5})$.

9.5. Let $t_1 = 1$ and $t_{n+1} = \frac{t_n^2 + 2}{2t_n}$ for $n \geq 1$. Assume that (t_n) converges and find the limit.

9.6. Let $x_1 = 1$ and $x_{n+1} = 3x_n^2$ for $n \geq 1$.

  (a) Show that if $a = \lim x_n$, then $a = \frac{1}{3}$ or $a = 0$.

  (b) Does $\lim x_n$ exist? Explain.

  (c) Discuss the apparent contradiction between parts (a) and (b).

9.7. Complete the proof of 9.7(c), i.e., give the standard argument needed to show that $\lim s_n = 0$.

9.8. Give the following when they exist. Otherwise assert "NOT EXIST."

  (a) $\lim n^3$ **(b)** $\lim(-n^3)$
  (c) $\lim(-n)^n$ **(d)** $\lim(1.01)^n$
  (e) $\lim n^n$

9.9. Suppose that there exists N_0 such that $s_n \leq t_n$ for all $n > N_0$.

  (a) Prove that if $\lim s_n = +\infty$, then $\lim t_n = +\infty$.

  (b) Prove that if $\lim t_n = -\infty$, then $\lim s_n = -\infty$.

  (c) Prove that if $\lim s_n$ and $\lim t_n$ exist, then $\lim s_n \leq \lim t_n$.

9.10. **(a)** Show that if $\lim s_n = +\infty$ and $k > 0$, then $\lim(ks_n) = +\infty$.

  (b) Show that $\lim s_n = +\infty$ if and only if $\lim(-s_n) = -\infty$.

  (c) Show that if $\lim s_n = +\infty$ and $k < 0$, then $\lim(ks_n) = -\infty$.

9.11. **(a)** Show that if $\lim s_n = +\infty$ and $\inf\{t_n : n \in \mathbb{N}\} > -\infty$, then $\lim(s_n + t_n) = +\infty$.

  (b) Show that if $\lim s_n = +\infty$ and $\lim t_n > -\infty$, then $\lim(s_n + t_n) = +\infty$.

  (c) Show that if $\lim s_n = +\infty$ and if (t_n) is a bounded sequence, then $\lim(s_n + t_n) = +\infty$.

9.12. Assume all $s_n \neq 0$ and that the limit $L = \lim \left|\frac{s_{n+1}}{s_n}\right|$ exists.

  (a) Show that if $L < 1$, then $\lim s_n = 0$. Hint: Select a so that $L < a < 1$ and obtain N so that $|s_{n+1}| < a|s_n|$ for $n \geq N$. Then show that $|s_n| < a^{n-N}|s_N|$ for $n > N$.

  (b) Show that if $L > 1$, then $\lim |s_n| = +\infty$. Hint: Apply (a) to the sequence $t_n = \frac{1}{|s_n|}$; see Theorem 9.10.

9.13. Show that

$$\lim_{n\to\infty} a^n = \begin{cases} 0 & \text{if } |a| < 1 \\ 1 & \text{if } a = 1 \\ +\infty & \text{if } a > 1 \\ \text{does not exist} & \text{if } a \le -1. \end{cases}$$

9.14. Let $p > 0$. Use Exercise 9.12 to show

$$\lim_{n\to\infty} \frac{a^n}{n^p} = \begin{cases} 0 & \text{if } |a| \le 1 \\ +\infty & \text{if } a > 1 \\ \text{does not exist} & \text{if } a < -1. \end{cases}$$

9.15. Show that $\lim_{n\to\infty} \frac{a^n}{n!} = 0$ for all $a \in \mathbb{R}$.

9.16. Use Theorems 9.9, 9.10 or Exercises 9.9–9.15 to prove the following:

(a) $\lim \frac{n^4+8n}{n^2+9} = +\infty$

(b) $\lim[\frac{2^n}{n^2} + (-1)^n] = +\infty$

(c) $\lim[\frac{3^n}{n^3} - \frac{3^n}{n!}] = +\infty$

9.17. Give a formal proof that $\lim n^2 = +\infty$ using only Definition 9.8.

9.18. (a) Verify $1 + a + a^2 + \cdots + a^n = \frac{1-a^{n+1}}{1-a}$ for $a \ne 1$.

(b) Find $\lim_{n\to\infty}(1 + a + a^2 + \cdots + a^n)$ for $|a| < 1$.

(c) Calculate $\lim_{n\to\infty}(1 + \frac{1}{3} + \frac{1}{9} + \frac{1}{27} + \cdots + \frac{1}{3^n})$.

(d) What is $\lim_{n\to\infty}(1 + a + a^2 + \cdots + a^n)$ for $a \ge 1$?

§10 Monotone Sequences and Cauchy Sequences

In this section we obtain two theorems [Theorems 10.2 and 10.11] that will allow us to conclude that certain sequences converge *without* knowing the limit in advance. These theorems are important because in practice the limits are not usually known in advance.

10.1 Definition.
A sequence (s_n) of real numbers is called a *nondecreasing sequence* if $s_n \le s_{n+1}$ for all n, and (s_n) is called a *nonincreasing sequence*

if $s_n \geq s_{n+1}$ for all n. Note that if (s_n) is nondecreasing, then $s_n \leq s_m$ whenever $n < m$. A sequence that is nondecreasing or nonincreasing will be called a *monotone sequence* or a *monotonic sequence*.

Example 1
The sequences defined by $a_n = 1 - \frac{1}{n}$, $b_n = n^3$ and $c_n = (1 + \frac{1}{n})^n$ are nondecreasing sequences, although this is not obvious for the sequence (c_n). The sequence $d_n = \frac{1}{n^2}$ is nonincreasing. The sequences $s_n = (-1)^n$, $t_n = \cos(\frac{n\pi}{3})$, $u_n = (-1)^n n$ and $v_n = \frac{(-1)^n}{n}$ are not monotonic sequences. Also $x_n = n^{1/n}$ is not monotonic, as can be seen by examining the first four values; see Example 1(d) in §7.

Of the sequences above, (a_n), (c_n), (d_n), (s_n), (t_n), (v_n) and (x_n) are bounded sequences. The remaining sequences, (b_n) and (u_n), are unbounded sequences.

10.2 Theorem.
All bounded monotone sequences converge.

Proof
Let (s_n) be a bounded nondecreasing sequence. Let S denote the set $\{s_n : n \in \mathbb{N}\}$, and let $u = \sup S$. Since S is bounded, u represents a real number. We show that $\lim s_n = u$. Let $\epsilon > 0$. Since $u - \epsilon$ is not an upper bound for S, there exists N such that $s_N > u - \epsilon$. Since (s_n) is nondecreasing, we have $s_N \leq s_n$ for all $n \geq N$. Of course, $s_n \leq u$ for all n, so $n > N$ implies $u - \epsilon < s_n \leq u$, which implies $|s_n - u| < \epsilon$. This shows that $\lim s_n = u$.

The proof for bounded nonincreasing sequences is left to Exercise 10.2. ∎

Note that the Completeness Axiom 4.4 is a vital ingredient in the proof of Theorem 10.2.

10.3 Discussion of Decimals.
We have not given much attention to the notion that real numbers are simply decimal expansions. This notion is substantially correct, but there are subtleties to be faced. For example, different decimal expansions can represent the same real number. The somewhat more

abstract developments of the set $\mathbb{R}$ of real numbers discussed in §6 turn out to be more satisfactory.

We restrict our attention to nonnegative decimal expansions and nonnegative real numbers. From our point of view, every nonnegative decimal expansion is shorthand for the limit of a bounded nondecreasing sequence of real numbers. Suppose we are given a decimal expansion $k.d_1 d_2 d_3 d_4 \cdots$ where k is a nonnegative integer and each d_j belongs to $\{0, 1, 2, 3, 4, 5, 6, 7, 8, 9\}$. Let

$$s_n = k + \frac{d_1}{10} + \frac{d_2}{10^2} + \cdots + \frac{d_n}{10^n}. \tag{1}$$

Then (s_n) is a nondecreasing sequence of real numbers, and (s_n) is bounded [by $k + 1$, in fact]. So by Theorem 10.2, (s_n) converges to a real number that we traditionally write as $k.d_1 d_2 d_3 d_4 \cdots$. For example, $3.3333 \cdots$ represents

$$\lim_{n \to \infty} \left(3 + \frac{3}{10} + \frac{3}{10^2} + \cdots + \frac{3}{10^n} \right).$$

To calculate this limit, we borrow the following fact about geometric series from Example 1 in §14:

$$\lim_{n \to \infty} a(1 + r + r^2 + \cdots + r^n) = \frac{a}{1 - r} \quad \text{for} \quad |r| < 1; \tag{2}$$

see also Exercise 9.18. In our case, $a = 3$ and $r = \frac{1}{10}$, so $3.3333 \cdots$ represents $\frac{3}{1 - \frac{1}{10}} = \frac{10}{3}$, as expected. Similarly, $0.9999 \cdots$ represents

$$\lim_{n \to \infty} \left(\frac{9}{10} + \frac{9}{10^2} + \cdots + \frac{9}{10^n} \right) = \frac{\frac{9}{10}}{1 - \frac{1}{10}} = 1.$$

Thus $0.9999 \cdots$ and $1.0000 \cdots$ are different decimal expansions that represent the same real number!

The converse of the preceding discussion also holds. That is, every nonnegative real number x has at least one decimal expansion. This will be proved, along with some related results, in the optional §16.

Unbounded monotone sequences also have limits.

10.4 Theorem.

  (i) *If (s_n) is an unbounded nondecreasing sequence, then $\lim s_n = +\infty$.*

  (ii) *If (s_n) is an unbounded nonincreasing sequence, then $\lim s_n = -\infty$.*

Proof

(i) Let (s_n) be an unbounded nondecreasing sequence. Let $M > 0$. Since the set $\{s_n : n \in \mathbb{N}\}$ is unbounded and it is bounded below by s_1, it must be unbounded above. Hence for some N in $\mathbb{N}$ we have $s_N > M$. Clearly $n > N$ implies $s_n \geq s_N > M$, so $\lim s_n = +\infty$.

  (ii) The proof is similar and is left to Exercise 10.5. ∎

10.5 Corollary.

If (s_n) is a monotone sequence, then the sequence either converges, diverges to $+\infty$, or diverges to $-\infty$. Thus $\lim s_n$ is always meaningful for monotone sequences.

Proof

Apply Theorems 10.2 and 10.4. ∎

Let (s_n) be a bounded sequence in $\mathbb{R}$; it may or may not converge. It is apparent from the definition of limit in 7.1 that the limiting behavior of (s_n) depends only on sets of the form $\{s_n : n > N\}$. For example, if $\lim s_n$ exists, clearly it must lie in the interval $[u_N, v_N]$ where

$$u_N = \inf\{s_n : n > N\} \quad \text{and} \quad v_N = \sup\{s_n : n > N\};$$

see Exercise 8.9. As N increases, the sets $\{s_n : n > N\}$ get smaller, so we have

$$u_1 \leq u_2 \leq u_3 \leq \cdots \quad \text{and} \quad v_1 \geq v_2 \geq v_3 \geq \cdots;$$

see Exercise 4.7(a). By Theorem 10.2 the limits $u = \lim_{N \to \infty} u_N$ and $v = \lim_{N \to \infty} v_N$ both exist, and $u \leq v$ since $u_N \leq v_N$ for all N. If $\lim s_n$ exists then, as noted above, $u_N \leq \lim s_n \leq v_N$ for all N, so we must have $u \leq \lim s_n \leq v$. The numbers u and v are useful whether $\lim s_n$ exists or not and are denoted $\liminf s_n$ and $\limsup s_n$, respectively.

10.6 Definition.
Let (s_n) be a sequence in $\mathbb{R}$. We define

$$\limsup s_n = \lim_{N\to\infty} \sup\{s_n : n > N\} \tag{1}$$

and

$$\liminf s_n = \lim_{N\to\infty} \inf\{s_n : n > N\}. \tag{2}$$

Note that in this definition we do not restrict (s_n) to be bounded. However, we adopt the following conventions. If (s_n) is not bounded above, $\sup\{s_n : n > N\} = +\infty$ for all N and we decree $\limsup s_n = +\infty$. Likewise, if (s_n) is not bounded below, $\inf\{s_n : n > N\} = -\infty$ for all N and we decree $\liminf s_n = -\infty$.

We emphasize that $\limsup s_n$ need not equal $\sup\{s_n : n \in \mathbb{N}\}$, but that $\limsup s_n \leq \sup\{s_n : n \in \mathbb{N}\}$. Some of the values s_n may be much larger than $\limsup s_n$; $\limsup s_n$ is the largest value that *infinitely many* s_n's can get close to. Similar remarks apply to $\liminf s_n$. These remarks will be clarified in Theorem 11.7 and §12, where we will give a thorough treatment of $\liminf$'s and $\limsup$'s. For now, we need a theorem that shows (s_n) has a limit if and only if $\liminf s_n = \limsup s_n$.

10.7 Theorem.
Let (s_n) be a sequence in $\mathbb{R}$.
  - **(i)** *If $\lim s_n$ is defined [as a real number, $+\infty$ or $-\infty$], then $\liminf s_n = \lim s_n = \limsup s_n$.*
  - **(ii)** *If $\liminf s_n = \limsup s_n$, then $\lim s_n$ is defined and $\lim s_n = \liminf s_n = \limsup s_n$.*

Proof
We use the notation $u_N = \inf\{s_n : n > N\}$, $v_N = \sup\{s_n : n > N\}$, $u = \lim u_N = \liminf s_n$ and $v = \lim v_N = \limsup s_n$.
  - **(i)** Suppose $\lim s_n = +\infty$. Let M be a positive real number. Then there is a natural number N so that

$$n > N \quad \text{implies} \quad s_n > M.$$

Then $u_N = \inf\{s_n : n > N\} \geq M$. It follows that $m > N$ implies $u_m \geq M$. In other words, the sequence (u_N) satisfies the condition defining $\lim u_N = +\infty$, i.e., $\liminf s_n = +\infty$. Likewise $\limsup s_n = +\infty$.

The case $\lim s_n = -\infty$ is handled in a similar manner.

Now suppose that $\lim s_n = s$ where s is a real number. Consider $\epsilon > 0$. There exists a natural number N such that $|s_n - s| < \epsilon$ for $n > N$. Thus $s_n < s + \epsilon$ for $n > N$, so

$$v_N = \sup\{s_n : n > N\} \le s + \epsilon.$$

Also, $m > N$ implies $v_m \le s + \epsilon$, so $\limsup s_n = \lim v_m \le s + \epsilon$. Since $\limsup s_n \le s + \epsilon$ for all $\epsilon > 0$, no matter how small, we conclude that $\limsup s_n \le s = \lim s_n$. A similar argument shows that $\lim s_n \le \liminf s_n$. Since $\liminf s_n \le \limsup s_n$, we infer that all three numbers are equal:

$$\liminf s_n = \lim s_n = \limsup s_n.$$

(ii) If $\liminf s_n = \limsup s_n = +\infty$ it is easy to show that $\lim s_n = +\infty$. And if $\liminf s_n = \limsup s_n = -\infty$ it is easy to show that $\lim s_n = -\infty$. We leave these two special cases to the reader.

Suppose, finally, that $\liminf s_n = \limsup s_n = s$ where s is a real number. We need to prove that $\lim s_n = s$. Let $\epsilon > 0$. Since $s = \lim v_N$ there exists a natural number N_0 such that

$$|s - \sup\{s_n : n > N_0\}| < \epsilon.$$

Thus $\sup\{s_n : n > N_0\} < s + \epsilon$, so

$$s_n < s + \epsilon \quad \text{for all} \quad n > N_0. \tag{1}$$

Similarly, since $s = \lim u_N$ there exists N_1 such that

$$|s - \inf\{s_n : n > N_1\}| < \epsilon,$$

hence $\inf\{s_n : n > N_1\} > s - \epsilon$, hence

$$s_n > s - \epsilon \quad \text{for all} \quad n > N_1. \tag{2}$$

From (1) and (2) we conclude

$$s - \epsilon < s_n < s + \epsilon \quad \text{for} \quad n > \max\{N_0, N_1\},$$

equivalently

$$|s_n - s| < \epsilon \quad \text{for} \quad n > \max\{N_0, N_1\}.$$

This proves that $\lim s_n = s$ as desired. ∎

If (s_n) converges, then $\lim \inf s_n = \lim \sup s_n$ by the theorem just proved, so for large N the numbers $\sup\{s_n : n > N\}$ and $\inf\{s_n : n > N\}$ must be close together. This implies that all the numbers in the set $\{s_n : n > N\}$ must be close to each other. This leads us to a concept of great theoretical importance that will be used throughout the book.

10.8 Definition.
A sequence (s_n) of real numbers is called a *Cauchy sequence* if

> for each $\epsilon > 0$ there exists a number N such that
>
> $m, n > N$ implies $|s_n - s_m| < \epsilon$. $\qquad\qquad$ (1)

Compare this definition with Definition 7.1.

10.9 Lemma.
Convergent sequences are Cauchy sequences.

Proof
Suppose that $\lim s_n = s$. The idea is that, since the terms s_n are close to s for large n, they also must be close to each other; indeed

$$|s_n - s_m| = |s_n - s + s - s_m| \leq |s_n - s| + |s - s_m|.$$

To be precise, let $\epsilon > 0$. Then there exists N such that

$$n > N \quad \text{implies} \quad |s_n - s| < \frac{\epsilon}{2}.$$

Clearly we may also write

$$m > N \quad \text{implies} \quad |s_m - s| < \frac{\epsilon}{2},$$

so

$$m, n > N \quad \text{implies} \quad |s_n - s_m| \leq |s_n - s| + |s - s_m| < \frac{\epsilon}{2} + \frac{\epsilon}{2} = \epsilon.$$

Thus (s_n) is a Cauchy sequence. ∎

10.10 Lemma.
Cauchy sequences are bounded.

Proof
The proof is similar to that of Theorem 9.1. Applying Definition 10.8 with $\epsilon = 1$ we obtain N in $\mathbb{N}$ so that

$$m, n > N \quad \text{implies} \quad |s_n - s_m| < 1.$$

In particular, $|s_n - s_{N+1}| < 1$ for $n > N$, so $|s_n| < |s_{N+1}| + 1$ for $n > N$. If $M = \max\{|s_{N+1}| + 1, |s_1|, |s_2|, \ldots, |s_N|\}$, then $|s_n| \le M$ for all $n \in \mathbb{N}$. ∎

The importance of the next theorem is the following consequence: To verify that a sequence converges it suffices to check that it is a Cauchy sequence, a property that does not involve the limit itself.

10.11 Theorem.
A sequence is a convergent sequence if and only if it is a Cauchy sequence.

Proof
The expression "if and only if" indicates that we have two assertions to verify: (i) convergent sequences are Cauchy sequences, and (ii) Cauchy sequences are convergent sequences. We already verified (i) in Lemma 10.9. To check (ii), consider a Cauchy sequence (s_n) and note that (s_n) is bounded by Lemma 10.10. By Theorem 10.7 we need only show

$$\liminf s_n = \limsup s_n. \tag{1}$$

Let $\epsilon > 0$. Since (s_n) is a Cauchy sequence, there exists N so that

$$m, n > N \quad \text{implies} \quad |s_n - s_m| < \epsilon.$$

In particular, $s_n < s_m + \epsilon$ for all $m, n > N$. This shows that $s_m + \epsilon$ is an upper bound for $\{s_n : n > N\}$, so $v_N = \sup\{s_n : n > N\} \le s_m + \epsilon$ for $m > N$. This, in turn, shows that $v_N - \epsilon$ is a lower bound for $\{s_m : m > N\}$, so $v_N - \epsilon \le \inf\{s_m : m > N\} = u_N$. Thus

$$\limsup s_n \le v_N \le u_N + \epsilon \le \liminf s_n + \epsilon.$$

Since this holds for all $\epsilon > 0$, we have $\limsup s_n \le \liminf s_n$. The opposite inequality always holds, so we have established (1). ∎

The proof of Theorem 10.11 uses Theorem 10.7, and Theorem 10.7 relies implicitly on the Completeness Axiom 4.4, since without the completeness axiom it is not clear that $\liminf s_n$ and $\limsup s_n$ are meaningful. The completeness axiom assures us that the expressions $\sup\{s_n : n > N\}$ and $\inf\{s_n : n > N\}$ in Definition 10.6 are meaningful, and Theorem 10.2 [which itself relies on the completeness axiom] assures us that the limits in Definition 10.6 also are meaningful.

Exercises on lim sup's and lim inf's appear in §§11 and 12.

Exercises

10.1. Which of the following sequences are nondecreasing? nonincreasing? bounded?

(a) $\frac{1}{n}$ (b) $\frac{(-1)^n}{n^2}$

(c) n^5 (d) $\sin(\frac{n\pi}{7})$

(e) $(-2)^n$ (f) $\frac{n}{3^n}$

10.2. Prove Theorem 10.2 for bounded nonincreasing sequences.

10.3. For a decimal expansion $k.d_1d_2d_3d_4\cdots$, let (s_n) be defined as in 10.3. Prove that $s_n < k+1$ for all $n \in \mathbb{N}$. Hint: $\frac{9}{10} + \frac{9}{10^2} + \cdots + \frac{9}{10^n} = 1 - \frac{1}{10^n}$ for all n.

10.4. Discuss why Theorems 10.2 and 10.11 would fail if we restricted our world of numbers to the set $\mathbb{Q}$ of rational numbers.

10.5. Prove Theorem 10.4(ii).

10.6. (a) Let (s_n) be a sequence such that

$$|s_{n+1} - s_n| < 2^{-n} \quad \text{for all} \quad n \in \mathbb{N}.$$

Prove that (s_n) is a Cauchy sequence and hence a convergent sequence.

(b) Is the result in (a) true if we only assume that $|s_{n+1} - s_n| < \frac{1}{n}$ for all $n \in \mathbb{N}$?

10.7. Let S be a bounded nonempty subset of $\mathbb{R}$ and suppose $\sup S \notin S$. Prove that there is a nondecreasing sequence (s_n) of points in S such that $\lim s_n = \sup S$.

10.8. Let (s_n) be a nondecreasing sequence of positive numbers and define $\sigma_n = \frac{1}{n}(s_1 + s_2 + \cdots + s_n)$. Prove that (σ_n) is a nondecreasing sequence.

10.9. Let $s_1 = 1$ and $s_{n+1} = (\frac{n}{n+1})s_n^2$ for $n \geq 1$.

  (a) Find s_2, s_3 and s_4.

  (b) Show that $\lim s_n$ exists.

  (c) Prove that $\lim s_n = 0$.

10.10. Let $s_1 = 1$ and $s_{n+1} = \frac{1}{3}(s_n + 1)$ for $n \geq 1$.

  (a) Find s_2, s_3 and s_4.

  (b) Use induction to show that $s_n > \frac{1}{2}$ for all n.

  (c) Show that (s_n) is a nonincreasing sequence.

  (d) Show that $\lim s_n$ exists and find $\lim s_n$.

10.11. Let $t_1 = 1$ and $t_{n+1} = [1 - \frac{1}{4n^2}] \cdot t_n$ for $n \geq 1$.

  (a) Show that $\lim t_n$ exists.

  (b) What do you think $\lim t_n$ is?

10.12. Let $t_1 = 1$ and $t_{n+1} = [1 - \frac{1}{(n+1)^2}] \cdot t_n$ for $n \geq 1$.

  (a) Show that $\lim t_n$ exists.

  (b) What do you think $\lim t_n$ is?

  (c) Use induction to show that $t_n = \frac{n+1}{2n}$.

  (d) Repeat part (b).

§11 Subsequences

11.1 Definition.
Suppose that $(s_n)_{n \in \mathbb{N}}$ is a sequence. A *subsequence* of this sequence is a sequence of the form $(t_k)_{k \in \mathbb{N}}$ where for each k there is a positive integer n_k such that

$$n_1 < n_2 < \cdots < n_k < n_{k+1} < \cdots \tag{1}$$

and

$$t_k = s_{n_k}. \tag{2}$$

Thus (t_k) is just a selection of some [possibly all] of the s_n's taken in order.

Here are some alternative ways to approach this concept. Note that (1) defines an infinite subset of $\mathbb{N}$, namely $\{n_1, n_2, n_3, \ldots\}$. Conversely, every infinite subset of $\mathbb{N}$ can be described by (1). Thus a subsequence of (s_n) is a sequence obtained by selecting, in order, an infinite subset of the terms.

For a more concise definition, recall that we can view the sequence $(s_n)_{n \in \mathbb{N}}$ as a function s with domain $\mathbb{N}$; see §7. For the subset $\{n_1, n_2, n_3, \ldots\}$, there is a natural function σ [lower case Greek sigma] given by $\sigma(k) = n_k$ for $k \in \mathbb{N}$. The function σ "selects" an infinite subset of $\mathbb{N}$, in order. The subsequence of s corresponding to σ is simply the composite function $t = s \circ \sigma$. That is,

$$t_k = t(k) = s \circ \sigma(k) = s(\sigma(k)) = s(n_k) = s_{n_k} \quad \text{for} \quad k \in \mathbb{N}. \tag{3}$$

Thus a sequence t is a subsequence of a sequence s if and only if $t = s \circ \sigma$ for some increasing function σ mapping $\mathbb{N}$ into $\mathbb{N}$. We will usually suppress the notation σ and often suppress the notation t also. Thus the phrase "a subsequence (s_{n_k}) of (s_n)" will refer to the subsequence defined by (1) and (2) or by (3), depending upon your point of view.

Example 1
Let (s_n) be the sequence defined by $s_n = n^2(-1)^n$. The positive terms of this sequence comprise a subsequence. In this case, the sequence (s_n) is

$$(-1, 4, -9, 16, -25, 36, -49, 64, \ldots)$$

and the subsequence is

$$(4, 16, 36, 64, 100, 144, \ldots).$$

More precisely, the subsequence is $(s_{n_k})_{k \in \mathbb{N}}$ where $n_k = 2k$ so that $s_{n_k} = (2k)^2(-1)^{2k} = 4k^2$. The selection function σ is given by $\sigma(k) = 2k$.

Example 2

Consider the sequence $a_n = \sin(\frac{n\pi}{3})$ and its subsequence (a_{n_k}) of nonnegative terms. The sequence $(a_n)_{n \in \mathbb{N}}$ is

$$(\frac{1}{2}\sqrt{3}, \frac{1}{2}\sqrt{3}, 0, -\frac{1}{2}\sqrt{3}, -\frac{1}{2}\sqrt{3}, 0, \frac{1}{2}\sqrt{3}, \frac{1}{2}\sqrt{3}, 0, -\frac{1}{2}\sqrt{3}, -\frac{1}{2}\sqrt{3}, 0, \ldots)$$

and the desired subsequence is

$$(\frac{1}{2}\sqrt{3}, \frac{1}{2}\sqrt{3}, 0, 0, \frac{1}{2}\sqrt{3}, \frac{1}{2}\sqrt{3}, 0, 0, \ldots).$$

It is evident that $n_1 = 1$, $n_2 = 2$, $n_3 = 3$, $n_4 = 6$, $n_5 = 7$, $n_6 = 8$, $n_7 = 9$, $n_8 = 12$, $n_9 = 13$, etc. We could obtain a general formula for n_k, but the project does not seem worth the effort.

Example 3

It can be shown that the set $\mathbb{Q}$ of rational numbers can be listed as a sequence (r_n), though it is tedious to specify an exact formula. Figure 11.1 suggests such a listing [with repetitions] where $r_1 = 0$, $r_2 = 1$, $r_3 = \frac{1}{2}$, $r_4 = -\frac{1}{2}$, $r_5 = -1$, $r_6 = -2$, $r_7 = -1$, etc. Readers familiar with some set theory will recognize this assertion as the fact that "$\mathbb{Q}$ is countable." This sequence has an amazing property: given any real number a there exists a subsequence (r_{n_k}) of (r_n) that converges to a, i.e., $\lim_{k \to \infty} r_{n_k} = a$. To see this, we will show how to define or construct step-by-step a subsequence (r_{n_k}) that satisfies

$$|r_{n_k} - a| < \frac{1}{k} \quad \text{for} \quad k \in \mathbb{N}. \tag{1}$$

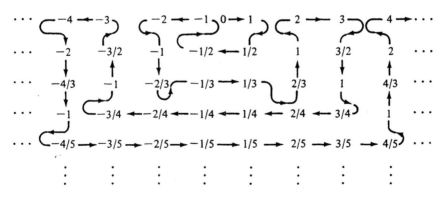

FIGURE 11.1

Specifically, we will assume $n_1, n_2, \ldots, n_k$ have been selected satisfying (1) and show how to select n_{k+1}. It is fairly evident that this will give us an infinite sequence $(n_k)_{k \in \mathbb{N}}$ and hence a subsequence (r_{n_k}) of (r_n) satisfying (1). To make this fully rigorous would require a technical lemma concerning step-by-step constructions whose proof depends in the end on Peano's axiom N5. For this reason, a construction of this sort is called an "inductive definition" or "definition by induction."

We now indicate the construction discussed above. Select n_1 so that $|r_{n_1} - a| < 1$; this is possible by the Denseness of $\mathbb{Q}$ 4.7. Suppose that $n_1, n_2, \ldots, n_k$ have been selected so that

$$n_1 < n_2 < \cdots < n_k \tag{2}$$

and

$$|r_{n_j} - a| < \frac{1}{j} \quad \text{for} \quad j = 1, 2, \ldots, k. \tag{3}$$

Since there are infinitely many rational numbers in the interval $(a - \frac{1}{k+1}, a + \frac{1}{k+1})$ by Exercise 4.11, there must exist an $n_{k+1} > n_k$ such that $r_{n_{k+1}}$ belongs to this interval. Then $|r_{n_{k+1}} - a| < \frac{1}{k+1}$ and hence (2) and (3) hold for $k+1$ in place of k. The procedure defines $(n_k)_{k \in \mathbb{N}}$ by induction. Since (3) holds, (1) holds and we conclude that $\lim_{k \to \infty} r_{n_k} = a$.

Example 4

Suppose that (s_n) is a sequence of positive numbers such that $\inf\{s_n : n \in \mathbb{N}\} = 0$. The sequence (s_n) need not converge or even be bounded, but it has a subsequence that converges monotonically to 0. We will again give an inductive construction. Since $\inf\{s_n : n \in \mathbb{N}\} = 0$, there exists $n_1 \in \mathbb{N}$ such that $s_{n_1} < 1$. Suppose that $n_1, n_2, \ldots, n_k$ have been selected so that

$$n_1 < n_2 < \cdots < n_k \tag{1}$$

and

$$s_{n_{j+1}} < \min\left\{s_{n_j}, \frac{1}{j+1}\right\} \quad \text{for} \quad j = 1, 2, \ldots, k-1. \tag{2}$$

Note that we are requiring $s_{n_{j+1}} < s_{n_j}$ so that our subsequence will be monotonic, and we are requiring $s_{n_{j+1}} < \frac{1}{j+1}$ to guarantee that it will converge to 0. Since $\min\{s_n : 1 \leq n \leq n_k\} > 0$, it follows that $\inf\{s_n : n > n_k\} = 0$. Thus there exists $n_{k+1} > n_k$ such that $s_{n_{k+1}} < \min\{s_{n_k}, \frac{1}{k+1}\}$. Hence (1) and (2) hold for $k + 1$ in place of k, and the construction continues by induction. As noted above, (2) shows that (s_{n_k}) converges monotonically to 0.

The next theorem is almost obvious.

11.2 Theorem.
If the sequence (s_n) converges, then every subsequence converges to the same limit.

Proof
Let (s_{n_k}) denote a subsequence of (s_n). Note that $n_k \geq k$ for all k. This is easy to prove by induction; in fact, $n_1 \geq 1$ and $n_k \geq k$ implies $n_{k+1} > n_k \geq k$ and hence $n_{k+1} \geq k + 1$.

Let $s = \lim s_n$ and let $\epsilon > 0$. There exists N so that $n > N$ implies $|s_n - s| < \epsilon$. Now $k > N$ implies $n_k > N$, which implies $|s_{n_k} - s| < \epsilon$. Thus

$$\lim_{k \to \infty} s_{n_k} = s.$$

∎

Our immediate goal is to prove the Bolzano-Weierstrass theorem which asserts that every bounded sequence has a convergent subsequence. First we prove a theorem about monotonic subsequences.

11.3 Theorem.
Every sequence (s_n) has a monotonic subsequence.

Proof
Let's say that the n-th term is *dominant* if it is greater than every term which follows it:

$$s_m < s_n \quad \text{for all} \quad m > n. \tag{1}$$

Case 1. Suppose that there are infinitely many dominant terms, and let (s_{n_k}) be any subsequence consisting solely of dominant terms. Then $s_{n_{k+1}} < s_{n_k}$ for all k by (1), so (s_{n_k}) is a decreasing sequence.

Case 2. Suppose that there are only finitely many dominant terms. Select n_1 so that s_{n_1} is beyond all the dominant terms of the sequence. Then

$$\text{given } N \geq n_1 \text{ there exists } m > N \text{ such that } \quad s_m \geq s_N. \qquad (2)$$

Applying (2) with $N = n_1$ we select $n_2 > n_1$ such that $s_{n_2} \geq s_{n_1}$. Suppose that $n_1, n_2, \ldots, n_k$ have been selected so that

$$n_1 < n_2 < \cdots < n_k \qquad (3)$$

and

$$s_{n_1} \leq s_{n_2} \leq \cdots \leq s_{n_k}. \qquad (4)$$

Applying (2) with $N = n_k$ we select $n_{k+1} > n_k$ such that $s_{n_{k+1}} \geq s_{n_k}$. Then (3) and (4) hold with $k+1$ in place of k, the procedure continues by induction, and we obtain a nondecreasing subsequence (s_{n_k}). ■

The elegant proof in Theorem 11.3 was brought to our attention by David M. Bloom and is based on a solution in D. J. Newman's beautiful book *A Problem Seminar*, Springer-Verlag, New York-Berlin-Heidelberg: 1982.

11.4 Corollary.
Let (s_n) be any sequence. There exists a monotonic subsequence whose limit is lim sup s_n, *and there exists a monotonic subsequence whose limit is* lim inf s_n.

Proof
For N in $\mathbb{N}$, let $v_N = \sup\{s_n : n > N\}$. Then $v_1 \geq v_2 \geq v_3 \geq \cdots$ and $v = \lim_N v_N = $ lim sup s_n; see Definition 10.6. Our task is to show that there is a monotonic subsequence of (s_n) that converges to v. If $v = -\infty$, then lim $s_n = -\infty$ by Theorem 10.7, the sequence (s_n) itself converges to lim sup s_n, and so a monotonic subsequence of (s_n) converges to lim sup s_n by Theorem 11.3. Henceforth we assume that $v \neq -\infty$.

First look at Case 1 of the previous proof. Then $s_{n_k} = \sup\{s_n : n \geq n_k\} = v_{n_k-1}$, so $\lim_{k \to \infty} s_{n_k} = \lim_N v_N = $ lim sup s_n, as desired.

Suppose now that there are only finitely many dominant terms, and let s_{m_0} be the last dominant term. We claim

$$\sup\{s_n : n > N\} = v_N = v \quad \text{for} \quad N > m_0. \qquad (1)$$

Otherwise, since (v_N) is a nonincreasing sequence, there is an integer $N > m_0$ so that $v_{N+1} < v_N$. So s_{N+1} must be bigger than

$v_{N+1} = \sup\{s_n : n \geq N + 2\}$, but this implies that s_{N+1} is dominant, contrary to the choice of m_0.

If infinitely many s_n equal v, simply select a subsequence of (s_n) consisting of terms equal to v. Otherwise, there exists $n_1 \geq m_0$ so that

$$s_n < v \quad \text{for all} \quad n \geq n_1. \tag{2}$$

Select a sequence (t_N) that increases to v. In fact, $t_N = v - \frac{1}{N}$ will do if v is finite, and $t_N = N$ will do if $v = +\infty$. The desired subsequence is obtained by induction, the first term being s_{n_1}. Note that, by (2), we have $s_{n_1} < v$. Assume that $n_1 < n_2 < \cdots < n_k$ have been selected so that

$$s_{n_1} < s_{n_2} < \cdots < s_{n_k} < v, \tag{3}$$

$$\text{and} \quad t_k < s_{n_k} \quad \text{for} \quad k \geq 2. \tag{4}$$

Using (1), we select $n_{k+1} > n_k$ so that $s_{n_{k+1}} > s_{n_k}$ and $s_{n_{k+1}} > t_{k+1}$. By (2), we also have $s_{n_{k+1}} < v$. The procedure continues by induction, and by (3) we obtain an increasing subsequence of (s_n). Also, by (4), we have $t_k < s_{n_k} < v$ for all k, so $v = \lim_{k \to \infty} t_k \leq \lim_{k \to \infty} s_{n_k} \leq v$, and the subsequence (s_{n_k}) converges to v as desired.

The assertion about $\liminf s_n$ has a similar proof, but it also can be derived from the first assertion; see Exercise 11.8. This revised proof is based on correspondence with Ray Hoobler, City College, CUNY. ∎

11.5 Bolzano-Weierstrass Theorem.
Every bounded sequence has a convergent subsequence.

Proof
If (s_n) is a bounded sequence, it has a monotonic subsequence by Theorem 11.3. The subsequence converges by Theorem 10.2. ∎

The Bolzano-Weierstrass theorem is very important and will be used at critical points in Chapter 3. Our proof, based on Theorem 11.3, is somewhat nonstandard for reasons we now discuss. Many of the notions introduced in this chapter make equally good sense in more general settings. For example, the ideas of convergent sequence, Cauchy sequence and bounded sequence all make sense for a sequence (s_n) where each s_n belongs to the plane. But the idea of a monotonic sequence does not carry over. It turns out that the Bolzano-Weierstrass theorem also holds in the plane and in many other settings [see Theorem 13.5], but clearly it would no longer be appropriate to prove it directly from an analogue of Theorem 11.3. Since the Bolzano-Weierstrass Theorem 11.5 generalizes to settings where Theorem 11.3 makes little sense, in applications we will emphasize 11.5 rather than 11.3.

We need one more notion, and then we will be able to tie our various concepts together in Theorem 11.7.

11.6 Definition.
Let (s_n) be a sequence in $\mathbb{R}$. A *subsequential limit* is any real number or symbol $+\infty$ or $-\infty$ that is the limit of some subsequence of (s_n).

When a sequence has a limit s, then all subsequences have limit s, so $\{s\}$ is the set of subsequential limits. The interesting case is when the original sequence does not have a limit. We return to some of the examples discussed after Definition 11.1.

Example 5
Consider (s_n) where $s_n = n^2(-1)^n$. The subsequence of even terms diverges to $+\infty$, and the subsequence of odd terms diverges to $-\infty$. All subsequences that have a limit diverge to $+\infty$ or $-\infty$, so that $\{-\infty, +\infty\}$ is exactly the set of subsequential limits of (s_n).

Example 6
Consider the sequence $a_n = \sin(\frac{n\pi}{3})$. This sequence takes each of the values $\frac{1}{2}\sqrt{3}$, 0 and $-\frac{1}{2}\sqrt{3}$ an infinite number of times. The only convergent subsequences are constant from some term on, and $\{-\frac{1}{2}\sqrt{3}, 0, \frac{1}{2}\sqrt{3}\}$ is the set of subsequential limits of (a_n). If $n_k = 3k$, then $a_{n_k} = 0$ for all $k \in \mathbb{N}$ and obviously $\lim_{k \to \infty} a_{n_k} = 0$. If $n_k = 6k + 1$, then $a_{n_k} = \frac{1}{2}\sqrt{3}$ for all k and $\lim_{k \to \infty} a_{n_k} = \frac{1}{2}\sqrt{3}$. And if $n_k = 6k + 4$, then $\lim_{k \to \infty} a_{n_k} = -\frac{1}{2}\sqrt{3}$.

Example 7
Let (r_n) be a list of all rational numbers. It was shown in Example 3 that every real number is a subsequential limit of (r_n). Also, $+\infty$

and $-\infty$ are subsequential limits; see Exercise 11.7. Consequently, $\mathbb{R} \cup \{-\infty, +\infty\}$ is the set of subsequential limits of (r_n).

Example 8
Let $b_n = n[1 + (-1)^n]$ for $n \in \mathbb{N}$. Then $b_n = 2n$ for even n and $b_n = 0$ for odd n. Thus $\{0, +\infty\}$ is the set of subsequential limits of (b_n).

11.7 Theorem.
Let (s_n) be any sequence in $\mathbb{R}$, and let S denote the set of subsequential limits of (s_n).
(i) S is nonempty.
(ii) $\sup S = \lim\sup s_n$ and $\inf S = \lim\inf s_n$.
(iii) $\lim s_n$ exists if and only if S has exactly one element, namely $\lim s_n$.

Proof
(i) is an immediate consequence of Corollary 11.4.

To prove (ii), consider any limit t of a subsequence (s_{n_k}) of (s_n). By Theorem 10.7 we have $t = \lim\inf s_{n_k} = \lim\sup s_{n_k}$. Since $\{s_{n_k} : k > N\} \subseteq \{s_n : n > N\}$ for each $N \in \mathbb{N}$, we have

$$\lim\inf s_n \leq \lim\inf s_{n_k} = t = \lim\sup s_{n_k} \leq \lim\sup s_n.$$

This inequality holds for all t in S; therefore

$$\lim\inf s_n \leq \inf S \leq \sup S \leq \lim\sup s_n.$$

Corollary 11.4 shows that $\lim\inf s_n$ and $\lim\sup s_n$ both belong to S. Therefore (ii) holds.

Assertion (iii) is simply a reformulation of Theorem 10.7. ∎

Theorem 11.7 and Corollary 11.4 show that $\lim\sup s_n$ is exactly the largest subsequential limit of (s_n), and $\lim\inf s_n$ is exactly the smallest subsequential limit of (s_n). This makes it easy to calculate $\lim\sup$'s and $\lim\inf$'s.

We return to the examples given before Theorem 11.7.

Example 9
If $s_n = n^2(-1)^n$, then $S = \{-\infty, +\infty\}$ as noted in Example 5. Therefore $\lim\sup s_n = \sup S = +\infty$ and $\lim\inf s_n = \inf S = -\infty$.

Example 10
If $a_n = \sin(\frac{n\pi}{3})$, then $S = \{-\frac{1}{2}\sqrt{3}, 0, \frac{1}{2}\sqrt{3}\}$ as observed in Example 6. Hence $\limsup a_n = \sup S = \frac{1}{2}\sqrt{3}$ and $\liminf a_n = \inf S = -\frac{1}{2}\sqrt{3}$.

Example 11
If (r_n) denotes a list of all rational numbers, then the set $\mathbb{R} \cup \{-\infty, +\infty\}$ is the set of subsequential limits of (r_n). Consequently we have $\limsup r_n = +\infty$ and $\liminf r_n = -\infty$.

Example 12
If $b_n = n[1 + (-1)^n]$, then $\limsup b_n = +\infty$ and $\liminf b_n = 0$; see Example 8.

The next result shows that the set S of subsequential limits always contains all limits of sequences *from* S. Such sets are called *closed sets*. Sets of this sort will be discussed further in the optional §13.

11.8 Theorem.
Let S denote the set of subsequential limits of a sequence (s_n). Suppose (t_n) is a sequence in $S \cap \mathbb{R}$ and that $t = \lim t_n$. Then t belongs to S.

Proof
Since a subsequence of (s_n) converges to t_1, there exists n_1 such that $|s_{n_1} - t_1| < 1$. Assume that $n_1, n_2, \ldots, n_k$ have been selected so that

$$n_1 < n_2 < \cdots < n_k \tag{1}$$

and

$$|s_{n_j} - t_j| < \frac{1}{j} \quad \text{for} \quad j = 1, 2, \ldots, k. \tag{2}$$

Since a subsequence of (s_n) converges to t_{k+1}, there exists $n_{k+1} > n_k$ such that $|s_{n_{k+1}} - t_{k+1}| < \frac{1}{k+1}$. Thus (1) and (2) hold for $k + 1$.

For the rest of the proof we need to consider cases. Suppose first that $t \in \mathbb{R}$, i.e., that t is not $+\infty$ or $-\infty$. Since

$$|s_{n_k} - t| \leq |s_{n_k} - t_k| + |t_k - t| < \frac{1}{k} + |t_k - t| \tag{3}$$

for all $k \in \mathbb{N}$, it follows easily that $\lim_{k \to \infty} s_{n_k} = t$, so t belongs to S. [To check that $\lim_{k \to \infty} s_{n_k} = t$, consider $\epsilon > 0$. There exists N so that $k > N$ implies $|t_k - t| < \frac{\epsilon}{2}$. If $k > \max\{N, \frac{2}{\epsilon}\}$, then $\frac{1}{k} < \frac{\epsilon}{2}$ and $|t_k - t| < \frac{\epsilon}{2}$, so $|s_{n_k} - t| < \epsilon$ by (3).]

Suppose next that $t = +\infty$. From (2) we have

$$s_{n_k} > t_k - \frac{1}{k} \quad \text{for} \quad k \in \mathbb{N}. \tag{4}$$

Since $\lim t_k = +\infty$ it follows easily that $\lim_{k\to\infty} s_{n_k} = +\infty$. Therefore $t = +\infty$ belongs to S. The case $t = -\infty$ is handled in a similar way. ∎

Exercises

11.1. Let $a_n = 3 + 2(-1)^n$ for $n \in \mathbb{N}$.

(a) List the first eight terms of the sequence (a_n).

(b) Give a subsequence that is constant [takes a single value]. Specify the selection function σ.

11.2. Consider the sequences defined as follows:

$$a_n = (-1)^n, \qquad b_n = \frac{1}{n}, \qquad c_n = n^2, \qquad d_n = \frac{6n+4}{7n-3}.$$

(a) For each sequence, give an example of a monotone subsequence.

(b) For each sequence, give its set of subsequential limits.

(c) For each sequence, give its lim sup and lim inf.

(d) Which of the sequences converges? diverges to $+\infty$? diverges to $-\infty$?

(e) Which of the sequences is bounded?

11.3. Repeat Exercise 11.2 for the sequences:

$$s_n = \cos(\frac{n\pi}{3}), \qquad t_n = \frac{3}{4n+1}, \qquad u_n = (-\frac{1}{2})^n, \qquad v_n = (-1)^n + \frac{1}{n}.$$

11.4. Repeat Exercise 11.2 for the sequences:

$$w_n = (-2)^n, \qquad x_n = 5^{(-1)^n}, \qquad y_n = 1 + (-1)^n, \qquad z_n = n\cos(\frac{n\pi}{4}).$$

11.5. Let (q_n) be an enumeration of all the rationals in the interval $(0, 1]$.

(a) Give the set of subsequential limits for (q_n).

(b) Give the values of $\lim \sup q_n$ and $\lim \inf q_n$.

11.6. Show that every subsequence of a subsequence of a given sequence is itself a subsequence of the given sequence. *Hint*: Define subsequences as in (3) of Definition 11.1.

11.7. Let (r_n) be an enumeration of the set $\mathbb{Q}$ of all rational numbers. Show that there exists a subsequence (r_{n_k}) such that $\lim_{k\to\infty} r_{n_k} = +\infty$.

11.8. (a) Use Definition 10.6 and Exercise 5.4 to prove that $\lim \inf s_n = -\lim \sup(-s_n)$.

(b) Let (t_k) be a monotonic subsequence of $(-s_n)$ converging to $\lim \sup(-s_n)$. Show that $(-t_k)$ is a monotonic subsequence of (s_n) converging to $\lim \inf s_n$. Observe that this completes the proof of Corollary 11.4.

11.9. (a) Show that the closed interval $[a, b]$ is a closed set.

(b) Is there a sequence (s_n) such that $(0,1)$ is its set of subsequential limits?

11.10. Let (s_n) be the sequence of numbers in Figure 11.2 listed in the indicated order.

(a) Find the set S of subsequential limits of (s_n).

(b) Determine $\lim \sup s_n$ and $\lim \inf s_n$.

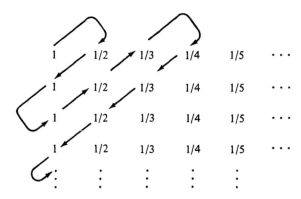

FIGURE 11.2

§12 lim sup's and lim inf's

Let (s_n) be any sequence of real numbers, and let S be the set of subsequential limits of (s_n). Recall that

$$\limsup s_n = \lim_{N \to \infty} \sup \{s_n : n > N\} = \sup S \qquad (*)$$

and

$$\liminf s_n = \lim_{N \to \infty} \inf \{s_n : n > N\} = \inf S. \qquad (**)$$

The first equalities in (*) and (**) are the definitions given in 10.6, and the second equalities are proved in Theorem 11.7. This section is designed to increase the students' familiarity with these concepts. Most of the material is given in the exercises. We illustrate the techniques by proving some results that will be needed later in the text.

12.1 Theorem.
If (s_n) converges to a positive real number s and (t_n) is any sequence, then

$$\limsup s_n t_n = s \cdot \limsup t_n.$$

Here we allow the conventions: $s \cdot (+\infty) = +\infty$ and $s \cdot (-\infty) = -\infty$ for $s > 0$.

Proof
We first show

$$\limsup s_n t_n \geq s \cdot \limsup t_n. \qquad (1)$$

We have three cases. Let $\beta = \limsup t_n$.

Case 1. Suppose β is finite.

By Corollary 11.4, there exists a subsequence (t_{n_k}) of (t_n) such that $\lim_{k \to \infty} t_{n_k} = \beta$. We also have $\lim_{k \to \infty} s_{n_k} = s$ [by Theorem 11.2], so $\lim_{k \to \infty} s_{n_k} t_{n_k} = s\beta$. Thus $(s_{n_k} t_{n_k})$ is a subsequence of $(s_n t_n)$ that converges to $s\beta$, and therefore $s\beta \leq \limsup s_n t_n$. [Recall that $\limsup s_n t_n$ is the largest possible limit of a subsequence of $(s_n t_n)$.] Thus (1) holds.

Case 2. Suppose $\beta = +\infty$.

There exists a subsequence (t_{n_k}) of (t_n) such that $\lim_{k\to\infty} t_{n_k} = +\infty$. Since $\lim_{k\to\infty} s_{n_k} = s > 0$, Theorem 9.9 shows that $\lim_{k\to\infty} s_{n_k} t_{n_k} = +\infty$. Hence $\limsup s_n t_n = +\infty$, so (1) clearly holds.

Case 3. Suppose $\beta = -\infty$.

Since $s > 0$, the right-hand side of (1) is equal to $s \cdot (-\infty) = -\infty$. Hence (1) is obvious in this case.

We have now established (1) in all cases. For the reversed inequality, we resort to a little trick. First note that we may ignore the first few terms of (s_n) and assume that all $s_n \neq 0$. Then we can write $\lim \frac{1}{s_n} = \frac{1}{s}$ by Lemma 9.5. Now we apply (1) with s_n replaced by $\frac{1}{s_n}$ and t_n replaced by $s_n t_n$:

$$\limsup t_n = \limsup (\frac{1}{s_n})(s_n t_n) \geq (\frac{1}{s}) \limsup s_n t_n,$$

i.e.,

$$\limsup s_n t_n \leq s \cdot \limsup t_n.$$

This inequality and (1) prove the theorem. ∎

The next theorem will be useful in dealing with infinite series; see the proof of the Ratio Test 14.8.

12.2 Theorem.

Let (s_n) be any sequence of nonzero real numbers. Then we have

$$\liminf \left| \frac{s_{n+1}}{s_n} \right| \leq \liminf |s_n|^{1/n} \leq \limsup |s_n|^{1/n} \leq \limsup \left| \frac{s_{n+1}}{s_n} \right|.$$

Proof

The middle inequality is obvious. The first and third inequalities have similar proofs. We will prove the third inequality and leave the first inequality to Exercise 12.11.

Let $\alpha = \limsup |s_n|^{1/n}$ and $L = \limsup |\frac{s_{n+1}}{s_n}|$. We need to prove that $\alpha \leq L$. This is obvious if $L = +\infty$, so we assume $L < +\infty$. To prove $\alpha \leq L$ it suffices to show

$$\alpha \leq L_1 \quad \text{for any} \quad L_1 > L. \tag{1}$$

Since

$$L = \limsup \left| \frac{s_{n+1}}{s_n} \right| = \lim_{N \to \infty} \sup \left\{ \left| \frac{s_{n+1}}{s_n} \right| : n > N \right\} < L_1,$$

there exists a natural number N such that

$$\sup \left\{ \left| \frac{s_{n+1}}{s_n} \right| : n \geq N \right\} < L_1.$$

Thus

$$\left| \frac{s_{n+1}}{s_n} \right| < L_1 \quad \text{for} \quad n \geq N. \tag{2}$$

Now for $n > N$ we can write

$$|s_n| = \left| \frac{s_n}{s_{n-1}} \right| \cdot \left| \frac{s_{n-1}}{s_{n-2}} \right| \cdots \left| \frac{s_{N+1}}{s_N} \right| \cdot |s_N|.$$

There are $n - N$ fractions here, so applying (2) we see that

$$|s_n| < L_1^{n-N} |s_N| \quad \text{for} \quad n > N.$$

Since L_1 and N are fixed in this argument, $a = L_1^{-N} |s_N|$ is a positive constant and we may write

$$|s_n| < L_1^n a \quad \text{for} \quad n > N.$$

Therefore we have

$$|s_n|^{1/n} < L_1 a^{1/n} \quad \text{for} \quad n > N.$$

Since $\lim_{n \to \infty} a^{1/n} = 1$ by Example 9.7(d), we conclude that $\alpha = \limsup |s_n|^{1/n} \leq L_1$; see Exercise 12.1. Consequently (1) holds as desired. ∎

12.3 Corollary.
If $\lim \left| \frac{s_{n+1}}{s_n} \right|$ exists [and equals L], then $\lim |s_n|^{1/n}$ exists [and equals L].

Proof
If $\lim \left| \frac{s_{n+1}}{s_n} \right| = L$, then all four values in Theorem 12.2 must equal L. Hence $\lim |s_n|^{1/n} = L$; see Theorem 10.7. ∎

Exercises

12.1. Let (s_n) and (t_n) be sequences and suppose that there exists N_0 such that $s_n \le t_n$ for all $n > N_0$. Show that $\liminf s_n \le \liminf t_n$ and $\limsup s_n \le \limsup t_n$. *Hint*: Use Definition 10.6 and Exercise 9.9(c).

12.2. Prove that $\limsup |s_n| = 0$ if and only if $\lim s_n = 0$.

12.3. Let (s_n) and (t_n) be the following sequences that repeat in cycles of four:

$$(s_n) = (0, 1, 2, 1, 0, 1, 2, 1, 0, 1, 2, 1, 0, 1, 2, 1, 0, \ldots)$$
$$(t_n) = (2, 1, 1, 0, 2, 1, 1, 0, 2, 1, 1, 0, 2, 1, 1, 0, 2, \ldots)$$

Find

(a) $\liminf s_n + \liminf t_n$,   **(b)** $\liminf(s_n + t_n)$,
(c) $\liminf s_n + \limsup t_n$,   **(d)** $\limsup(s_n + t_n)$,
(e) $\limsup s_n + \limsup t_n$,   **(f)** $\liminf(s_n t_n)$,
(g) $\limsup(s_n t_n)$

12.4. Show that $\limsup(s_n + t_n) \le \limsup s_n + \limsup t_n$ for bounded sequences (s_n) and (t_n). *Hint*: First show

$$\sup\{s_n + t_n : n > N\} \le \sup\{s_n : n > N\} + \sup\{t_n : n > N\}.$$

Then apply Exercise 9.9(c).

12.5. Use Exercises 11.8(a) and 12.4 to prove

$$\liminf(s_n + t_n) \ge \liminf s_n + \liminf t_n$$

for bounded sequences (s_n) and (t_n).

12.6. Let (s_n) be a bounded sequence, and let k be a nonnegative real number.

(a) Prove that $\limsup(ks_n) = k \cdot \limsup s_n$.

(b) Do the same for $\liminf$. *Hint*: Use Exercise 11.8(a).

(c) What happens in (a) and (b) if $k < 0$?

12.7. Prove that if $\limsup s_n = +\infty$ and $k > 0$, then $\limsup(ks_n) = +\infty$.

12.8. Let (s_n) and (t_n) be bounded sequences of nonnegative numbers. Prove that $\limsup s_n t_n \le (\limsup s_n)(\limsup t_n)$.

12.9. **(a)** Prove that if $\lim s_n = +\infty$ and $\liminf t_n > 0$, then $\lim s_n t_n = +\infty$.

(b) Prove that if $\limsup s_n = +\infty$ and $\liminf t_n > 0$, then $\limsup s_n t_n = +\infty$.

(c) Observe that Exercise 12.7 is the special case of (b) where $t_n = k$ for all $n \in \mathbb{N}$.

12.10. Prove that (s_n) is bounded if and only if $\lim \sup |s_n| < +\infty$.

12.11. Prove the first inequality in Theorem 12.2.

12.12. Let (s_n) be a sequence of nonnegative numbers, and for each n define $\sigma_n = \frac{1}{n}(s_1 + s_2 + \cdots + s_n)$.

(a) Show that

$$\lim \inf s_n \leq \lim \inf \sigma_n \leq \lim \sup \sigma_n \leq \lim \sup s_n.$$

Hint: For the last inequality, show first that $M > N$ implies

$$\sup\{\sigma_n : n > M\} \leq \frac{1}{M}(s_1 + s_2 + \cdots + s_N) + \sup\{s_n : n > N\}.$$

(b) Show that if $\lim s_n$ exists, then $\lim \sigma_n$ exists and $\lim \sigma_n = \lim s_n$.

12.13. Let (s_n) be a bounded sequence in $\mathbb{R}$. Let A be the set of $a \in \mathbb{R}$ such that $\{n \in \mathbb{N} : s_n < a\}$ is finite, i.e., all but finitely many s_n are $\geq a$. Let B be the set of $b \in \mathbb{R}$ such that $\{n \in \mathbb{N} : s_n > b\}$ is finite. Prove that $\sup A = \lim \inf s_n$ and $\inf B = \lim \sup s_n$.

12.14. Calculate (a) $\lim (n!)^{1/n}$, (b) $\lim \frac{1}{n}(n!)^{1/n}$.

§13 * Some Topological Concepts in Metric Spaces

In this book we are restricting our attention to analysis on $\mathbb{R}$. Accordingly, we have taken full advantage of the order properties of $\mathbb{R}$ and studied such important notions as lim sup's and lim inf's. In §3 we briefly introduced a distance function on $\mathbb{R}$. Most of our analysis could have been based on the notion of distance, in which case it becomes easy and natural to work in a more general setting. For example, analysis on the k-dimensional Euclidean spaces $\mathbb{R}^k$ is important, but these spaces do not have the useful natural ordering that $\mathbb{R}$ has, unless of course $k = 1$.

13.1 Definition.
Let S be a set, and suppose d is a function defined for all pairs (x, y) of elements from S satisfying
  D1. $d(x, x) = 0$ for all $x \in S$ and $d(x, y) > 0$ for distinct x, y in S.
  D2. $d(x, y) = d(y, x)$ for all $x, y \in S$.
  D3. $d(x, z) \le d(x, y) + d(y, z)$ for all $x, y, z \in S$.
Such a function d is called a *distance function* or a *metric* on S. A *metric space* S is a set S together with a metric on it. Properly speaking, the metric space is the pair (S, d) since a set S may well have more than one metric on it; see Exercise 13.1.

Example 1
As in Definition 3.4, let $\mathrm{dist}(a, b) = |a - b|$ for $a, b \in \mathbb{R}$. Then dist is a metric on $\mathbb{R}$. Note that Corollary 3.6 gives D3 in this case. As remarked there, the inequality

$$\mathrm{dist}(a, c) \le \mathrm{dist}(a, b) + \mathrm{dist}(b, c)$$

is called the triangle inequality. In fact, for any metric d, property D3 is called the *triangle inequality*.

Example 2
The space of all k-tuples

$$\pmb{x} = (x_1, x_2, \ldots, x_k) \quad \text{where} \quad x_j \in \mathbb{R} \quad \text{for} \quad j = 1, 2, \ldots, k,$$

is called *k-dimensional Euclidean space* and written $\mathbb{R}^k$. As noted in Exercise 13.1, $\mathbb{R}^k$ has several metrics on it. The most familiar metric is the one that gives the ordinary distance in the plane $\mathbb{R}^2$ or in 3-space $\mathbb{R}^3$:

$$d(\pmb{x}, \pmb{y}) = \left[\sum_{j=1}^{k} (x_j - y_j)^2 \right]^{1/2}.$$

[The summation notation $\sum$ is explained in 14.1.] Obviously this function d satisfies properties D1 and D2. The triangle inequality D3 is not so obvious. For a proof, see for example [33], §6.1, or [36], 1.37.

13.2 Definition.

A sequence (s_n) in a metric space (S, d) *converges to* s in S if $\lim_{n \to \infty} d(s_n, s) = 0$. A sequence (s_n) in S is a *Cauchy sequence* if for each $\epsilon > 0$ there exists an N such that

$$m, n > N \quad \text{implies} \quad d(s_m, s_n) < \epsilon.$$

The metric space (S, d) is said to be *complete* if every Cauchy sequence in S converges to some element in S.

Since the Completeness Axiom 4.4 deals with least upper bounds, the word "complete" now appears to have two meanings. However, these two uses of the term are very closely related and both reflect the property that the space is complete, i.e., has no gaps. Theorem 10.11 asserts that the metric space $(\mathbb{R}, \text{dist})$ is a complete metric space, and the proof uses the Completeness Axiom 4.4. We could just as well have taken as an axiom the completeness of $(\mathbb{R}, \text{dist})$ as a metric space and proved the least upper bound property in 4.4 as a theorem. We did not do so because the concept of least upper bound in $\mathbb{R}$ seems to us more fundamental than the concept of Cauchy sequence.

We will prove that $\mathbb{R}^k$ is complete. But we have a notational problem, since we like subscripts for sequences and for coordinates of points in $\mathbb{R}^k$. When there is a conflict, we will write $(x^{(n)})$ for a sequence instead of (x_n). In this case,

$$x^{(n)} = (x_1^{(n)}, x_2^{(n)}, \ldots, x_k^{(n)}).$$

Unless otherwise specified, *the metric in $\mathbb{R}^k$ is always as given in Example 2.*

13.3 Lemma.

A sequence $(x^{(n)})$ in $\mathbb{R}^k$ converges if and only if for each $j = 1, 2, \ldots, k$, the sequence $(x_j^{(n)})$ converges in $\mathbb{R}$. A sequence $(x^{(n)})$ in $\mathbb{R}^k$ is a Cauchy sequence if and only if each sequence $(x_j^{(n)})$ is a Cauchy sequence in $\mathbb{R}$.

Proof

The proof of the first assertion is left to Exercise 13.2. For the second assertion, we first observe for x, y in $\mathbb{R}^k$ and $j = 1, 2, \ldots, k$,

$$|x_j - y_j| \leq d(x, y) \leq \sqrt{k} \max\{|x_j - y_j| : j = 1, 2, \ldots, k\}. \tag{1}$$

Suppose $(x^{(n)})$ is a Cauchy sequence in $\mathbb{R}^k$, and consider fixed j. If $\epsilon > 0$, there exists N such that

$$m, n > N \quad \text{implies} \quad d(x^{(m)}, x^{(n)}) < \epsilon.$$

From (1) we see that

$$m, n > N \quad \text{implies} \quad |x_j^{(m)} - x_j^{(n)}| < \epsilon,$$

so $(x_j^{(n)})$ is a Cauchy sequence in $\mathbb{R}$.

Now suppose each sequence $(x_j^{(n)})$ is a Cauchy sequence in $\mathbb{R}$. Let $\epsilon > 0$. For $j = 1, 2, \ldots, k$, there exist N_j such that

$$m, n > N_j \quad \text{implies} \quad |x_j^{(m)} - x_j^{(n)}| < \frac{\epsilon}{\sqrt{k}}.$$

If $N = \max\{N_1, N_2, \ldots, N_k\}$, then by (1)

$$m, n > N \quad \text{implies} \quad d(x^{(m)}, x^{(n)}) < \epsilon,$$

i.e., $(x^{(n)})$ is a Cauchy sequence in $\mathbb{R}^k$. ∎

13.4 Theorem.
Euclidean k-space $\mathbb{R}^k$ is complete.

Proof
Consider a Cauchy sequence $(x^{(n)})$ in $\mathbb{R}^k$. By Lemma 13.3, $(x_j^{(n)})$ is a Cauchy sequence in $\mathbb{R}$ for each j. Hence by Theorem 10.11, $(x_j^{(n)})$ converges to some real number x_j. By Lemma 13.3 again, the sequence $(x^{(n)})$ converges, in fact to $x = (x_1, x_2, \ldots, x_k)$. ∎

We now can prove the Bolzano-Weierstrass theorem for $\mathbb{R}^k$; compare Theorem 11.5. A set S in $\mathbb{R}^k$ is *bounded* if there exists $M > 0$ such that

$$\max\{|x_j| : j = 1, 2, \ldots, k\} \leq M \quad \text{for all} \quad x \in S.$$

13.5 Bolzano-Weierstrass Theorem.
Every bounded sequence in $\mathbb{R}^k$ has a convergent subsequence.

Proof
Let $(x^{(n)})$ be a bounded sequence in $\mathbb{R}^k$. Then each sequence $(x_j^{(n)})$ is bounded in $\mathbb{R}$. By Theorem 11.5, we may replace $(x^{(n)})$ by a subsequence such that $(x_1^{(n)})$ converges. By the same theorem, we may

replace $(\pmb{x}^{(n)})$ by a subsequence of the subsequence such that $(x_2^{(n)})$ converges. Of course, $(x_1^{(n)})$ still converges by Theorem 11.2. Repeating this argument k times, we obtain a sequence $(\pmb{x}^{(n)})$ so that each sequence $(x_j^{(n)})$ converges, $j = 1, 2, \ldots, k$. This sequence represents a subsequence of the original sequence, and it converges in $\mathbb{R}^k$ by Lemma 13.3. ∎

13.6 Definition.
Let (S, d) be a metric space. Let E be a subset of S. An element $s_0 \in E$ is *interior to* E if for some $r > 0$ we have

$$\{s \in S : d(s, s_0) < r\} \subseteq E.$$

We write E° for the set of points in E that are interior to E. The set E is *open in* S if every point in E is interior to E, i.e., if $E = E^\circ$.

13.7 Discussion.
One can show [Exercise 13.4]
  (i) S is open in S [trivial].
  (ii) The empty set $\varnothing$ is open in S [trivial].
  (iii) The union of *any* collection of open sets is open.
  (iv) The intersection of *finitely many* open sets is again an open set.

Our study of $\mathbb{R}^k$ and the exercises suggest that metric spaces are fairly general and useful objects. When one is interested in convergence of certain objects [such as points or functions], there is often a metric that assists in the study of the convergence. But sometimes no metric will work and yet there is still some sort of convergence notion. Frequently the appropriate vehicle is what is called a *topology*. This is a set S for which certain subsets are decreed to be *open sets*. In general, all that is required is that the family of open sets satisfies (i)–(iv) above. In particular, the open sets defined by a metric form a topology. We will not pursue this abstract theory. However, because of this abstract theory, concepts that can be defined in terms of open sets [see Definitions 13.8, 13.11 and 22.1] are called *topological*, hence the title of this section.

13.8 Definition.
Let (S, d) be a metric space. A subset E of S is *closed* if its complement $S \setminus E$ is an open set. In other words, E is closed if $E = S \setminus U$ where U is an open set.

Because of (iii) in Discussion 13.7, the intersection of *any* collection of closed sets in closed [Exercise 13.5]. The *closure* E^- of a set E is the intersection of all closed sets containing E. The *boundary* of E is the set $E^- \setminus E^\circ$; points in this set are called *boundary points of E*.

To get a feel for these notions, we state some easy facts and leave the proofs as exercises.

13.9 Proposition.
Let E be a subset of a metric space (S, d).
  (a) *The set E is closed if and only if $E = E^-$.*
  (b) *The set E is closed if and only if it contains the limit of every convergent sequence of points in E.*
  (c) *An element is in E^- if and only if it is the limit of some sequence of points in E.*
  (d) *A point is in the boundary of E if and only if it belongs to the closure of both E and its complement.*

Example 3
In $\mathbb{R}$, open intervals (a, b) are open sets. Closed intervals $[a, b]$ are closed sets. The interior of $[a, b]$ is (a, b). The boundary of both (a, b) and $[a, b]$ is the two-element set $\{a, b\}$.

Every open set in $\mathbb{R}$ is the union of a disjoint sequence of open intervals [Exercise 13.7]. A closed set in $\mathbb{R}$ need not be the union of a disjoint sequence of closed intervals and points; such a set appears in Example 5.

Example 4
In $\mathbb{R}^k$, open balls $\{x : d(x, x_0) < r\}$ are open sets, and closed balls $\{x : d(x, x_0) \leq r\}$ are closed sets. The boundary of each of these sets is $\{x : d(x, x_0) = r\}$. In the plane $\mathbb{R}^2$, the sets

$$\{(x_1, x_2) : x_1 > 0\} \quad \text{and} \quad \{(x_1, x_2) : x_1 > 0 \text{ and } x_2 > 0\}$$

are open. If $>$ is replaced by $\geq$, we obtain closed sets. Many sets are neither open nor closed, for example

$$\{(x_1, x_2) : x_1 > 0 \text{ and } x_2 \geq 0\}.$$

13.10 Theorem.
Let (F_n) be a decreasing sequence [i.e., $F_1 \supseteq F_2 \supseteq \cdots$] of closed bounded nonempty sets in $\mathbb{R}^k$. Then $F = \cap_{n=1}^{\infty} F_n$ is also closed, bounded and nonempty.

Proof
Clearly F is closed and bounded. It is the nonemptiness that needs proving! For each n, select an element x_n in F_n. By the Bolzano-Weierstrass theorem 13.5, a subsequence $(x_{n_m})_{m=1}^{\infty}$ of (x_n) converges to some element x_0 in $\mathbb{R}^k$. To show $x_0 \in F$, it suffices to show $x_0 \in F_{n_0}$ with n_0 fixed. If $m \geq n_0$, then $n_m \geq n_0$, so $x_{n_m} \in F_{n_m} \subseteq F_{n_0}$. Hence the sequence $(x_{n_m})_{m=n_0}^{\infty}$ consists of points in F_{n_0} and converges to x_0. Thus x_0 belongs to F_{n_0} by (b) of Proposition 13.9. ∎

Example 5
Here is a famous nonempty closed set in $\mathbb{R}$ called the *Cantor set*. Pictorially, $F = \cap_{n=1}^{\infty} F_n$ where F_n are sketched in Figure 13.1. The Cantor set has some remarkable properties. The sum of the lengths of the intervals comprising F_n is $\left(\frac{2}{3}\right)^{n-1}$ and this tends to 0 as $n \to \infty$. Yet the intersection F is so large that it cannot be written as a sequence; in set-theoretic terms it is "uncountable." The interior of

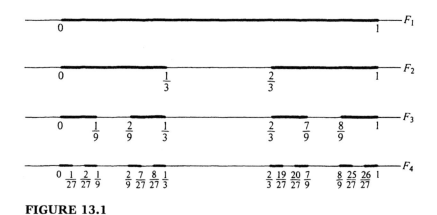

FIGURE 13.1

F is the empty set, so F equals its boundary. For more details, see [36], 2.44, or [23], 6.62.

13.11 Definition.
Let (S, d) be a metric space. A family $\mathcal{U}$ of open sets is said to be an *open cover for a set* E if each point of E belongs to at least one set in $\mathcal{U}$, i.e.,

$$E \subseteq \bigcup \{U : U \in \mathcal{U}\}.$$

A *subcover* of $\mathcal{U}$ is any subfamily of $\mathcal{U}$ that also covers E. A cover or subcover is *finite* if it contains only finitely many sets; the sets themselves may be infinite.

A set E is *compact* if every open cover of E has a finite subcover of E.

This rather abstract definition is very important in advanced analysis; see, for example, [22]. In $\mathbb{R}^k$, compact sets are nicely characterized, as follows.

13.12 Heine-Borel Theorem.
A subset E of $\mathbb{R}^k$ is compact if and only if it is closed and bounded.

Proof
Suppose that E is compact. For each $m \in \mathbb{N}$, let U_m consist of all x in $\mathbb{R}^k$ such that

$$\max\{|x_j| : j = 1, 2, \ldots, k\} < m.$$

The family $\mathcal{U} = \{U_m : m \in \mathbb{N}\}$ is an open cover of E [it covers $\mathbb{R}^k$!], so a finite subfamily of $\mathcal{U}$ covers E. If U_{m_0} is the largest member of the subfamily, then $E \subseteq U_{m_0}$. It follows that E is bounded. To show that E is closed, consider any point x_0 in $\mathbb{R}^k \setminus E$. For $m \in \mathbb{N}$, let

$$V_m = \left\{ x \in \mathbb{R}^k : d(x, x_0) > \frac{1}{m} \right\}.$$

Then each V_m is open in $\mathbb{R}^k$ and $\mathcal{V} = \{V_m : m \in \mathbb{N}\}$ covers E since $\bigcup_{m=1}^{\infty} V_m = \mathbb{R}^k \setminus \{x_0\}$. Since E can be covered by finitely many V_m, for some m_0 we have

$$E \subseteq \left\{ x \in \mathbb{R}^k : d(x, x_0) > \frac{1}{m_0} \right\}.$$

Thus $\{x \in \mathbb{R}^k : d(x, x_0) < \frac{1}{m_0}\} \subseteq \mathbb{R}^k \setminus E$, so that x_0 is interior to $\mathbb{R}^k \setminus E$. Since x_0 in $\mathbb{R}^k \setminus E$ was arbitrary, $\mathbb{R}^k \setminus E$ is an open set. Hence E is a closed set.

Now suppose that E is closed and bounded. Since E is bounded, E is a subset of some set F having the form

$$F = \{x \in \mathbb{R}^k : |x_j| \le m \quad \text{for} \quad j = 1, 2, \ldots, k\}.$$

As noted in Exercise 13.12, it suffices to prove that F is compact. We do so in the next proposition after some preparation. ■

The set F in the last proof is a *k-cell* because it has the following form. There exist closed intervals $[a_1, b_1], [a_2, b_2], \ldots, [a_k, b_k]$ so that

$$F = \{x \in \mathbb{R}^k : x_j \in [a_j, b_j] \quad \text{for} \quad j = 1, 2, \ldots, k\}.$$

The *diameter* of F is

$$\delta = \left[\sum_{j=1}^{k} (b_j - a_j)^2 \right]^{1/2};$$

that is, $\delta = \sup\{d(x, y) : x, y \in F\}$. Using midpoints $c_j = \frac{1}{2}(a_j + b_j)$ of $[a_j, b_j]$, we see that F is a union of 2^k *k*-cells each having diameter $\frac{\delta}{2}$. If this remark is not clear, consider first the cases $k = 2$ and $k = 3$.

13.13 Proposition.
Every k-cell F in $\mathbb{R}^k$ is compact.

Proof
Assume F is not compact. Then there exists an open cover $\mathcal{U}$ of F, no finite subfamily of which covers F. Let δ denote the diameter of F. As noted above, F is a union of 2^k *k*-cells having diameter $\frac{\delta}{2}$. At least one of these 2^k *k*-cells, which we denote by F_1, cannot be covered by finitely many sets from $\mathcal{U}$. Likewise, F_1 contains a *k*-cell F_2 of diameter $\frac{\delta}{4}$ which cannot be covered by finitely many sets from $\mathcal{U}$. Continuing in this fashion, we obtain a sequence (F_n) of *k*-cells such that

$$F_1 \supseteq F_2 \supseteq F_3 \supseteq \cdots; \tag{1}$$

$$F_n \text{ has diameter } \delta \cdot 2^{-n}; \tag{2}$$

$$F_n \text{ cannot be covered by finitely many sets from } \mathcal{U}. \tag{3}$$

By Theorem 13.10, the intersection $\bigcap_{n=1}^{\infty} F_n$ contains a point x_0. This point belongs to some set U_0 in $\mathcal{U}$. Since U_0 is open, there exists $r > 0$ so that

$$\{x \in \mathbb{R}^k : d(x, x_0) < r\} \subseteq U_0.$$

It follows that $F_n \subseteq U_0$ provided $\delta \cdot 2^{-n} < r$, but this contradicts (3) in a dramatic way. ∎

Since $\mathbb{R} = \mathbb{R}^1$, the preceding results apply to $\mathbb{R}$.

Exercises

13.1. For points x, y in $\mathbb{R}^k$, let

$$d_1(x, y) = \max\{|x_j - y_j| : j = 1, 2, \ldots, k\}$$

and

$$d_2(x, y) = \sum_{j=1}^{k} |x_j - y_j|.$$

(a) Show that d_1 and d_2 are metrics for $\mathbb{R}^k$.

(b) Show that d_1 and d_2 are complete.

13.2. (a) Prove (1) in Lemma 13.3.

(b) Prove the first assertion in Lemma 13.3.

13.3. Let B be the set of all bounded sequences $x = (x_1, x_2, \ldots)$, and define $d(x, y) = \sup\{|x_j - y_j| : j = 1, 2, \ldots\}$.

(a) Show that d is a metric for B.

(b) Does $d^*(x, y) = \sum_{j=1}^{\infty} |x_j - y_j|$ define a metric for B?

13.4. Prove (iii) and (iv) in Discussion 13.7.

13.5. (a) Verify one of DeMorgan's Laws for sets:

$$\bigcap\{S \setminus U : U \in \mathcal{U}\} = S \setminus \bigcup\{U : U \in \mathcal{U}\}.$$

(b) Show that the intersection of any collection of closed sets is a closed set.

13.6. Prove Proposition 13.9.

13.7. Show that every open set in $\mathbb{R}$ is the disjoint union of a finite or infinite sequence of open intervals.

13.8. (a) Verify the assertions in Example 3.

(b) Verify the assertions in Example 4.

13.9. Find the closures of the following sets:

(a) $\{\frac{1}{n} : n \in \mathbb{N}\}$,

(b) $\mathbb{Q}$, the set of rational numbers,

(c) $\{r \in \mathbb{Q} : r^2 < 2\}$.

13.10. Show that the interior of each of the following sets is the empty set.

(a) $\{\frac{1}{n} : n \in \mathbb{N}\}$,

(b) $\mathbb{Q}$, the set of rational numbers,

(c) the Cantor set in Example 5.

13.11. Let E be a subset of $\mathbb{R}^k$. Show that E is compact if and only if every sequence in E has a subsequence that converges to a point *in E*.

13.12. Let (S, d) be any metric space.

(a) Show that if E is a closed subset of a compact set F, then E is also compact.

(b) Show that the finite union of compact sets in S is compact.

13.13. Let E be a compact nonempty subset of $\mathbb{R}$. Show that $\sup E$ and $\inf E$ belong to E.

13.14. Let E be a compact nonempty subset of $\mathbb{R}^k$, and let $\delta = \sup\{d(x,y) : x, y \in E\}$. Show that E contains points x_0, y_0 such that $d(x_0, y_0) = \delta$.

13.15. Let (B, d) be as in Exercise 13.3, and let F consist of all $x \in B$ such that $\sup\{|x_j| : j = 1, 2, \ldots\} \le 1$.

(a) Show that F is closed and bounded. [A set F in a metric space (S, d) is *bounded* if there exist $s_0 \in S$ and $r > 0$ such that $F \subseteq \{s \in S : d(s, s_0) \le r\}$.]

(b) Show that F is not compact. *Hint:* For each x in F, let $U(x) = \{y \in B : d(y, x) < 1\}$, and consider the cover $\mathcal{U}$ of F consisting of all $U(x)$. For each $n \in \mathbb{N}$, let $x^{(n)}$ be defined so that $x_n^{(n)} = -1$ and $x_j^{(n)} = 1$ for $j \ne n$. Show that distinct $x^{(n)}$ cannot belong to the same member of $\mathcal{U}$.

§14 Series

Our thorough treatment of sequences now allows us to quickly obtain the basic properties of infinite series.

14.1 Summation Notation.
The notation $\sum_{k=m}^{n} a_k$ is shorthand for the sum $a_m + a_{m+1} + \cdots + a_n$. The symbol "$\sum$" instructs us to sum and the decorations "$k = m$" and "n" tell us to sum the summands obtained by successively substituting $m, m+1, \ldots, n$ for k. For example, $\sum_{k=2}^{5} \frac{1}{k^2+k}$ is shorthand for

$$\frac{1}{2^2+2} + \frac{1}{3^2+3} + \frac{1}{4^2+4} + \frac{1}{5^2+5} = \frac{1}{6} + \frac{1}{12} + \frac{1}{20} + \frac{1}{30}$$

and $\sum_{k=0}^{n} 2^{-k}$ is shorthand for $1 + 1/2 + 1/4 + \cdots + 1/2^n$.

The symbol $\sum_{n=m}^{\infty} a_n$ is shorthand for $a_m + a_{m+1} + a_{m+2} + \cdots$, although we have not yet assigned meaning to such an infinite sum. We now do so.

14.2 Infinite Series.
To assign meaning to $\sum_{n=m}^{\infty} a_n$, we consider the sequences $(s_n)_{n=m}^{\infty}$ of *partial sums*:

$$s_n = a_m + a_{m+1} + \cdots + a_n = \sum_{k=m}^{n} a_k.$$

The infinite series $\sum_{n=m}^{\infty} a_n$ is said to *converge* provided the sequence (s_n) of partial sums converges to a real number S, in which case we define $\sum_{n=m}^{\infty} a_n = S$. Thus

$$\sum_{n=m}^{\infty} a_n = S \quad \text{means} \quad \lim s_n = S \quad \text{or} \quad \lim_{n \to \infty} \left(\sum_{k=m}^{n} a_k \right) = S.$$

A series that does not converge is said to *diverge*. We say that $\sum_{n=m}^{\infty} a_n$ *diverges to* $+\infty$ and we write $\sum_{n=m}^{\infty} a_n = +\infty$ provided $\lim s_n = +\infty$; a similar remark applies to $-\infty$. The symbol $\sum_{n=m}^{\infty} a_n$ has no meaning unless the series converges or diverges to $+\infty$ or $-\infty$. Often we will be concerned with properties of infinite series but not their exact values or precisely where the summation begins, in which case we may write $\sum a_n$ rather than $\sum_{n=m}^{\infty} a_n$.

If the terms a_n of an infinite series $\sum a_n$ are all nonnegative, then the partial sums (s_n) form a nondecreasing sequence, so Theorems 10.2 and 10.4 show that $\sum a_n$ either converges or diverges to $+\infty$. In particular, $\sum |a_n|$ is meaningful for any sequence (a_n) whatever. The series $\sum a_n$ is said to *converge absolutely* or to be *absolutely convergent* if $\sum |a_n|$ converges. Absolutely convergent series are convergent, as we shall see in 14.7.

Example 1
A series of the form $\sum_{n=0}^{\infty} ar^n$ for constants a and r is called a *geometric series*. These are the easiest series to sum. For $r \neq 1$, the partial sums s_n are given by

$$\sum_{k=0}^{n} ar^k = a\frac{1 - r^{n+1}}{1 - r}. \tag{1}$$

This identity can be verified by mathematical induction or by multiplying both sides by $1 - r$, in which case the right hand side equals $a - ar^{n+1}$ and the left side becomes

$$(1 - r)\sum_{k=0}^{n} ar^k = \sum_{k=0}^{n} ar^k - \sum_{k=0}^{n} ar^{k+1}$$
$$= a + ar + ar^2 + \cdots + ar^n$$
$$-(ar + ar^2 + \cdots + ar^n + ar^{n+1})$$
$$= a - ar^{n+1}.$$

For $|r| < 1$, we have $\lim_{n\to\infty} r^{n+1} = 0$ by Example 7(b) in §9, so from (1) we have $\lim_{n\to\infty} s_n = \frac{a}{1-r}$. This proves

$$\sum_{n=0}^{\infty} ar^n = \frac{a}{1 - r} \quad \text{if} \quad |r| < 1. \tag{2}$$

If $a \neq 0$ and $|r| \geq 1$, then the sequence (ar^n) does not converge to 0, so the series $\sum ar^n$ diverges by Corollary 14.5 below.

Example 2
Formula (2) of Example 1 and the next result are very important and both should be used whenever possible, even though we will not prove (1) below until the next section. Consider a fixed positive

real number p. Then

$$\sum_{n=1}^{\infty} \frac{1}{n^p} \quad \text{converges if and only if} \quad p > 1. \tag{1}$$

In particular, for $p \leq 1$, we can write $\sum 1/n^p = +\infty$. The exact value of the series for $p > 1$ is not easy to determine. Here are some remarkable formulas that can be shown by techniques [Fourier series or complex variables, to name two possibilities] that will not be covered in this text.

$$\sum_{n=1}^{\infty} \frac{1}{n^2} = \frac{\pi^2}{6} = 1.6449\cdots, \tag{2}$$

$$\sum_{n=1}^{\infty} \frac{1}{n^4} = \frac{\pi^4}{90} = 1.0823\cdots. \tag{3}$$

Similar formulas hold for $\sum_{n=1}^{\infty} \frac{1}{n^p}$ when p is any even integer, but no such elegant formulas are known for p odd. In particular, no such formula is known for $\sum_{n=1}^{\infty} \frac{1}{n^3}$ though of course this series converges and can be approximated as closely as desired.

It is worth emphasizing that it is often easier to prove that a limit exists or that a series converges than to determine its exact value. In the next section we will show without much difficulty that $\sum \frac{1}{n^p}$ converges for all $p > 1$, but it is a lot harder to show that the sum is $\frac{\pi^2}{6}$ when $p = 2$ and no one knows exactly what the sum is for $p = 3$.

14.3 Definition.
We say that a series $\sum a_n$ satisfies the *Cauchy criterion* if its sequence (s_n) of partial sums is a Cauchy sequence [see Definition 10.8]:

> for each $\epsilon > 0$ there exists a number N such that
> $m, n > N$ implies $|s_n - s_m| < \epsilon.$ (1)

Nothing is lost in this definition if we impose the restriction $n > m$. Moreover, it is only a notational matter to work with $m - 1$ where $m \leq n$ instead of m where $m < n$. Therefore (1) is equivalent to

> for each $\epsilon > 0$ there exists a number N such that
> $n \geq m > N$ implies $|s_n - s_{m-1}| < \epsilon.$ (2)

Since $s_n - s_{m-1} = \sum_{k=m}^{n} a_k$, condition (2) can be rewritten

for each $\epsilon > 0$ there exists a number N such that

$$n \geq m > N \text{ implies } \left| \sum_{k=m}^{n} a_k \right| < \epsilon. \tag{3}$$

We will usually use version (3) of the *Cauchy criterion*. Theorem 10.11 implies the following.

14.4 Theorem.
A series converges if and only if it satisfies the Cauchy criterion.

14.5 Corollary.
If a series $\sum a_n$ converges, then $\lim a_n = 0$.

Proof
Since the series converges, (3) in Definition 14.3 holds. In particular, (3) in 14.3 holds for $n = m$; i.e., for each $\epsilon > 0$ there exists a number N such that $n > N$ implies $|a_n| < \epsilon$. Thus $\lim a_n = 0$. ∎

The converse of Corollary 14.5 does not hold as the example $\sum 1/n = +\infty$ shows.

We next give several tests to assist us in determining whether a series converges. The first test is elementary but useful.

14.6 Comparison Test.
Let $\sum a_n$ be a series where $a_n \geq 0$ for all n.
  (i) If $\sum a_n$ converges and $|b_n| \leq a_n$ for all n, then $\sum b_n$ converges.
  (ii) If $\sum a_n = +\infty$ and $b_n \geq a_n$ for all n, then $\sum b_n = +\infty$.

Proof
  (i) For $n \geq m$ we have

$$\left| \sum_{k=m}^{n} b_k \right| \leq \sum_{k=m}^{n} |b_k| \leq \sum_{k=m}^{n} a_k;$$

the first inequality follows from the triangle inequality [Exercise 3.6(b)]. Since $\sum a_n$ converges, it satisfies the Cauchy criterion 14.3(3). It follows from (1) that $\sum b_n$ also satisfies the Cauchy criterion, and hence $\sum b_n$ converges.

(ii) Let (s_n) and (t_n) be the sequences of partial sums for $\sum a_n$ and $\sum b_n$, respectively. Since $b_n \geq a_n$ for all n, we obviously have $t_n \geq s_n$ for all n. Since $\lim s_n = +\infty$, we conclude that $\lim t_n = +\infty$, i.e., $\sum b_n = +\infty$. ■

14.7 Corollary.
Absolutely convergent series are convergent.

Proof
Suppose that $\sum b_n$ is absolutely convergent. This means that $\sum a_n$ converges where $a_n = |b_n|$ for all n. Then $|b_n| \leq a_n$ trivially, so $\sum b_n$ converges by 14.6(i). ■

We next state the Ratio Test which is popular because it is often easy to use. But it has defects: It isn't as general as the Root Test. An important result concerning the radius of convergence of a power series uses the Root Test. Finally, the Ratio Test is worthless if some of the a_n's equal 0. To review lim sup's and lim inf's, see 10.6, 10.7, 11.7 and §12.

14.8 Ratio Test.
A series $\sum a_n$ of nonzero terms
  (i) *converges absolutely if* $\limsup |a_{n+1}/a_n| < 1$,
  (ii) *diverges if* $\liminf |a_{n+1}/a_n| > 1$.
  (iii) *Otherwise* $\liminf |a_{n+1}/a_n| \leq 1 \leq \limsup |a_{n+1}/a_n|$ *and the test gives no information.*

We give the proof after the proof of the Root Test.
Remember that if $\lim |a_{n+1}/a_n|$ exists, then it is equal to both $\limsup |a_{n+1}/a_n|$ and $\liminf |a_{n+1}/a_n|$ and hence the Ratio Test will give information unless, or course, the limit $\lim |a_{n+1}/a_n|$ equals 1.

14.9 Root Test.
Let $\sum a_n$ be a series and let $\alpha = \limsup |a_n|^{1/n}$. The series $\sum a_n$
  (i) *converges absolutely if* $\alpha < 1$,
  (ii) *diverges if* $\alpha > 1$.
  (iii) *Otherwise* $\alpha = 1$ *and the test gives no information.*

Proof

(i) Suppose $\alpha < 1$, and select $\epsilon > 0$ so that $\alpha + \epsilon < 1$. Then by Definition 10.6 there is a natural number N such that

$$\alpha - \epsilon < \sup\{|a_n|^{1/n} : n > N\} < \alpha + \epsilon.$$

In particular, we have $|a_n|^{1/n} < \alpha + \epsilon$ for $n > N$, so

$$|a_n| < (\alpha + \epsilon)^n \quad \text{for} \quad n > N.$$

Since $0 < \alpha + \epsilon < 1$, the geometric series $\sum_{n=N+1}^{\infty} (\alpha + \epsilon)^n$ converges, and the Comparison Test shows that the series $\sum_{n=N+1}^{\infty} a_n$ also converges. Then clearly $\sum a_n$ converges; see Exercise 14.9.

(ii) If $\alpha > 1$, then by Corollary 11.4 a subsequence of $|a_n|^{1/n}$ has limit $\alpha > 1$. It follows that $|a_n| > 1$ for infinitely many choices of n. In particular, the sequence (a_n) cannot possibly converge to 0, so the series $\sum a_n$ cannot converge by Corollary 14.5.

(iii) For each of the series $\sum \frac{1}{n}$ and $\sum \frac{1}{n^2}$, α turns out to equal 1 as can be seen by applying 9.7(c). Since $\sum \frac{1}{n}$ diverges and $\sum \frac{1}{n^2}$ converges, the equality $\alpha = 1$ does not guarantee either convergence or divergence of the series. ∎

Proof of the Ratio Test

Let $\alpha = \limsup |a_n|^{1/n}$. By Theorem 12.2 we have

$$\liminf \left| \frac{a_{n+1}}{a_n} \right| \leq \alpha \leq \limsup \left| \frac{a_{n+1}}{a_n} \right|. \tag{1}$$

If $\limsup |a_{n+1}/a_n| < 1$, then $\alpha < 1$ and the series converges by the Root Test. If $\liminf |a_{n+1}/a_n| > 1$, then $\alpha > 1$ and the series diverges by the Root Test. Assertion 14.8(iii) is verified by again examining the series $\sum 1/n$ and $\sum 1/n^2$. ∎

Inequality (1) in the proof of the Ratio Test shows that the Root Test is superior to the Ratio Test in the following sense: Whenever the Root Test gives no information [i.e., $\alpha = 1$] the Ratio Test will surely also give no information. On the other hand, Example 8 below gives a series for which the Ratio Test gives no information but which converges by the Root Test. Nevertheless, the tests usually fail together as the next remark shows.

14.10 Remark.
If the terms a_n are nonzero and if $\lim |a_{n+1}/a_n| = 1$, then $\alpha = \limsup |a_n|^{1/n} = 1$ by Corollary 12.3, so neither the Ratio Test nor the Root Test gives information concerning the convergence of $\sum a_n$.

We have three tests for convergence of a series [Comparison, Ratio, Root] and we will obtain two more in the next section. There is no clearcut strategy advising us which test to try first. However, if the form of a given series $\sum a_n$ does not suggest a particular strategy, and if the ratios a_{n+1}/a_n are easy to calculate, one may as well try the Ratio Test first.

Example 3
Consider the series

$$\sum_{n=2}^{\infty} \left(-\frac{1}{3}\right)^n = \frac{1}{9} - \frac{1}{27} + \frac{1}{81} - \frac{1}{243} + \cdots. \tag{1}$$

This is a geometric series and has the form $\sum_{n=0}^{\infty} ar^n$ if we write it as $(1/9) \sum_{n=0}^{\infty} (-1/3)^n$. Here $a = 1/9$ and $r = -1/3$, so by (2) of Example 1 the sum is $(1/9)/[1 - (-1/3)] = 1/12$.

The series (1) can also be shown to converge by the Comparison Test, since $\sum 1/3^n$ converges by the Ratio Test or by the Root Test. In fact, if $a_n = (-1/3)^n$, then $\lim |a_{n+1}/a_n| = \limsup |a_n|^{1/n} = 1/3$. Of course, none of these tests will give us the exact value of the series (1).

Example 4
Consider the series

$$\sum \frac{n}{n^2 + 3}. \tag{1}$$

If $a_n = \dfrac{n}{n^2 + 3}$, then

$$\frac{a_{n+1}}{a_n} = \frac{n+1}{(n+1)^2 + 3} \cdot \frac{n^2 + 3}{n} = \frac{n+1}{n} \cdot \frac{n^2 + 3}{n^2 + 2n + 4},$$

so $\lim |a_{n+1}/a_n| = 1$. As noted in 14.10, neither the Ratio Test nor the Root Test gives any information in this case. Before trying the Comparison Test we need to decide whether we *believe* the series converges or not. Since a_n is approximately $1/n$ for large n and since $\sum(1/n)$ diverges, we expect the series (1) to diverge. Now

$$\frac{n}{n^2 + 3} \geq \frac{n}{n^2 + 3n^2} = \frac{n}{4n^2} = \frac{1}{4n}.$$

Since $\sum(1/n)$ diverges, $\sum(1/4n)$ also diverges [its partial sums are $s_n/4$ where $s_n = \sum_{k=1}^{n}(1/k)$], so (1) diverges by the Comparison Test.

Example 5
Consider the series

$$\sum \frac{1}{n^2 + 1}.$$ (1)

As the reader should check, neither the Ratio Rest nor the Root Test gives any information. The nth term is approximately $\frac{1}{n^2}$ and in fact $\frac{1}{n^2+1} \leq \frac{1}{n^2}$. Since $\sum \frac{1}{n^2}$ converges, the series (1) converges by the Comparison Test.

Example 6
Consider the series

$$\sum \frac{n}{3^n}.$$ (1)

If $a_n = n/3^n$, then $a_{n+1}/a_n = (n + 1)/(3n)$, so $\lim |a_{n+1}/a_n| = 1/3$. Hence the series (1) converges by the Ratio Test. In this case, applying the Root Test is not much more difficult provided we recall $\lim n^{1/n} = 1$. It is also possible to show that (1) converges by comparing it with a suitable geometric series.

Example 7
Consider the series

$$\sum a_n \text{ where } a_n = \left[\frac{2}{(-1)^n - 3}\right]^n.$$ (1)

The form of a_n suggests the Root Test. Since $|a_n|^{1/n} = 1$ for even n and $|a_n|^{1/n} = 1/2$ for odd n, we have $\alpha = \limsup |a_n|^{1/n} = 1$. So the Root Test gives no information and the Ratio Test cannot help either. On the other hand, if we had been alert, we would have observed that $a_n = 1$ for even n, so (a_n) cannot converge to 0. Therefore the series (1) diverges by Corollary 14.5.

Example 8

Consider the series

$$\sum_{n=0}^{\infty} 2^{(-1)^n - n} = 2 + \frac{1}{4} + \frac{1}{2} + \frac{1}{16} + \frac{1}{8} + \frac{1}{64} + \cdots. \tag{1}$$

Let $a_n = 2^{(-1)^n - n}$. Since $a_n \leq \frac{1}{2^{n-1}}$ for all n, we can quickly conclude that the series converges by the Comparison Test. But our real interest in this series is that it illustrates the difference between the Ratio Test and the Root Test. Since $a_{n+1}/a_n = 1/8$ for even n and $a_{n+1}/a_n = 2$ for odd n, we have

$$\frac{1}{8} = \liminf \left| \frac{a_{n+1}}{a_n} \right| < 1 < \limsup \left| \frac{a_{n+1}}{a_n} \right| = 2.$$

Hence the Ratio Test gives no information.

Note that $(a_n)^{1/n} = 2^{\frac{1}{n} - 1}$ for even n and $(a_n)^{1/n} = 2^{-\frac{1}{n} - 1}$ for odd n. Since $\lim 2^{\frac{1}{n}} = \lim 2^{-\frac{1}{n}} = 1$ by Example 7(d) in §9, we conclude that $\lim(a_n)^{1/n} = \frac{1}{2}$. Therefore $\alpha = \limsup(a_n)^{1/n} = \frac{1}{2} < 1$ and the series (1) converges by the Root Test.

Example 9

Consider the series

$$\sum \frac{(-1)^n}{\sqrt{n}}. \tag{1}$$

Since $\lim \sqrt{n/(n+1)} = 1$, neither the Ratio Test nor the Root Test gives any information. Since $\sum \frac{1}{\sqrt{n}}$ diverges, we will not be able to use the Comparison Test 14.6(i) to show that (1) converges. Since the terms of the series (1) are not all nonnegative, we will not be able to use the Comparison Test 14.6(ii) to show that (1) diverges. It turns out that this series converges by the Alternating Series Test 15.3, which we have deferred to the next section.

Exercises

14.1. Determine which of the following series converge. Justify your answers.

(a) $\sum \frac{n^4}{2^n}$ 　　　　　　　　　　　(b) $\sum \frac{2^n}{n!}$

(c) $\sum \frac{n^2}{3^n}$

(d) $\sum \frac{n!}{n^4+3}$

(e) $\sum \frac{\cos^2 n}{n^2}$

(f) $\sum_{n=2}^{\infty} \frac{1}{\log n}$

14.2. Repeat Exercise 14.1 for the following.

(a) $\sum \frac{n-1}{n^2}$

(b) $\sum (-1)^n$

(c) $\sum \frac{3n}{n^3}$

(d) $\sum \frac{n^3}{3^n}$

(e) $\sum \frac{n^2}{n!}$

(f) $\sum \frac{1}{n^n}$

(g) $\sum \frac{n}{2^n}$

14.3. Repeat Exercise 14.1 for the following.

(a) $\sum \frac{1}{\sqrt{n!}}$

(b) $\sum \frac{2+\cos n}{3^n}$

(c) $\sum \frac{1}{2^n+n}$

(d) $\sum (\frac{1}{2})^n (50 + \frac{2}{n})$

(e) $\sum \sin(\frac{n\pi}{9})$

(f) $\sum \frac{(100)^n}{n!}$

14.4. Repeat Exercise 14.1 for the following.

(a) $\sum_{n=2}^{\infty} \frac{1}{[n+(-1)^n]^2}$

(b) $\sum [\sqrt{n+1} - \sqrt{n}]$

(c) $\sum \frac{n!}{n^n}$

14.5. Suppose that $\sum a_n = A$ and $\sum b_n = B$ where A and B are real numbers. Use limit theorems from § 9 to quickly prove the following.

(a) $\sum (a_n + b_n) = A + B$.

(b) $\sum k a_n = kA$ for $k \in \mathbb{R}$.

(c) Is $\sum a_n b_n = AB$ a reasonable conjecture? Discuss.

14.6. **(a)** Prove that if $\sum |a_n|$ converges and (b_n) is a bounded sequence, then $\sum a_n b_n$ converges. *Hint*: Use Theorem 14.4.

(b) Observe that Corollary 14.7 is a special case of part (a).

14.7. Prove that if $\sum a_n$ is a convergent series of nonnegative numbers and $p > 1$, then $\sum a_n^p$ converges.

14.8. Show that if $\sum a_n$ and $\sum b_n$ are convergent series of nonnegative numbers, then $\sum \sqrt{a_n b_n}$ converges. *Hint*: Show that $\sqrt{a_n b_n} \le a_n + b_n$ for all n.

14.9. The convergence of a series does not depend on any finite number of the terms, though of course the value of the limit does. More precisely, consider series $\sum a_n$ and $\sum b_n$ and suppose that the set $\{n \in \mathbb{N} : a_n \ne b_n\}$ is finite. Then the series both converge or else they both diverge. Prove this. *Hint*: This is almost obvious from Theorem 14.4.

14.10. Find a series $\sum a_n$ which diverges by the Root Test but for which the Ratio Test gives no information. Compare Example 8.

14.11. Let (a_n) be a sequence of nonzero real numbers such that the sequence $(\frac{a_{n+1}}{a_n})$ of ratios is a constant sequence. Show that $\sum a_n$ is a geometric series.

14.12. Let $(a_n)_{n\in\mathbb{N}}$ be a sequence such that $\liminf |a_n| = 0$. Prove that there is a subsequence $(a_{n_k})_{k\in\mathbb{N}}$ such that $\sum_{k=1}^{\infty} a_{n_k}$ converges.

14.13. We have seen that it is often a lot harder to find the value of an infinite sum than to show that it exists. Here are some sums that can be handled.

(a) Calculate $\sum_{n=1}^{\infty} (\frac{2}{3})^n$ and $\sum_{n=1}^{\infty} (-\frac{2}{3})^n$.

(b) Prove $\sum_{n=1}^{\infty} \frac{1}{n(n+1)} = 1$. *Hint*: Note that $\sum_{k=1}^{n} \frac{1}{k(k+1)} = \sum_{k=1}^{n} [\frac{1}{k} - \frac{1}{k+1}]$.

(c) Prove that $\sum_{n=1}^{\infty} \frac{n-1}{2^{n+1}} = \frac{1}{2}$. *Hint*: Note that $\frac{k-1}{2^{k+1}} = \frac{k}{2^k} - \frac{k+1}{2^{k+1}}$.

(d) Use (c) to calculate $\sum_{n=1}^{\infty} \frac{n}{2^n}$.

14.14. Prove that $\sum_{n=2}^{\infty} \frac{1}{n}$ diverges by comparing with the series $\sum_{n=2}^{\infty} a_n$ where (a_n) is the sequence

$$\left(\frac{1}{2}, \frac{1}{4}, \frac{1}{4}, \frac{1}{8}, \frac{1}{8}, \frac{1}{8}, \frac{1}{8}, \frac{1}{16}, \frac{1}{16}, \frac{1}{16}, \frac{1}{16}, \frac{1}{16}, \frac{1}{16}, \frac{1}{16}, \frac{1}{16}, \frac{1}{32}, \frac{1}{32}, \cdots \right).$$

§15 Alternating Series and Integral Tests

Sometimes one can check convergence or divergence of series by comparing the partial sums with familiar integrals. We illustrate.

Example 1

We show that $\sum \frac{1}{n} = +\infty$.

Consider the picture of the function $f(x) = \frac{1}{x}$ in Figure 15.1. For $n \geq 1$ it is evident that

$$\sum_{k=1}^{n} \frac{1}{k} = \text{Sum of the areas of the first } n \text{ rectangles in Figure 15.1}$$

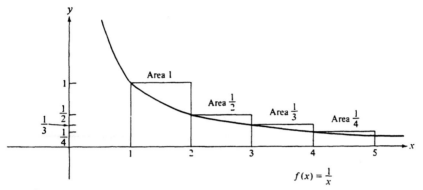

FIGURE 15.1

$\ge$ Area under the curve $\dfrac{1}{x}$ between 1 and $n + 1$

$$= \int_1^{n+1} \frac{1}{x}\, dx = \log(n + 1).$$

Since $\lim_{n\to\infty} \log(n + 1) = +\infty$, we conclude that $\sum_{n=1}^{\infty} \frac{1}{n} = +\infty$.

The series $\sum \frac{1}{n}$ diverges very slowly. In Example 7 of §16, we observe that $\sum_{n=1}^{N} \frac{1}{n}$ is approximately $\log_e N + 0.5772$. Thus for $N = 1,000$ the sum is approximately 7.485, and for $N = 1,000,000$ the sum is approximately 14.393.

Another proof that $\sum \frac{1}{n}$ diverges was indicated in Exercise 14.14. However, an integral test is useful to establish the next result.

Example 2
We show that $\sum \frac{1}{n^2}$ converges.
Consider the graph of $f(x) = \frac{1}{x^2}$ in Figure 15.2. Then we have

$$\sum_{k=1}^{n} \frac{1}{k^2} = \text{Sum of the areas of the first } n \text{ rectangles}$$

$$\le 1 + \int_1^n \frac{1}{x^2}\, dx = 2 - \frac{1}{n} < 2$$

for all $n \ge 1$. Thus the partial sums form an increasing sequence that is bounded above by 2. Therefore $\sum_{n=1}^{\infty} \frac{1}{n^2}$ converges and its sum is less than or equal to 2. Actually, we have already mentioned [without proof!] that the sum is $\frac{\pi^2}{6} = 1.6449\cdots$.

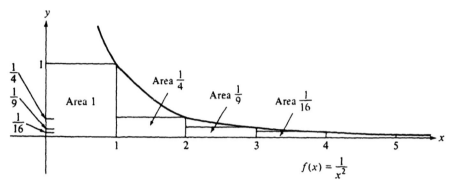

FIGURE 15.2

Note that in estimating $\sum_{k=1}^{n} \frac{1}{k^2}$ we did not simply write $\sum_{k=1}^{n} \frac{1}{k^2} \le \int_{0}^{n} \frac{1}{x^2}\,dx$, even though this is true, because this integral is infinite. We were after a *finite* upper bound for the partial sums.

The techniques just illustrated can be used to prove the following theorem.

15.1 Theorem.
$\sum \frac{1}{n^p}$ converges if and only if $p > 1$.

Proof
Supply your own picture and observe that if $p > 1$, then

$$\sum_{k=1}^{n} \frac{1}{k^p} \le 1 + \int_{1}^{n} \frac{1}{x^p}\,dx = 1 + \frac{1}{p-1}(1 - \frac{1}{n^{p-1}}) < 1 + \frac{1}{p-1} = \frac{p}{p-1}.$$

Consequently $\sum_{n=1}^{\infty} \frac{1}{n^p} \le \frac{p}{p-1} < +\infty$.

Suppose that $0 < p \le 1$. Then $\frac{1}{n} \le \frac{1}{n^p}$ for all n. Since $\sum \frac{1}{n}$ diverges, we see that $\sum \frac{1}{n^p}$ diverges by the Comparison Test. ∎

15.2 Integral Tests.
Here are the conditions under which an integral test is advisable:
  (a) The tests in §14 do not seem to apply.
  (b) The terms of the series $\sum a_n$ are nonnegative.
  (c) There is a nice nonincreasing function f on $[1, \infty)$ such that $f(n) = a_n$ for all n [f is *nonincreasing* if $x < y$ implies $f(x) \ge f(y)$].

(d) The integral of f is easy to calculate or estimate.

If $\lim_{n \to \infty} \int_1^n f(x)\, dx = +\infty$, then the series will diverge just as in Example 1. If $\lim_{n \to \infty} \int_1^n f(x)\, dx < +\infty$, then the series will converge just as in Example 2. The interested reader may formulate and prove the general result [Exercise 15.8].

The following result is a bit tricky to prove, but it enables us to conclude that series like $\sum \frac{(-1)^n}{\sqrt{n}}$ converge even though they do not converge absolutely. See Example 9 in §14.

15.3 Alternating Series Theorem.
If $a_1 \geq a_2 \geq \cdots \geq a_n \geq \cdots \geq 0$ and $\lim a_n = 0$, then the alternating series $\sum (-1)^n a_n$ converges.

The series $\sum (-1)^n a_n$ is called an *alternating series* because the signs of the terms alternate between $+$ and $-$.

Proof
It suffices to show that the series satisfies the Cauchy criterion 14.3(3). This will follow easily from

$$n \geq m > N \quad \text{implies} \quad \left| \sum_{k=m}^{n} (-1)^k a_k \right| \leq a_N, \tag{1}$$

since for each $\epsilon > 0$ there exists $N \in \mathbb{N}$ such that $a_N < \epsilon$.

To prove (1), we fix $n \geq m$ and define

$$A = a_m - a_{m+1} + a_{m+2} - a_{m+3} + \cdots \pm a_n$$

so that

$$\sum_{k=m}^{n} (-1)^k a_k = (-1)^m A. \tag{2}$$

If $n - m$ is odd, the last term of A is $-a_n$, so

$$A = [a_m - a_{m+1}] + [a_{m+2} - a_{m+3}] + \cdots + [a_{n-1} - a_n] \geq 0$$

and also

$$A = a_m - [a_{m+1} - a_{m+2}] - [a_{m+3} - a_{m+4}] - \cdots - [a_{n-2} - a_{n-1}] - a_n \leq a_m.$$

Remember that the numbers in brackets are nonnegative, since (a_n) is nonincreasing. If $n - m$ is even, the last term of A is $+a_n$, so

$$A = [a_m - a_{m+1}] + [a_{m+2} - a_{m+3}] + \cdots + [a_{n-2} - a_{n-1}] + a_n \geq 0$$

and

$$A = a_m - [a_{m+1} - a_{m+2}] - [a_{m+3} - a_{m+4}] - \cdots - [a_{n-1} - a_n] \leq a_m.$$

In either case we have $0 \leq A \leq a_m$. Hence from (2) we see that

$$\left| \sum_{k=m}^{n} (-1)^k a_k \right| = A \leq a_m.$$

Assertion (1) now follows since $n \geq m > N$ implies

$$\left| \sum_{k=m}^{n} (-1)^k a_k \right| \leq a_m \leq a_N.$$

∎

Exercises

15.1. Determine which of the following series converge. Justify your answers.
   (a) $\sum \frac{(-1)^n}{n}$ (b) $\sum \frac{(-1)^n n!}{2^n}$

15.2. Repeat Exercise 15.1 for the following.
   (a) $\sum [\sin(\frac{n\pi}{6})]^n$ (b) $\sum [\sin(\frac{n\pi}{7})]^n$

15.3. Show that $\sum_{n=2}^{\infty} \frac{1}{n(\log n)^p}$ converges if and only if $p > 1$.

15.4. Determine which of the following series converge. Justify your answers.
   (a) $\sum_{n=2}^{\infty} \frac{1}{\sqrt{n} \log n}$ (b) $\sum_{n=2}^{\infty} \frac{\log n}{n}$
   (c) $\sum_{n=4}^{\infty} \frac{1}{n(\log n)(\log \log n)}$ (d) $\sum_{n=2}^{\infty} \frac{\log n}{n^2}$

15.5. Why didn't we use the Comparison Test to prove Theorem 15.1 for $p > 1$?

15.6. (a) Give an example of a divergent series $\sum a_n$ for which $\sum a_n^2$ converges.

   (b) Observe that if $\sum a_n$ is a convergent series of nonnegative terms, then $\sum a_n^2$ also converges. See Exercise 14.7.

(c) Give an example of a convergent series $\sum a_n$ for which $\sum a_n^2$ diverges.

15.7. (a) Prove that if (a_n) is a nonincreasing sequence of real numbers and if $\sum a_n$ converges, then $\lim n a_n = 0$. *Hint*: Consider $|a_{N+1} + a_{N+2} + \cdots + a_n|$ for suitable N.

(b) Use (a) to give another proof that $\sum \frac{1}{n}$ diverges.

15.8. Formulate and prove a general integral test as advised in 15.2.

§16 * Decimal Expansions of Real Numbers

We begin by recalling the brief discussion of decimals in Discussion 10.3. There we considered a decimal expansion $k.d_1 d_2 d_3 \cdots$ where k is a nonnegative integer and each digit d_j belongs to $\{0, 1, 2, 3, 4, 5, 6, 7, 8, 9\}$. This expansion represents the real number

$$k + \sum_{j=1}^{\infty} \frac{d_j}{10^j} = k + \sum_{j=1}^{\infty} d_j \cdot 10^{-j}$$

which we also can write as

$$\lim_{n \to \infty} s_n \quad \text{where} \quad s_n = k + \sum_{j=1}^{n} d_j \cdot 10^{-j}.$$

Thus *every such decimal expansion represents a nonnegative real number.* We will prove the converse after we formalize the process of long division. The development here is based on some suggestions by Karl Stromberg.

16.1 Long Division.
Let's first consider positive integers a and b where $a < b$. We analyze the familiar long division process which gives a decimal expansion for $\frac{a}{b}$. Figure 16.1 shows the first few steps where $a = 3$ and $b = 7$. If we name the digits $d_1, d_2, d_3, \ldots$ and the remainders $r_1, r_2, r_3, \ldots$, then so far $d_1 = 4$, $d_2 = 2$ and $r_1 = 2$, $r_2 = 6$. At the next step we divide 7 into $60 = 10 \cdot r_2$ and obtain $60 = 7 \cdot 8 + 4$. The quotient 8 becomes the

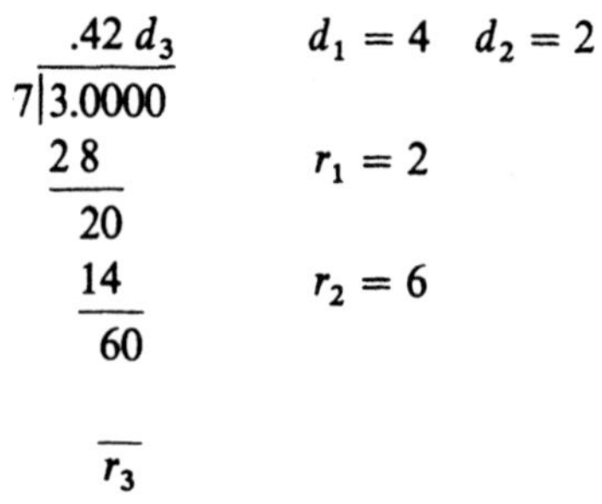

$$.42\ d_3 \qquad d_1 = 4 \quad d_2 = 2$$
$$\overline{7\,|\,3.0000}$$
$$28 \qquad\qquad r_1 = 2$$
$$\overline{20}$$
$$14 \qquad\qquad r_2 = 6$$
$$\overline{60}$$

$$\overline{}$$
$$r_3$$

FIGURE 16.1

third digit d_3, we place the product 56 under 60, subtract and obtain a new remainder $4 = r_3$. That is, we are calculating the remainder obtained by dividing 60 by 7. Next we multiply the remainder $r_3 = 4$ by 10 and repeat the process. At each stage

$$d_n \in \{0, 1, 2, 3, 4, 5, 6, 7, 8, 9\}$$
$$r_n = 10 \cdot r_{n-1} - 7 \cdot d_n$$
$$0 \le r_n < 7.$$

These results hold for $n = 1, 2, \ldots$ if we set $r_0 = 3$. In general, we set $r_0 = a$ and obtain

$$d_n \in \{0, 1, 2, 3, 4, 5, 6, 7, 8, 9\} \tag{1}$$
$$r_n = 10 \cdot r_{n-1} - b \cdot d_n \tag{2}$$
$$0 \le r_n < b. \tag{3}$$

We next show that this construction is well defined in general and that the decimal expansion represents $\frac{a}{b}$. In what follows we *do not need to assume that a and b are integers; a* and *b* will represent positive numbers. The only noticable change in our construction will be that the "remainders" r_n will not necessarily be integers. We also do not assume that $a < b$, so the first step will be a little different than in our example. The first step will provide us with the integer part of $\frac{a}{b}$.

Let $\mathbb{Z}^+ = \mathbb{N} \cup \{0\}$. By the Archimedean property 4.6, we have $a \le nb$ for some positive integer n. Hence $\{m \in \mathbb{Z}^+ : mb \le a\}$ is finite. This set is also nonempty, since it contains 0, so we can define

$$k = \max\{m \in \mathbb{Z}^+ : mb \le a\}.$$

Thus $kb \le a < (k+1)b$. Let $r_0 = a - kb$ and note that $0 \le r_0 < b$. Next define

$$d_1 = \max\{d \in \mathbb{Z}^+ : db \le 10 \cdot r_0\}$$

and

$$r_1 = 10 \cdot r_0 - d_1 b.$$

Note that $d_1 \le 9$, because $10 \cdot b \le 10 \cdot r_0$ would imply $b \le r_0$, a contradiction. Also note that $d_1 b \le 10 \cdot r_0 < (d_1 + 1)b$, so $0 \le r_1 = 10 \cdot r_0 - d_1 b < b$. Thus the following holds for $n = 1$:

$$d_n \in \{0, 1, 2, 3, 4, 5, 6, 7, 8, 9\} \tag{1}$$
$$r_n = 10 \cdot r_{n-1} - d_n b \tag{2}$$
$$0 \le r_n < b. \tag{3}$$

Suppose that $d_1, d_2, \ldots, d_n \in \mathbb{Z}^+$ and $r_0, r_1, \ldots, r_n$ have been defined satisfying (1)–(3). Next define

$$d_{n+1} = \max\{d \in \mathbb{Z}^+ : db \le 10 \cdot r_n\}$$

and

$$r_{n+1} = 10 \cdot r_n - d_{n+1} b.$$

Then $d_{n+1} \le 9$ since $10 \cdot b \le 10 \cdot r_n$ would imply $b \le r_n$, violating (3). Hence (1) holds for $n+1$ and (2) is obvious for $n+1$ by our definition of r_{n+1}. Finally $d_{n+1} b \le 10 \cdot r_n < (d_{n+1} + 1)b$ implies $0 \le r_{n+1} < b$, so (3) holds for $n + 1$. The construction of the sequences (d_n) and (r_n) satisfying (1)–(3) is completed by an appeal to the principle of induction.

To see that the decimal expansion $k.d_1 d_2 d_3 \cdots$ represents $\frac{a}{b}$, we observe that (2) implies

$$r_n \cdot 10^{-n} = r_{n-1} \cdot 10^{-n+1} - d_n \cdot 10^{-n} \cdot b$$

for $n \ge 1$. Transposing and changing n to j, we obtain

$$d_j \cdot 10^{-j} \cdot b = r_{j-1} \cdot 10^{-j+1} - r_j \cdot 10^{-j}$$

for $j \geq 1$. When we sum from $j = 1$ to $j = n$, most of the terms on the right side cancel [it's called a telescoping sum]. Hence the partial sums s_n for the decimal expansion satisfy

$$s_n \cdot b = \left[k + \sum_{j=1}^{n} d_j \cdot 10^{-j} \right] \cdot b = kb + r_0 - r_n \cdot 10^{-n}.$$

In view of (3), we have $\lim_n [r_n \cdot 10^{-n}] = 0$, so $\lim_n s_n = k + \frac{r_0}{b}$. Recall that $r_0 = a - kb$; hence

$$\lim_{n \to \infty} s_n = k + \frac{a - kb}{b} = \frac{a}{b}.$$

Thus $k.d_1 d_2 d_3 \cdots$ is a decimal expansion for $\frac{a}{b}$.

16.2 Theorem.
Every nonnegative real number x has at least one decimal expansion.

Proof
Let $a = x$ and $b = 1$ in 16.1 above. ∎

As noted in Discussion 10.3, $1.000 \cdots$ and $.999 \cdots$ are decimal expansions for the same real number. That is, the series

$$1 + \sum_{j=1}^{\infty} 0 \cdot 10^{-j} \qquad \text{and} \qquad \sum_{j=1}^{\infty} 9 \cdot 10^{-j}$$

have the same value, namely 1. Similarly, $2.75000 \cdots$ and $2.74999 \cdots$ are both decimal expansions for $\frac{11}{4}$ [Exercise 16.1]. The next theorem shows that this is essentially the only way a number can have distinct decimal expansions.

16.3 Theorem.
A real number x has exactly one decimal expansion or else x has two decimal expansions, one ending in a sequence of all 0's and the other ending in a sequence of all 9's.

Proof
We assume $x \geq 0$. If x has decimal expansions $k.000 \cdots$ with $k > 0$, then it has one other decimal expansion, namely $(k-1).999 \cdots$. If x has decimal expansion $k.d_1 d_2 d_3 \cdots d_r 000 \cdots$ where $d_r \neq 0$, then it

has one other decimal expansion $k.d_1d_2d_3 \cdots (d_r - 1)9999 \cdots$. The reader can easily check these claims [Exercise 16.2].

Now suppose that x has two distinct decimal expansions $k.d_1d_2d_3 \cdots$ and $\ell.e_1e_2e_3 \cdots$. Suppose that $k < \ell$. If any $d_j < 9$, then by Exercise 16.3 we have

$$x < k + \sum_{j=1}^{\infty} 9 \cdot 10^{-j} = k + 1 \leq \ell \leq x,$$

a contradiction. It follows that $x = k + 1 = \ell$ and its decimal expansions must be $k.999 \cdots$ and $(k+1).000 \cdots$. In the remaining case, we have $k = \ell$. Let

$$m = \min\{j : d_j \neq e_j\}.$$

We may assume that $d_m < e_m$. If $d_j < 9$ for any $j > m$, then by Exercise 16.3,

$$x < k + \sum_{j=1}^{m} d_j \cdot 10^{-j} + \sum_{j=m+1}^{\infty} 9 \cdot 10^{-j} = k + \sum_{j=1}^{m} d_j \cdot 10^{-j} + 10^{-m}$$

$$= k + \sum_{j=1}^{m-1} e_j \cdot 10^{-j} + d_m \cdot 10^{-m} + 10^{-m} \leq k + \sum_{j=1}^{m} e_j \cdot 10^{-j} \leq x,$$

a contradiction. Thus $d_j = 9$ for $j > m$. Likewise, if $e_j > 0$ for any $j > m$, then

$$x > k + \sum_{j=1}^{m} e_j \cdot 10^{-j} = k + \sum_{j=1}^{m-1} d_j \cdot 10^{-j} + e_m \cdot 10^{-m}$$

$$\geq k + \sum_{j=1}^{m-1} d_j \cdot 10^{-j} + d_m \cdot 10^{-m} + 10^{-m}$$

$$= k + \sum_{j=1}^{m} d_j \cdot 10^{-j} + \sum_{j=m+1}^{\infty} 9 \cdot 10^{-j} \geq x,$$

a contradiction. So in this case, $d_j = 9$ for $j > m$, $e_m = d_m + 1$ and $e_j = 0$ for $j > m$. ∎

16.4 Definition.
An expression of the form

$$k.d_1d_2 \cdots d_\ell \overline{d_{\ell+1} \cdots d_{\ell+r}}$$

represents the decimal expansion in which the block $d_{\ell+1} \cdots d_{\ell+r}$ is repeated indefinitely:

$$k.d_1 d_2 \cdots d_\ell d_{\ell+1} \cdots d_{\ell+r} d_{\ell+1} \cdots d_{\ell+r} d_{\ell+1} \cdots d_{\ell+r} d_{\ell+1} \cdots d_{\ell+r} \cdots.$$

We call such an expansion a *repeating decimal*.

Example 1
Every integer is a repeating decimal. For example, $17 = 17.\overline{0} = 17.000\cdots$. Another simple example is

$$.\overline{8} = .888\cdots = \sum_{j=1}^{\infty} 8 \cdot 10^{-j} = \frac{8}{10} \sum_{j=0}^{\infty} 10^{-j} = \frac{8}{10} \cdot \frac{10}{9} = \frac{8}{9}.$$

Example 2
The expression $3.9\overline{67}$ represents the repeating decimal $3.9676767\cdots$. We evaluate this as follows:

$$3.9\overline{67} = 3 + 9 \cdot 10^{-1} + 6 \cdot 10^{-2} + 7 \cdot 10^{-3} + 6 \cdot 10^{-4} + 7 \cdot 10^{-5} + \cdots$$

$$= 3 + 9 \cdot 10^{-1} + 67 \cdot 10^{-3} \sum_{j=0}^{\infty} (10^{-2})^j$$

$$= 3 + 9 \cdot 10^{-1} + 67 \cdot 10^{-3} \left(\frac{100}{99} \right) = 3 + \frac{9}{10} + \frac{67}{990}$$

$$= \frac{3928}{990} = \frac{1964}{495}.$$

Thus the repeating decimal $3.9\overline{67}$ represents the rational number $\frac{1964}{495}$. Any repeating decimal can be evaluated as a rational number in this way, as we'll show in the next theorem.

Example 3
We find the decimal expansion for $\frac{11}{7}$. By the usual long division process in 16.1, we find

$$\frac{11}{7} = 1.571\,428\,571\,428\,571\,428\,571\,428\,571\,428\,571\cdots,$$

i.e., $\frac{11}{7} = 1.\overline{571\,428}$. To check this, observe

$$1.\overline{571428} = 1 + 571428 \cdot 10^{-6} \sum_{j=0}^{\infty} (10^{-6})^j = 1 + \frac{571428}{999999}$$

$$= 1 + \frac{4}{7} = \frac{11}{7}.$$

Many books give the next theorem as an exercise, probably to avoid the complicated notation.

16.5 Theorem.
A real number x is rational if and only if its decimal expansion is repeating. [Theorem 16.3 shows that if x has two decimal expansions, they are both repeating.]

Proof
First assume $x \geq 0$ has a repeating decimal expansion $x = k.d_1 d_2 \cdots d_\ell \overline{d_{\ell+1} \cdots d_{\ell+r}}$. Then

$$x = k + \sum_{j=1}^{\ell} d_j \cdot 10^{-j} + 10^{-\ell} y$$

where

$$y = .\overline{d_{\ell+1} \cdots d_{\ell+r}},$$

so it suffices to show such y are rational. To simplify the notation, we write

$$y = .\overline{e_1 e_2 \cdots e_r}.$$

A little computation shows that

$$y = \sum_{j=1}^{r} e_j \cdot 10^{-j} \left[\sum_{j=0}^{\infty} (10^{-r} y^j) \right] = \sum_{j=1}^{r} e_j \cdot 10^{-j} \frac{10^r}{10^r - 1}.$$

In fact, if we write $e_1 e_2 \cdots e_r$ for the usual *decimal* $\sum_{j=0}^{r-1} e_j \cdot 10^{r-1-j}$ *not the product*, then $y = \frac{e_1 e_2 \cdots e_r}{10^r - 1}$; see Example 3.

Next consider any positive rational, say $\frac{a}{b}$ where $a, b \in \mathbb{N}$. We may assume that $a < b$. As we saw in 16.1, $\frac{a}{b}$ is given by the decimal expansion $.d_1 d_2 d_3 \cdots$ where $r_0 = a$,

$$d_n \in \{0, 1, 2, 3, 4, 5, 6, 7, 8, 9\} \tag{1}$$
$$r_n = 10 \cdot r_{n-1} - d_n b \tag{2}$$
$$0 \leq r_n < b, \tag{3}$$

for $n \geq 1$. Since a and b are integers, each r_n is an integer. Thus (3) can be written

$$r_n \in \{0, 1, 2, \ldots, b - 1\} \quad \text{for} \quad n \geq 0. \tag{4}$$

This set has b elements, so the first $b + 1$ remainders r_n cannot all be distinct. That is, there exist integers $m \geq 0$ and $p > 0$ so that

$$0 \leq m < m + p \leq b \quad \text{and} \quad r_m = r_{m+p}.$$

From the construction giving (1)–(3) it is clear that given r_{n-1}, the integers r_n and d_n are uniquely determined. Thus

$$r_j = r_k \quad \text{implies} \quad r_{j+1} = r_{k+1} \quad \text{and} \quad d_{j+1} = d_{k+1}.$$

Since $r_m = r_{m+p}$, we conclude that $r_{m+1} = r_{m+1+p}$ and $d_{m+1} = d_{m+1+p}$. A simple induction shows that the statement

$$\text{"}r_n = r_{n+p} \quad \text{and} \quad d_n = d_{n+p}\text{"}$$

holds for all integers $n \geq m + 1$. Thus the decimal expansion of $\frac{a}{b}$ is periodic with period p after the first m digits. That is,

$$\frac{a}{b} = .d_1 d_2 \cdots d_m \overline{d_{m+1} \cdots d_{m+p}}. \qquad \blacksquare$$

Example 4
An expansion such as

.1010010001000010000010000001000000010000000010000000001100 $\cdots$

must represent an irrational number, since it cannot be a repeating decimal: we've arranged for arbitrarily long blocks of 0's.

Example 5
We do not know the complete decimal expansions of $\sqrt{2}$, $\sqrt{3}$ and many other familiar irrational numbers, but we know that they cannot be repeating by virtue of the last theorem.

Example 6
We have claimed that π and e are irrational. These facts and many others are proved in a fascinating book by Ivan Niven [30]. Here is

the proof that

$$e = \sum_{k=0}^{\infty} \frac{1}{k!}$$

is irrational. Assume that $e = \frac{a}{b}$ where $a, b \in \mathbb{N}$. Then both $b!e$ and $b! \sum_{k=0}^{b} \frac{1}{k!}$ must be integers, so the difference

$$b! \sum_{k=b+1}^{\infty} \frac{1}{k!}$$

must be a positive integer. On the other hand, this last number is less than

$$\frac{1}{b+1} + \frac{1}{(b+1)^2} + \frac{1}{(b+1)^3} + \cdots = \frac{1}{b} \leq 1,$$

a contradiction.

Example 7
There is a famous number introduced by Euler over 200 years ago that arises in the study of the gamma function. It is known as *Euler's constant* and is defined by

$$\gamma = \lim_{n \to \infty} \left[\sum_{k=1}^{n} \frac{1}{k} - \log_e n \right].$$

Even though

$$\lim_{n \to \infty} \sum_{k=1}^{n} \frac{1}{k} = +\infty \quad \text{and} \quad \lim_{n \to \infty} \log_e n = +\infty,$$

the limit defining γ exists and is finite [Exercise 16.9]. In fact, γ is approximately .577216. The amazing fact is that no one knows whether γ is rational or not. Most mathematicians believe γ is irrational. This is because it is "easier" for a number to be irrational, since repeating decimal expansions must be regular. The remark in Exercise 16.8 hints at another reason it is easier for a number to be irrational.

Exercises

16.1. **(a)** Show that $2.74\bar{9}$ and $2.75\bar{0}$ are both decimal expansions for $\frac{11}{4}$.

(b) Which of these expansions arises from the long division process described in 16.1?

16.2. Verify the claims in the first paragraph of the proof of Theorem 16.3.

16.3. Suppose that $\sum a_n$ and $\sum b_n$ are convergent series of nonnegative numbers. Show that if $a_n \le b_n$ for all n and if $a_n < b_n$ for at least one n, then $\sum a_n < \sum b_n$.

16.4. Write the following repeating decimals as rationals, i.e., as fractions of integers.

(a) $.2$ (b) $.0\overline{2}$
(c) $.\overline{02}$ (d) $3.\overline{14}$
(e) $.\overline{10}$ (f) $.1\overline{492}$

16.5. Find the decimal expansions of the following rational numbers.

(a) $\frac{1}{8}$ (b) $\frac{1}{16}$
(c) $\frac{2}{3}$ (d) $\frac{7}{9}$
(e) $\frac{6}{11}$ (f) $\frac{22}{7}$

16.6. Find the decimal expansions of $\frac{1}{7}$, $\frac{2}{7}$, $\frac{3}{7}$, $\frac{4}{7}$, $\frac{5}{7}$ and $\frac{6}{7}$. Note the interesting pattern.

16.7. Is $.1234567891011121314151617181920212223242526\cdots$ rational?

16.8. Let (s_n) be a sequence of numbers in $(0, 1)$. Each s_n has a decimal expansion $.d_1^{(n)} d_2^{(n)} d_3^{(n)} \cdots$. For each n, let $e_n = 6$ if $d_n^{(n)} \ne 6$ and $e_n = 7$ if $d_n^{(n)} = 6$. Show that $e_1 e_2 e_3 \cdots$ is the decimal expansion for some number y in $(0, 1)$ and that $y \ne s_n$ for all n. *Remark*: This shows that the elements of $(0, 1)$ cannot be listed as a sequence. In set-theoretic parlance, $(0, 1)$ is "uncountable." Since the set $\mathbb{Q} \cap (0, 1)$ can be listed as a sequence, there must be a lot of irrational numbers in $(0, 1)$!

16.9. Let $\gamma_n = \left(\sum_{k=1}^{n} \frac{1}{k} \right) - \log_e n = \sum_{k=1}^{n} \frac{1}{k} - \int_1^n \frac{1}{t} dt$.

(a) Show that (γ_n) is a decreasing sequence. *Hint*: Look at $\gamma_n - \gamma_{n+1}$.

(b) Show that $0 < \gamma_n \le 1$ for all n.

(c) Observe that $\gamma = \lim_n \gamma_n$ exists and is finite.

3

CHAPTER

Continuity

Most of the calculus involves the study of continuous functions. In this chapter we study continuous and uniformly continuous functions.

§17 Continuous Functions

Recall that the salient features of a function f are:

(a) the set on which f is defined, called the *domain of f* and written dom(f);

(b) the assignment, rule or formula specifying the value $f(x)$ of f at each x in dom(f).

We will be concerned with functions f such that dom(f) $\subseteq \mathbb{R}$ and such that f is a *real-valued function*, i.e., $f(x) \in \mathbb{R}$ for all $x \in$ dom(f). Properly speaking, the symbol f represents the function while $f(x)$ represents the value of the function at x. However, a function is often given by specifying its values and without mentioning its domain. In this case, the domain is understood to be the *natural domain*: the largest subset of $\mathbb{R}$ on which the function is a well defined real-valued function. Thus "the function $f(x) = \frac{1}{x}$" is shorthand for "the function f given by $f(x) = \frac{1}{x}$ with natural domain $\{x \in \mathbb{R} : x \neq 0\}$." Similarly, the natural domain of $g(x) = \sqrt{4 - x^2}$ is $[-2, 2]$ and the natural domain of $\csc x = \frac{1}{\sin x}$ is the set of real numbers x not of the form $n\pi, n \in \mathbb{Z}$.

In keeping with the approach in this book, we will define continuity in terms of sequences. We then show that our definition is equivalent to the usual ϵ-δ definition.

17.1 Definition.
Let f be a real-valued function whose domain is a subset of $\mathbb{R}$. The function f is *continuous at x_0* in dom(f) if, for every sequence (x_n) in dom(f) converging to x_0, we have $\lim_n f(x_n) = f(x_0)$. If f is continuous at each point of a set $S \subseteq$ dom(f), then f is said to be *continuous on S*. The function f is said to be *continuous* if it is continuous on dom(f).

Our definition implies that the values $f(x)$ are close to $f(x_0)$ when the values x are close to x_0. The next theorem says this in another way. In fact, condition (1) of the next theorem is the ϵ-δ definition of continuity given in many calculus books.

17.2 Theorem.
Let f be a real-valued function whose domain is a subset of $\mathbb{R}$. Then f is continuous at $x_0 \in$ dom(f) if and only if

$$\text{for each } \epsilon > 0 \text{ there exists } \delta > 0 \text{ such that} \tag{1}$$
$$x \in \text{dom}(f) \text{ and } |x - x_0| < \delta \text{ imply } |f(x) - f(x_0)| < \epsilon.$$

Proof
Suppose that (1) holds, and consider a sequence (x_n) in dom(f) such that $\lim x_n = x_0$. We need to prove that $\lim f(x_n) = f(x_0)$. Let $\epsilon > 0$. By (1), there exists $\delta > 0$ such that

$$x \in \text{dom}(f) \quad \text{and} \quad |x - x_0| < \delta \quad \text{imply} \quad |f(x) - f(x_0)| < \epsilon.$$

Since $\lim x_n = x_0$, there exists N so that

$$n > N \quad \text{implies} \quad |x_n - x_0| < \delta.$$

It follows that

$$n > N \quad \text{implies} \quad |f(x_n) - f(x_0)| < \epsilon.$$

This proves that $\lim f(x_n) = f(x_0)$.

Now assume that f is continuous at x_0, but that (1) fails. Then there exists $\epsilon > 0$ so that the implication

"$x \in \text{dom}(f)$ and $|x - x_0| < \delta$ imply $|f(x) - f(x_0)| < \epsilon$"

fails for each $\delta > 0$. In particular, the implication

"$x \in \text{dom}(f)$ and $|x - x_0| < \dfrac{1}{n}$ imply $|f(x) - f(x_0)| < \epsilon$"

fails for each $n \in \mathbb{N}$. So for each $n \in \mathbb{N}$ there exists x_n in $\text{dom}(f)$ such that $|x_n - x_0| < \frac{1}{n}$ and yet $|f(x_n) - f(x_0)| \geq \epsilon$. Thus we have $\lim x_n = x_0$, but we cannot have $\lim f(x_n) = f(x_0)$ since $|f(x_n) - f(x_0)| \geq \epsilon$ for all n. This shows that f cannot be continuous at x_0, contrary to our assumption. ∎

As the next example illustrates, it is sometimes easier to work with the sequential definition of continuity in Definition 17.1 than the ϵ-δ property in Theorem 17.2. However, it is important to get comfortable with the ϵ-δ property, partly because the definition of uniform continuity is more closely related to the ϵ-δ property than the sequential definition.

Example 1
Let $f(x) = 2x^2 + 1$ for $x \in \mathbb{R}$. Prove that f is continuous on $\mathbb{R}$ by
    (a) using the definition,
    (b) using the ϵ-δ property of Theorem 17.2.

Solution
(a) Suppose that $\lim x_n = x_0$. Then we have

$$\lim f(x_n) = \lim[2x_n^2 + 1] = 2[\lim x_n]^2 + 1 = 2x_0^2 + 1 = f(x_0)$$

where the second equality is an application of the limit theorems 9.2–9.4. Hence f is continuous at each x_0 in $\mathbb{R}$.

(b) Let x_0 be in $\mathbb{R}$ and let $\epsilon > 0$. We want to show $|f(x) - f(x_0)| < \epsilon$ provided $|x - x_0|$ is sufficiently small, i.e., less than some δ. We observe that

$$|f(x) - f(x_0)| = |2x^2 + 1 - (2x_0^2 + 1)| = |2x^2 - 2x_0^2|$$
$$= 2|x - x_0| \cdot |x + x_0|.$$

We need to get a bound for $x + x_0|$ that does not depend on x. We notice that if $|x - x_0| < 1$, say, then $|x| < |x_0| + 1$ and hence $|x + x_0| \leq |x| + |x_0| < 2|x_0| + 1$. Thus we have

$$|f(x) - f(x_0)| < 2|x - x_0|(2|x_0| + 1)$$

provided $|x - x_0| < 1$. To arrange for $2|x - x_0|(2|x_0| + 1) < \epsilon$, it suffices to have $|x - x_0| < \frac{\epsilon}{2(2|x_0|+1)}$ and also $|x - x_0| < 1$. So we put

$$\delta = \min\left\{1, \frac{\epsilon}{2(2|x_0| + 1)}\right\}.$$

The work above shows that $|x - x_0| < \delta$ implies $|f(x) - f(x_0)| < \epsilon$, as desired. ☐

The reason that solution (a) in Example 1 is so much easier is that the careful analysis was done in proving the limit theorems in §9.

Example 2
Let $f(x) = x^2 \sin(\frac{1}{x})$ for $x \neq 0$ and $f(0) = 0$. The graph of f in Figure 17.1 *looks* continuous. Prove that f is continuous at 0.

Solution
Let $\epsilon > 0$. Clearly $|f(x) - f(0)| = |f(x)| \leq x^2$ for all x. Since we want this to be less than ϵ, we set $\delta = \sqrt{\epsilon}$. Then $|x - 0| < \delta$ implies $x^2 < \delta^2 = \epsilon$, so

$$|x - 0| < \delta \quad \text{implies} \quad |f(x) - f(0)| < \epsilon.$$

Hence the ϵ-δ property holds and f is continuous at 0. ☐

In Example 2 it would have been equally easy to show that if $\lim x_n = 0$ then $\lim f(x_n) = 0$. The function f in Example 2 is also continuous at the other points of $\mathbb{R}$; see Example 4.

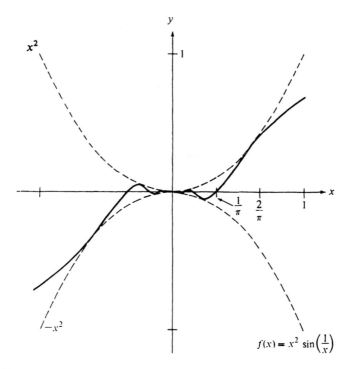

$$f(x) = x^2 \sin\left(\frac{1}{x}\right)$$

FIGURE 17.1

Example 3
Let $f(x) = \frac{1}{x} \sin(\frac{1}{x^2})$ for $x \neq 0$ and $f(0) = 0$; see Figure 17.2. Show that f is *discontinuous*, i.e., not continuous, at 0.

Solution
It suffices for us to find a sequence (x_n) converging to 0 such that $f(x_n)$ does not converge to $f(0) = 0$. So we will arrange for $\frac{1}{x_n} \sin(\frac{1}{x_n^2}) = \frac{1}{x_n}$ where $x_n \to 0$. Thus we want $\sin(\frac{1}{x_n^2}) = 1$, $\frac{1}{x_n^2} = 2\pi n + \frac{\pi}{2}$, $x_n^2 = \frac{1}{2\pi n + \frac{\pi}{2}}$ or $x_n = \frac{1}{\sqrt{2\pi n + \frac{\pi}{2}}}$. Then $\lim x_n = 0$ while $\lim f(x_n) = \lim \frac{1}{x_n} = +\infty$. ☐

Let f be a real-valued function. For k in $\mathbb{R}$, kf signifies the function defined by $(kf)(x) = kf(x)$ for $x \in \text{dom}(f)$. Also $|f|$ is the function defined by $|f|(x) = |f(x)|$ for $x \in \text{dom}(f)$. Thus if f is given by $f(x) = \sqrt{x} - 4$ for $x \geq 0$, then $3f$ is given by $(3f)(x) = 3\sqrt{x} - 12$ for $x \geq 0$, and $|f|$ is given by $|f|(x) = |\sqrt{x} - 4|$ for $x \geq 0$. Here is an easy theorem.

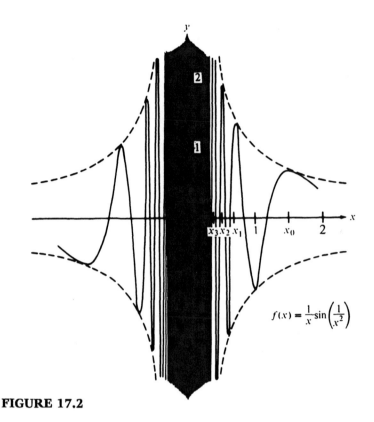

$$f(x) = \frac{1}{x}\sin\left(\frac{1}{x^2}\right)$$

FIGURE 17.2

17.3 Theorem.
Let f be a real-valued function with $\mathrm{dom}(f) \subseteq \mathbb{R}$. If f is continuous at x_0 in $\mathrm{dom}(f)$, then $|f|$ and kf, $k \in \mathbb{R}$, are continuous at x_0.

Proof
Consider a sequence (x_n) in $\mathrm{dom}(f)$ converging to x_0. Since f is continuous at x_0, we have $\lim f(x_n) = f(x_0)$. Theorem 9.2 shows that $\lim kf(x_n) = kf(x_0)$. This proves that kf is continuous at x_0.

To prove that $|f|$ is continuous at x_0, we need to prove that $\lim |f(x_n)| = |f(x_0)|$. This follows from the inequality

$$||f(x_n)| - |f(x_0)|| \le |f(x_n) - f(x_0)|; \qquad (1)$$

see Exercise 3.5. [In detail, consider $\epsilon > 0$. Since $\lim f(x_n) = f(x_0)$, there exists N such that $n > N$ implies $|f(x_n) - f(x_0)| < \epsilon$. So by (1),

$n > N$ implies

$$||f(x_n)| - |f(x_0)|| < \epsilon;$$

thus $\lim |f(x_n)| = |f(x_0)|.]$ ∎

We remind readers that if f and g are real-valued functions, then we can combine f and g to obtain new functions:

$$(f + g)(x) = f(x) + g(x); \qquad fg(x) = f(x)g(x);$$
$$(f/g)(x) = \frac{f(x)}{g(x)}; \qquad g \circ f(x) = g(f(x));$$
$$\max(f, g)(x) = \max\{f(x), g(x)\}; \qquad \min(f, g)(x) = \min\{f(x), g(x)\}.$$

The function $g \circ f$ is called the *composition of g and f*. Each of these new functions is defined exactly where they make sense. Thus the domains of $f + g$, fg, $\max(f, g)$ and $\min(f, g)$ are $\text{dom}(f) \cap \text{dom}(g)$, the domain of f/g is the set $\text{dom}(f) \cap \{x \in \text{dom}(g) : g(x) \neq 0\}$, and the domain of $g \circ f$ is $\{x \in \text{dom}(f) : f(x) \in \text{dom}(g)\}$. Note that $f + g = g + f$ and $fg = gf$ but that in general $f \circ g \neq g \circ f$.

These new functions are continuous if f and g are continuous.

17.4 Theorem.
Let f and g be real-valued functions that are continuous at x_0 in $\mathbb{R}$. Then
    (i) $f + g$ *is continuous at x_0;*
    (ii) fg *is continuous at x_0;*
    (iii) f/g *is continuous at x_0 if $g(x_0) \neq 0$.*

Proof
We are given that $x_0 \in \text{dom}(f) \cap \text{dom}(g)$. Let (x_n) be a sequence in $\text{dom}(f) \cap \text{dom}(g)$ converging to x_0. Then we have $\lim f(x_n) = f(x_0)$ and $\lim g(x_n) = g(x_0)$. Theorem 9.3 shows that

$$\lim(f + g)(x_n) = \lim[f(x_n) + g(x_n)] = \lim f(x_n) + \lim g(x_n)$$
$$= f(x_0) + g(x_0) = (f + g)(x_0),$$

so $f + g$ is continuous at x_0. Likewise, Theorem 9.4 implies that fg is continuous at x_0.

To handle f/g we assume $g(x_0) \neq 0$ and consider a sequence (x_n) in $\text{dom}(f) \cap \{x \in \text{dom}(g) : g(x) \neq 0\}$ converging to x_0. Then

Theorem 9.6 shows that

$$\lim\left(\frac{f}{g}\right)(x_n) = \lim\frac{f(x_n)}{g(x_n)} = \frac{f(x_0)}{g(x_0)} = \left(\frac{f}{g}\right)(x_0);$$

so f/g is continuous at x_0. ∎

17.5 Theorem.

If f is continuous at x_0 and g is continuous at $f(x_0)$, then the composite function $g \circ f$ is continuous at x_0.

Proof
We are given that $x_0 \in \text{dom}(f)$ and that $f(x_0) \in \text{dom}(g)$. Let (x_n) be a sequence in $\{x \in \text{dom}(f) : f(x) \in \text{dom}(g)\}$ converging to x_0. Since f is continuous at x_0, we have $\lim f(x_n) = f(x_0)$. Since the sequence $(f(x_n))$ converges to $f(x_0)$ and g is continuous at $f(x_0)$, we also have $\lim g(f(x_n)) = g(f(x_0))$; that is, $\lim g \circ f(x_n) = g \circ f(x_0)$. Hence $g \circ f$ is continuous at x_0. ∎

Example 4
For this example, let us accept as known that polynomial functions and the functions $\sin x$, $\cos x$ and e^x are continuous on $\mathbb{R}$. Then $4e^x$ and $|\sin x|$ are continuous on $\mathbb{R}$ by Theorem 17.3. The function $\sin x + 4e^x + x^3$ is continuous on $\mathbb{R}$ by (i) of Theorem 17.4. The function $x^4 \sin x$ is continuous on $\mathbb{R}$ by (ii) of Theorem 17.4, and (iii) of Theorem 17.4 shows that $\tan x = \frac{\sin x}{\cos x}$ is continuous wherever $\cos x \neq 0$, i.e., at all x not of the form $n\pi + \frac{\pi}{2}$, $n \in \mathbb{Z}$. Theorem 17.5 tells us that $e^{\sin x}$ and $\cos(e^x)$ are continuous on $\mathbb{R}$; for example, $\cos(e^x) = g \circ f(x)$ where $f(x) = e^x$ and $g(x) = \cos x$. Several applications of Theorems 17.3–17.5 will show that $x^2 \sin(\frac{1}{x})$ and $\frac{1}{x}\sin(\frac{1}{x^2})$ are continuous at all nonzero x in $\mathbb{R}$.

Example 5
Let f and g be continuous at x_0 in $\mathbb{R}$. Prove that $\max(f,g)$ is continuous at x_0.

Solution
First observe that

$$\max(f,g) = \frac{1}{2}(f+g) + \frac{1}{2}|f - g|.$$

This equation holds because $\max\{a, b\} = \frac{1}{2}(a+b) + \frac{1}{2}|a-b|$ is true for all $a, b \in \mathbb{R}$, a fact which is easily checked by considering the cases $a \geq b$ and $a < b$. By Theorem 17.4(i), $f + g$ and $f - g$ are continuous at x_0. Hence $|f - g|$ is continuous at x_0 by Theorem 17.3. Then $\frac{1}{2}(f + g)$ and $\frac{1}{2}|f - g|$ are continuous at x_0 by Theorem 17.3, and another application of Theorem 17.4(i) shows that $\max(f, g)$ is continuous at x_0. □

Exercises

17.1. Let $f(x) = \sqrt{4 - x}$ for $x \leq 4$ and $g(x) = x^2$ for all $x \in \mathbb{R}$.

(a) Give the domains of $f + g$, fg, $f \circ g$ and $g \circ f$.

(b) Find the values $f \circ g(0)$, $g \circ f(0)$, $f \circ g(1)$, $g \circ f(1)$, $f \circ g(2)$ and $g \circ f(2)$.

(c) Are the functions $f \circ g$ and $g \circ f$ equal?

(d) Are $f \circ g(3)$ and $g \circ f(3)$ meaningful?

17.2. Let $f(x) = 4$ for $x \geq 0$, $f(x) = 0$ for $x < 0$, and $g(x) = x^2$ for all x. Thus $\text{dom}(f) = \text{dom}(g) = \mathbb{R}$.

(a) Determine the following functions: $f + g$, fg, $f \circ g$, $g \circ f$. Be sure to specify their domains.

(b) Which of the functions $f, g, f+g, fg, f\circ g, g\circ f$ is continuous?

17.3. Accept on faith that the following familiar functions are continuous on their domains: $\sin x$, $\cos x$, e^x, 2^x, $\log_e x$ for $x > 0$, x^p for $x > 0$ [p any real number]. Use these facts and theorems in this section to prove that the following functions are also continuous.

(a) $\log_e(1 + \cos^4 x)$

(b) $[\sin^2 x + \cos^6 x]^\pi$

(c) 2^{x^2}

(d) 8^x

(e) $\tan x$ for $x \neq$ odd multiple of $\frac{\pi}{2}$

(f) $x \sin(\frac{1}{x})$ for $x \neq 0$

(g) $x^2 \sin(\frac{1}{x})$ for $x \neq 0$

(h) $\frac{1}{x}\sin(\frac{1}{x^2})$ for $x \neq 0$

17.4. Prove that the function $\sqrt{x}$ is continuous on its domain $[0, \infty)$.
Hint: Apply Example 5 in §8.

17.5. (a) Prove that if $m \in \mathbb{N}$, then the function $f(x) = x^m$ is continuous on $\mathbb{R}$.

(b) Prove that every *polynomial function* $p(x) = a_0 + a_1 x + \cdots + a_n x^n$ is continuous on $\mathbb{R}$.

17.6. A *rational function* is a function f of the form p/q where p and q are polynomial functions. The domain of f is $\{x \in \mathbb{R} : q(x) \neq 0\}$. Prove that every rational function is continuous. *Hint:* Use Exercise 17.5.

17.7. (a) Observe that if k is in $\mathbb{R}$, then the function $g(x) = kx$ is continuous by Exercise 17.5.

(b) Prove that $|x|$ is a continuous function on $\mathbb{R}$.

(c) Use (a) and (b) and Theorem 17.5 to give another proof of Theorem 17.3.

17.8. Let f and g be real-valued functions.

(a) Show that $\min(f, g) = \frac{1}{2}(f + g) - \frac{1}{2}|f - g|$.

(b) Show that $\min(f, g) = -\max(-f, -g)$.

(c) Use (a) or (b) to prove that if f and g are continuous at x_0 in $\mathbb{R}$, then $\min(f, g)$ is continuous at x_0.

17.9. Prove that each of the following functions is continuous at x_0 by verifying the ϵ-δ property of Theorem 17.2.

(a) $f(x) = x^2$, $x_0 = 2$;

(b) $f(x) = \sqrt{x}$, $x_0 = 0$;

(c) $f(x) = x \sin(\frac{1}{x})$ for $x \neq 0$ and $f(0) = 0$, $x_0 = 0$;

(d) $g(x) = x^3$, x_0 arbitrary.
Hint for (d): $x^3 - x_0^3 = (x - x_0)(x^2 + x_0 x + x_0^2)$.

17.10. Prove that the following functions are discontinuous at the indicated points. You may use either Definition 17.1 or the ϵ-δ property in Theorem 17.2.

(a) $f(x) = 1$ for $x > 0$ and $f(x) = 0$ for $x \leq 0$, $x_0 = 0$;

(b) $g(x) = \sin(\frac{1}{x})$ for $x \neq 0$ and $g(0) = 0$, $x_0 = 0$;

(c) $\text{sgn}(x) = -1$ for $x < 0$, $\text{sgn}(x) = 1$ for $x > 0$, and $\text{sgn}(0) = 0$, $x_0 = 0$;

(d) $P(x) = 15$ for $0 \le x < 1$ and $P(x) = 15+13n$ for $n \le x < n+1$, x_0 a positive integer.
The function sgn is called the *signum function*; note that $\text{sgn}(x) = \frac{x}{|x|}$ for $x \ne 0$. The definition of P, the postage-stamp function *circa* 1979, means P takes the value 15 on the interval $[0, 1)$, the value 28 on the interval $[1, 2)$, the value 41 on the interval $[2, 3)$, etc.

17.11. Let f be a real-valued function with $\text{dom}(f) \subseteq \mathbb{R}$. Prove that f is continuous at x_0 if and only if, for every *monotonic* sequence (x_n) in $\text{dom}(f)$ converging to x_0, we have $\lim f(x_n) = f(x_0)$. Hint: Don't forget Theorem 11.3. *(NOT ON EXAM)*

17.12. **(a)** Let f be a continuous real-valued function with domain (a, b). Show that if $f(r) = 0$ for each rational number r in (a, b), then $f(x) = 0$ for all $x \in (a, b)$.

(b) Let f and g be continuous real-valued functions on (a, b) such that $f(r) = g(r)$ for each rational number r in (a, b). Prove that $f(x) = g(x)$ for all $x \in (a, b)$.

17.13. **(a)** Let $f(x) = 1$ for rational numbers x and $f(x) = 0$ for irrational numbers. Show that f is discontinuous at every x in $\mathbb{R}$.

(b) Let $h(x) = x$ for rational numbers x and $h(x) = 0$ for irrational numbers. Show that h is continuous at $x = 0$ and at no other point.

17.14. For each rational number x, write x as $\frac{p}{q}$ where p, q are integers with no common factors and $q > 0$, and then define $f(x) = \frac{1}{q}$. Also define $f(x) = 0$ for all $x \in \mathbb{R} \setminus \mathbb{Q}$. Thus $f(x) = 1$ for each integer, $f(\frac{1}{2}) = f(-\frac{1}{2}) = f(\frac{3}{2}) = \cdots = \frac{1}{2}$, etc. Show that f is continuous at each point of $\mathbb{R} \setminus \mathbb{Q}$ and discontinuous at each point of $\mathbb{Q}$.

17.15. Let f be a real-valued function whose domain is a subset of $\mathbb{R}$. Show that f is continuous at x_0 in $\text{dom}(f)$ if and only if, for every sequence (x_n) in $\text{dom}(f) \setminus \{x_0\}$ that converges to x_0, we have $\lim f(x_n) = f(x_0)$.

§18 Properties of Continuous Functions

A real-valued function f is said to be *bounded* if $\{f(x) : x \in \text{dom}(f)\}$ is a bounded set, i.e., if there exists a real number M such that $|f(x)| \leq M$ for all $x \in \text{dom}(f)$.

18.1 Theorem.
Let f be a continuous real-valued function on a closed interval $[a, b]$. Then f is a bounded function. Moreover, f assumes its maximum and minimum values on $[a, b]$; that is, there exist $x_0, y_0 \in [a, b]$ such that $f(x_0) \leq f(x) \leq f(y_0)$ for all $x \in [a, b]$.

Proof
Assume that f is not bounded on $[a, b]$. Then to each $n \in \mathbb{N}$ there corresponds an $x_n \in [a, b]$ such that $|f(x_n)| > n$. By the Bolzano-Weierstrass theorem 11.5, (x_n) has a subsequence (x_{n_k}) that converges to some real number x_0. The number x_0 also must belong to the closed interval $[a, b]$, as noted in Exercise 8.9. Since f is continuous at x_0, we have $\lim_{k \to \infty} f(x_{n_k}) = f(x_0)$, but we also have $\lim_{k \to \infty} |f(x_{n_k})| = +\infty$, which is a contradiction. It follows that f is bounded.

Now let $M = \sup\{f(x) : x \in [a, b]\}$; M is finite by the preceding paragraph. For each $n \in \mathbb{N}$ there exists $y_n \in [a, b]$ such that $M - \frac{1}{n} < f(y_n) \leq M$. Hence we have $\lim f(y_n) = M$. By the Bolzano-Weierstrass theorem, there is a subsequence (y_{n_k}) of (y_n) converging to a limit y_0 in $[a, b]$. Since f is continuous at y_0, we have $f(y_0) = \lim_{k \to \infty} f(y_{n_k})$. Since $(f(y_{n_k}))_{k \in \mathbb{N}}$ is a subsequence of $(f(y_n))_{n \in \mathbb{N}}$, Theorem 11.2 shows that $\lim_{k \to \infty} f(y_{n_k}) = \lim_{n \to \infty} f(y_n) = M$ and therefore $f(y_0) = M$. Thus f assumes its maximum at y_0.

The last paragraph applies to the function $-f$, so $-f$ assumes its maximum at some $x_0 \in [a, b]$. It follows easily that f assumes its minimum at x_0; see Exercise 8.1. ∎

Theorem 18.1 is used all the time, at least implicitly, in solving maximum-minimum problems in calculus because it is taken for granted that the problems have solutions, namely that a continuous function on a closed interval actually takes on a maximum and a minimum. If the domain is not a closed interval, one must be careful; see Exercise 18.3.

Theorem 18.1 is false if the closed interval $[a, b]$ is replaced by an open interval. For example, $f(x) = \frac{1}{x}$ is continuous but unbounded on $(0, 1)$. The function x^2 is continuous and bounded on $(-1, 1)$, but it does not have a maximum value on $(-1, 1)$.

18.2 Intermediate Value Theorem.
If f is a continuous real-valued function on an interval I, then f has the intermediate value property on I: Whenever $a, b \in I$, $a < b$ and y lies between $f(a)$ and $f(b)$ [i.e., $f(a) < y < f(b)$ or $f(b) < y < f(a)$], there exists at least one $x \in (a, b)$ such that $f(x) = y$.

Proof
We assume $f(a) < y < f(b)$; the other case is similar. Let $S = \{x \in [a, b] : f(x) < y\}$; see Figure 18.1. Since a belongs to S, S is nonempty, so $x_0 = \sup S$ represents a number in $[a, b]$. For each $n \in \mathbb{N}$, $x_0 - \frac{1}{n}$ is not an upper bound for S, so there exists $s_n \in S$ such that $x_0 - \frac{1}{n} < s_n \le x_0$. Thus $\lim s_n = x_0$ and, since $f(s_n) < y$ for all n, we have

$$f(x_0) = \lim f(s_n) \le y. \tag{1}$$

Let $t_n = \min\{b, x_0 + \frac{1}{n}\}$. Since $x_0 \le t_n \le x_0 + \frac{1}{n}$ we have $\lim t_n = x_0$. Each t_n belongs to $[a, b]$ but not to S, so $f(t_n) \ge y$ for all n. Therefore

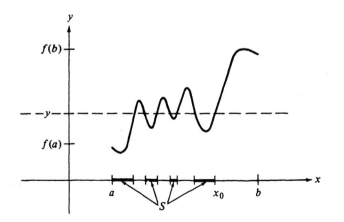

FIGURE 18.1

we have

$$f(x_0) = \lim f(t_n) \geq y;$$

this and (1) imply that $f(x_0) = y$. ∎

18.3 Corollary.

If f is a continuous real-valued function on an interval I, then the set $f(I) = \{f(x) : x \in I\}$ is also an interval or a single point.

Proof

The set $J = f(I)$ has the property:

$$y_0, y_1 \in J \quad \text{and} \quad y_0 < y < y_1 \quad \text{imply} \quad y \in J. \tag{1}$$

If $\inf J < \sup J$, then such a set J must be an interval. In fact, we will show that

$$\inf J < y < \sup J \quad \text{implies} \quad y \in J, \tag{2}$$

so J is an interval with endpoints $\inf J$ and $\sup J$; $\inf J$ and $\sup J$ may or may not belong to J and they may or may not be finite.

To prove (2) from (1), consider $\inf J < y < \sup J$. Then there exist y_0, y_1 in J so that $y_0 < y < y_1$. Thus $y \in J$ by (1). ∎

Example 1

Let f be a continuous function mapping $[0, 1]$ into $[0, 1]$. In other words, dom(f) = $[0, 1]$ and $f(x) \in [0, 1]$ for all $x \in [0, 1]$. Show that f has a *fixed point*, i.e., a point $x_0 \in [0, 1]$ such that $f(x_0) = x_0$; x_0 is left "fixed" by f.

Solution

The graph of f lies in the unit square; see Figure 18.2. Our assertion is equivalent to the assertion that the graph of f crosses the $y = x$ line, which is almost obvious.

A rigorous proof involves a little trick. We consider $g(x) = f(x) - x$ which is also a continuous function on $[0, 1]$. Since $g(0) = f(0) - 0 = f(0) \geq 0$ and $g(1) = f(1) - 1 \leq 1 - 1 = 0$, the Intermediate Value theorem shows that $g(x_0) = 0$ for some $x_0 \in [0, 1]$. Then obviously we have $f(x_0) = x_0$. □

Example 2

Show that if $y > 0$ and $m \in \mathbb{N}$, then y has a positive mth root.

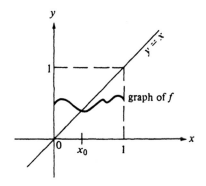

FIGURE 18.2

Solution
The function $f(x) = x^m$ is continuous [Exercise 17.5]. There exists $b > 0$ so that $y \le b^m$; in fact, if $y \le 1$ let $b = 1$ and if $y > 1$ let $b = y$. Thus $f(0) < y \le f(b)$ and the Intermediate Value theorem implies that $f(x) = y$ for some x in $(0, b]$. So $y = x^m$ and x is an mth root of y. $\qquad\qquad\square$

Let us analyze the function $f(x) = x^m$ in Example 2 more closely. It is a *strictly increasing function* on $[0, \infty)$:

$$0 \le x_1 < x_2 \quad \text{implies} \quad x_1^m < x_2^m.$$

Therefore f is one-to-one and each nonnegative y has exactly one nonnegative mth root. This assures us that the notation $y^{1/m}$ is unambiguous. In fact, $f^{-1}(y) = y^{1/m}$ is the *inverse function* of f since $f^{-1} \circ f(x) = x$ for $x \in \text{dom}(f)$ and $f \circ f^{-1}(y) = y$ for $y \in \text{dom}(f^{-1})$. Since $f(x) = x^m$ is continuous, the function $f^{-1}(y) = y^{1/m}$ is continuous on $[0, \infty)$ by the next theorem. Note that for $m = 2$ this result appears in Exercise 17.4.

18.4 Theorem.
Let f be a continuous strictly increasing function on some interval I. Then $f(I)$ is an interval J by Corollary 18.3 and f^{-1} represents a function with domain J. The function f^{-1} is a continuous strictly increasing function on J.

Proof

The function f^{-1} is easily shown to be strictly increasing. Since f^{-1} maps J onto I, the next theorem shows that f^{-1} is continuous. ■

18.5 Theorem.
Let g be a strictly increasing function on an interval J such that $g(J)$ is an interval I. Then g is continuous on J.

Proof

Consider x_0 in J. We assume x_0 is not an endpoint of J; tiny changes in the proof are needed otherwise. Then $g(x_0)$ is not an endpoint of I, so there exists $\epsilon_0 > 0$ such that $(g(x_0) - \epsilon_0, g(x_0) + \epsilon_0) \subseteq I$.

Let $\epsilon > 0$. Since we only need to verify the ϵ-δ property of Theorem 17.2 for small ϵ, we may assume that $\epsilon < \epsilon_0$. Then there exist $x_1, x_2 \in J$ such that $g(x_1) = g(x_0) - \epsilon$ and $g(x_2) = g(x_0) + \epsilon$. Clearly we have $x_1 < x_0 < x_2$. Also, if $x_1 < x < x_2$, then $g(x_1) < g(x) < g(x_2)$, hence $g(x_0) - \epsilon < g(x) < g(x_0) + \epsilon$, and hence $|g(x) - g(x_0)| < \epsilon$. Now if we put $\delta = \min\{x_2 - x_0, x_0 - x_1\}$, then $|x - x_0| < \delta$ implies $x_1 < x < x_2$ and hence $|g(x) - g(x_0)| < \epsilon$. ■

Theorem 18.5 provides a partial converse to the Intermediate Value theorem, since it tells us that a strictly increasing function with the intermediate value property is continuous. However, Exercise 18.12 shows that a function can have the intermediate value property without being continuous.

18.6 Theorem.
Let f be a one-to-one continuous function on an interval I. Then f is strictly increasing [$x_1 < x_2$ implies $f(x_1) < f(x_2)$] or strictly decreasing [$x_1 < x_2$ implies $f(x_1) > f(x_2)$].

Proof
First we show

$$\text{if } a < b < c \text{ in } I, \text{ then } f(b) \text{ lies between } f(a) \text{ and } f(c). \quad (1)$$

If not, then $f(b) > \max\{f(a), f(c)\}$, say. Select y so that $f(b) > y > \max\{f(a), f(c)\}$. By the Intermediate Value theorem 18.2 applied to $[a, b]$ and to $[b, c]$, there exist $x_1 \in (a, b)$ and $x_2 \in (b, c)$ such that $f(x_1) = f(x_2) = y$. This contradicts the one-to-one property of f.

Now select any $a_0 < b_0$ in I and suppose, say, that $f(a_0) < f(b_0)$. We will show that f is strictly increasing on I. By (1) we have

$$f(x) < f(a_0) \qquad \text{for} \quad x < a_0 \qquad \text{[since } x < a_0 < b_0\text{]},$$
$$f(a_0) < f(x) < f(b_0) \quad \text{for} \quad a_0 < x < b_0,$$
$$f(b_0) < f(x) \qquad \text{for} \quad x > b_0 \qquad \text{[since } a_0 < b_0 < x\text{]}.$$

In particular,

$$f(x) < f(a_0) \quad \text{for all} \quad x < a_0, \tag{2}$$
$$f(a_0) < f(x) \quad \text{for all} \quad x > a_0. \tag{3}$$

Consider any $x_1 < x_2$ in I. If $x_1 \leq a_0 \leq x_2$, then $f(x_1) < f(x_2)$ by (2) and (3). If $x_1 < x_2 < a_0$, then $f(x_1) < f(a_0)$ by (2), so by (1) we have $f(x_1) < f(x_2)$. Finally, if $a_0 < x_1 < x_2$, then $f(a_0) < f(x_2)$, so that $f(x_1) < f(x_2)$. ∎

Exercises

18.1. Let f be as in Theorem 18.1. Show that if the function $-f$ assumes its maximum at $x_0 \in [a, b]$, then f assumes its minimum at x_0.

18.2. Reread the proof of Theorem 18.1 with $[a, b]$ replaced by (a, b). Where does it break down? Discuss.

18.3. Use calculus to find the maximum and minimum of $f(x) = x^3 - 6x^2 + 9x + 1$ on $[0, 5)$.

18.4. Let $S \subseteq \mathbb{R}$ and suppose there exists a sequence (x_n) in S that converges to a number $x_0 \notin S$. Show that there exists an unbounded continuous function on S.

18.5. **(a)** Let f and g be continuous functions on $[a, b]$ such that $f(a) \geq g(a)$ and $f(b) \leq g(b)$. Prove that $f(x_0) = g(x_0)$ for at least one x_0 in $[a, b]$.

   (b) Show that Example 1 can be viewed as a special case of part (a).

18.6. Prove that $x = \cos x$ for some x in $(0, \frac{\pi}{2})$.

18.7. Prove that $x2^x = 1$ for some x in $(0, 1)$.

18.8. Suppose that f is a real-valued continuous function on $\mathbb{R}$ and that $f(a)f(b) < 0$ for some $a, b \in \mathbb{R}$. Prove that there exists x between a and b such that $f(x) = 0$.

18.9. Prove that a polynomial function f of odd degree has at least one real root. *Hint*: It may help to consider first the case of a cubic, i.e., $f(x) = a_0 + a_1 x + a_2 x^2 + a_3 x^3$ where $a_3 \neq 0$.

18.10. Suppose that f is continuous on $[0, 2]$ and that $f(0) = f(2)$. Prove that there exist x, y in $[0, 2]$ such that $|y - x| = 1$ and $f(x) = f(y)$. *Hint*: Consider $g(x) = f(x + 1) - f(x)$ on $[0, 1]$.

18.11. **(a)** Show that if f is strictly increasing on an interval I, then $-f$ is strictly decreasing on I.

(b) State and prove Theorems 18.4 and 18.5 for strictly decreasing functions.

18.12. Let $f(x) = \sin(\frac{1}{x})$ for $x \neq 0$ and let $f(0) = 0$.

(a) Observe that f is discontinuous at 0 by Exercise 17.10(b).

(b) Show that f has the intermediate value property on $\mathbb{R}$.

§19 Uniform Continuity

Let f be a real-valued function whose domain is a subset of $\mathbb{R}$. Theorem 17.2 tells us that f is continuous on a set $S \subseteq \mathrm{dom}(f)$ if and only if

$$\text{for each } x_0 \in S \text{ and } \epsilon > 0 \text{ there is } \delta > 0 \text{ so that} \atop x \in \mathrm{dom}(f), |x - x_0| < \delta \text{ imply } |f(x) - f(x_0)| < \epsilon. \qquad (*)$$

The choice of δ depends on $\epsilon > 0$ and on the point x_0 in S.

Example 1
We verify (*) for the function $f(x) = \frac{1}{x^2}$ on $(0, \infty)$. Let $x_0 > 0$ and $\epsilon > 0$. We need to show that $|f(x) - f(x_0)| < \epsilon$ for $|x - x_0|$ sufficiently small. Note that

$$f(x) - f(x_0) = \frac{1}{x^2} - \frac{1}{x_0^2} = \frac{x_0^2 - x^2}{x^2 x_0^2} = \frac{(x_0 - x)(x_0 + x)}{x^2 x_0^2}. \qquad (1)$$

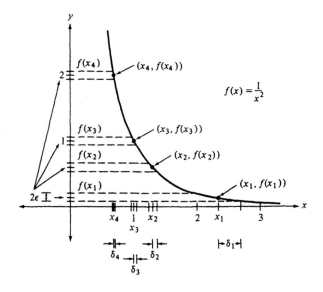

FIGURE 19.1

If $|x - x_0| < \frac{x_0}{2}$, then we have $|x| > \frac{x_0}{2}$, $|x| < \frac{3x_0}{2}$ and $|x_0 + x| < \frac{5x_0}{2}$. These observations and (1) show that if $|x - x_0| < \frac{x_0}{2}$, then

$$|f(x) - f(x_0)| < \frac{|x_0 - x| \cdot \frac{5x_0}{2}}{(\frac{x_0}{2})^2 x_0^2} = \frac{10|x_0 - x|}{x_0^3}.$$

Thus if we set $\delta = \min\{\frac{x_0}{2}, \frac{x_0^3 \epsilon}{10}\}$, then

$$|x - x_0| < \delta \quad \text{implies} \quad |f(x) - f(x_0)| < \epsilon.$$

This establishes (*) for f on $(0, \infty)$. Note that δ depends on both ϵ and x_0. Even if ϵ is fixed, δ gets small when x_0 is small. This shows that *our* choice of δ depends on x_0 as well as ϵ, though this may be because we obtained δ via sloppy estimates. As a matter of fact, in this case δ *must* depend on x_0 as well as ϵ; see Example 3. Figure 19.1 shows how a fixed ϵ requires smaller and smaller δ as x_0 approaches 0. [In the figure, δ_1 signifies a δ that works for x_1 and ϵ, δ_2 signifies a δ that works for x_2 and ϵ, etc.]

It turns out to be very useful to know when the δ in condition (*) can be chosen to depend only on $\epsilon > 0$ and S, so that δ does not depend on the particular point x_0. Such functions are said to be uniformly continuous on S. In the definition, the points x and x_0 play a symmetric role and so we will call them x and y.

19.1 Definition.
Let f be a real-valued function defined on a set $S \subseteq \mathbb{R}$. Then f is *uniformly continuous on S* if

> for each $\epsilon > 0$ there exists $\delta > 0$ such that
> $x, y \in S$ and $|x - y| < \delta$ imply $|f(x) - f(y)| < \epsilon$. (1)

We will say that f is *uniformly continuous* if f is uniformly continuous on $\text{dom}(f)$.

Note that if a function is uniformly continuous on its domain, then it must be continuous on its domain. This should be obvious; if it isn't, Theorem 17.2 and Definition 19.1 should be carefully scrutinized. Note also that uniform continuity is a property concerning a function *and a set* [on which it is defined]. It makes no sense to speak of a function being uniformly continuous at a point.

Example 2
We show that $f(x) = \frac{1}{x^2}$ is uniformly continuous on any set of the form $[a, \infty)$ where $a > 0$. Here a is fixed. Let $\epsilon > 0$. We need to show that there exists $\delta > 0$ such that

$$x \ge a, \quad y \ge a \quad \text{and} \quad |x - y| < \delta \quad \text{imply} \quad |f(x) - f(y)| < \epsilon. \quad (1)$$

As in formula (1) of Example 1, we have

$$f(x) - f(y) = \frac{(y - x)(y + x)}{x^2 y^2}.$$

If we can show that $\frac{y+x}{x^2 y^2}$ is bounded on $[a, \infty)$ by a constant M, then we will take $\delta = \frac{\epsilon}{M}$. But we have

$$\frac{y + x}{x^2 y^2} = \frac{1}{x^2 y} + \frac{1}{xy^2} \le \frac{1}{a^3} + \frac{1}{a^3} = \frac{2}{a^3},$$

so we set $\delta = \frac{\epsilon a^3}{2}$. It is now straightforward to verify (1). In fact, $x \ge a$, $y \ge a$ and $|x - y| < \delta$ imply

$$|f(x) - f(y)| = \frac{|y - x| \cdot |y + x|}{x^2 y^2} < \delta\left(\frac{1}{x^2 y} + \frac{1}{xy^2}\right) \le \frac{2\delta}{a^3} = \epsilon.$$

We have shown that f is uniformly continuous on $[a, \infty)$ since δ depends only on ϵ and the set $[a, \infty)$.

Example 3
The function $f(x) = \frac{1}{x^2}$ is not uniformly continuous on the set $(0, \infty)$ or even on the set $(0, 1)$.

We will prove this by directly violating the definition of uniform continuity. The squeamish reader may skip this demonstration and wait for the easy proof in Example 6. We will show that (1) in Definition 19.1 fails for $\epsilon = 1$; that is

$$\text{for each } \delta > 0 \text{ there exist } x, y \text{ in } (0, 1) \text{ such that} \qquad (1)$$
$$|x - y| < \delta \text{ and yet } |f(x) - f(y)| \geq 1.$$

[Actually, for this function, (1) in 19.1 fails for all $\epsilon > 0$.] To show (1) it suffices to take $y = x + \frac{\delta}{2}$ and arrange for

$$\left| f(x) - f\left(x + \frac{\delta}{2}\right) \right| \geq 1. \qquad (2)$$

[The motivation for this maneuver is to go from two unknowns, x and y, in (1) to one unknown, x, in (2).] By (1) in Example 1, (2) is equivalent to

$$1 \leq \frac{(x + \frac{\delta}{2} - x)(x + \frac{\delta}{2} + x)}{x^2(x + \frac{\delta}{2})^2} = \frac{\delta(2x + \frac{\delta}{2})}{2x^2(x + \frac{\delta}{2})^2}. \qquad (3)$$

It suffices to prove (1) for $\delta < \frac{1}{2}$. To obtain (3), let us try $x = \delta$ for no particular reason. Then

$$\frac{\delta(2\delta + \frac{\delta}{2})}{2\delta^2(\delta + \frac{\delta}{2})^2} = \frac{\frac{5\delta^2}{2}}{\frac{9\delta^4}{2}} = \frac{5}{9\delta^2} \geq \frac{5}{9(\frac{1}{2})^2} = \frac{20}{9} > 1.$$

We were lucky! To summarize, we have shown that if $0 < \delta < \frac{1}{2}$, then $|f(\delta) - f(\delta + \frac{\delta}{2})| > 1$, so (1) holds with $x = \delta$ and $y = \delta + \frac{\delta}{2}$.

Example 4
Is the function $f(x) = x^2$ uniformly continuous on $[-7, 7]$? To see if it is, consider $\epsilon > 0$. Note that $|f(x) - f(y)| = |x^2 - y^2| = |x - y| \cdot |x + y|$. Since $|x + y| \leq 14$ for x, y in $[-7, 7]$, we have

$$|f(x) - f(y)| \leq 14|x - y| \quad \text{for} \quad x, y \in [-7, 7].$$

Thus if $\delta = \frac{\epsilon}{14}$, then

$$x, y \in [-7, 7] \quad \text{and} \quad |x - y| < \delta \quad \text{imply} \quad |f(x) - f(y)| < \epsilon.$$

We have shown that f is uniformly continuous on $[-7, 7]$. A similar proof would work for $f(x) = x^2$ on any closed interval. However, these results are not accidents as the next important theorem shows.

19.2 Theorem.
If f is continuous on a closed interval $[a, b]$, then f is uniformly continuous on $[a, b]$.

Proof
Assume that f is not uniformly continuous on $[a, b]$. Then there exists $\epsilon > 0$ such that for each $\delta > 0$ the implication

$$\text{``}|x - y| < \delta \quad \text{implies} \quad |f(x) - f(y)| < \epsilon\text{''}$$

fails. That is, for each $\delta > 0$ there exist $x, y \in [a, b]$ such that $|x-y| < \delta$ and yet $|f(x) - f(y)| \geq \epsilon$. Then for each $n \in \mathbb{N}$ there exist x_n, y_n in $[a, b]$ such that $|x_n - y_n| < \frac{1}{n}$ and yet $|f(x_n) - f(y_n)| \geq \epsilon$. By the Bolzano-Weierstrass theorem 11.5, a subsequence (x_{n_k}) of (x_n) converges. Moreover, if $x_0 = \lim_{k\to\infty} x_{n_k}$, then x_0 belongs to $[a, b]$; see Exercise 8.9. Clearly we also have $x_0 = \lim_{k\to\infty} y_{n_k}$. Since f is continuous at x_0, we have

$$f(x_0) = \lim_{k\to\infty} f(x_{n_k}) = \lim_{k\to\infty} f(y_{n_k}),$$

so

$$\lim_{k\to\infty} [f(x_{n_k}) - f(y_{n_k})] = 0.$$

Since $|f(x_{n_k}) - f(y_{n_k})| \geq \epsilon$ for all k, we have a contradiction. We conclude that f is uniformly continuous on $[a, b]$. ∎

The preceding proof used only two properties of $[a, b]$: (a) boundedness, so that the Bolzano-Weierstrass theorem applies, and (b) a convergent sequence in $[a, b]$ must converge to an element in $[a, b]$. As noted prior to Theorem 11.8, sets with property (b) are called *closed sets*. Hence Theorem 19.2 has the following generalization. *If f is continuous on a closed and bounded set S, then f is uniformly continuous on S.* See also Theorems 21.4 and 13.12 that appear in optional sections.

Example 5
In view of Theorem 19.2, the following functions are uniformly continuous on the indicated sets: x^{73} on $[-13, 13]$, $\sqrt{x}$ on $[0, 400]$, $x^{17} \sin(e^x) - e^{4x} \cos 2x$ on $[-8\pi, 8\pi]$, and $\frac{1}{x^6}$ on $[\frac{1}{4}, 44]$.

19.3 Discussion.
Example 5 illustrates the power of Theorem 19.2, but it still may not be clear why uniform continuity is worth studying. One of the important applications of uniform continuity concerns the integrability of continuous functions on closed intervals. To see the relevance of uniform continuity, consider a continuous nonnegative real-valued function f on $[0, 1]$. For $n \in \mathbb{N}$ and $i = 0, 1, 2, \ldots, n-1$, let

$$M_{i,n} = \sup \left\{ f(x) : x \in \left[\frac{i}{n}, \frac{i+1}{n}\right] \right\} \quad \text{and} \quad m_{i,n} = \inf \left\{ f(x) : x \in \left[\frac{i}{n}, \frac{i+1}{n}\right] \right\}.$$

Then the sum of the areas of the rectangles in Figure 19.2(a) equals

$$U_n = \frac{1}{n} \sum_{i=0}^{n-1} M_{i,n}$$

and the sum of the areas of the rectangles in Figure 19.2(b) equals

$$L_n = \frac{1}{n} \sum_{i=0}^{n-1} m_{i,n}.$$

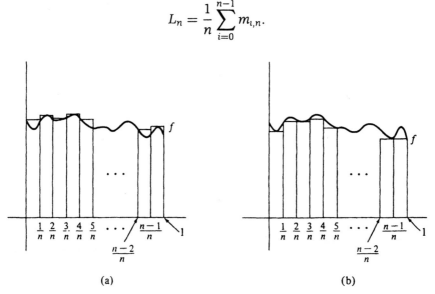

(a)   (b)

FIGURE 19.2

The function f would turn out to be Riemann integrable provided the numbers U_n and L_n are close together for large n, i.e., if

$$\lim_{n \to \infty} (U_n - L_n) = 0; \tag{1}$$

see Exercise 32.6. Moreover, we would have $\int_0^1 f(x)dx = \lim U_n = \lim L_n$. Relation (1) may appear obvious from Figure 19.2, but uniform continuity is needed to prove it. First note that

$$0 \le U_n - L_n = \frac{1}{n} \sum_{i=0}^{n-1} (M_{i,n} - m_{i,n})$$

for all n. Let $\epsilon > 0$. By Theorem 19.2, f is uniformly continuous on $[0, 1]$, so there exists $\delta > 0$ such that

$$x, y \in [0, 1] \quad \text{and} \quad |x - y| < \delta \quad \text{imply} \quad |f(x) - f(y)| < \epsilon. \tag{2}$$

Select N so that $\frac{1}{N} < \delta$. Consider $n > N$; for $i = 0, 1, 2, \ldots, n - 1$, Theorem 18.1 shows that there exist x_i, y_i in $[\frac{i}{n}, \frac{i+1}{n}]$ satisfying $f(x_i) = m_{i,n}$ and $f(y_i) = M_{i,n}$. Since $|x_i - y_i| \le \frac{1}{n} < \frac{1}{N} < \delta$, (2) shows that

$$M_{i,n} - m_{i,n} = f(y_i) - f(x_i) < \epsilon,$$

so

$$0 \le U_n - L_n = \frac{1}{n} \sum_{i=0}^{n-1} (M_{i,n} - m_{i,n}) < \frac{1}{n} \sum_{i=0}^{n-1} \epsilon = \epsilon.$$

This proves (1) as desired.

The next two theorems show that uniformly continuous functions have nice properties.

19.4 Theorem.
If f is uniformly continuous on a set S and (s_n) is a Cauchy sequence in S, then $(f(s_n))$ is a Cauchy sequence.

Proof
Let (s_n) be a Cauchy sequence in S and let $\epsilon > 0$. Since f is uniformly continuous on S, there exists $\delta > 0$ so that

$$x, y \in S \quad \text{and} \quad |x - y| < \delta \quad \text{imply} \quad |f(x) - f(y)| < \epsilon. \tag{1}$$

Since (s_n) is a Cauchy sequence, there exists N so that

$$m, n > N \quad \text{implies} \quad |s_n - s_m| < \delta.$$

From (1) we see that

$$m, n > N \quad \text{implies} \quad |f(s_n) - f(s_m)| < \epsilon.$$

This proves that $(f(s_n))$ is also a Cauchy sequence. ∎

Example 6
We show that $f(x) = \frac{1}{x^2}$ is not uniformly continuous on $(0, 1)$. Let $s_n = \frac{1}{n}$ for $n \in \mathbb{N}$. Then (s_n) is obviously a Cauchy sequence in $(0, 1)$. Since $f(s_n) = n^2$, $(f(s_n))$ is not a Cauchy sequence. Therefore f cannot be uniformly continuous on $(0, 1)$ by Theorem 19.4.

The next theorem involves extensions of functions. We say that a function $\tilde{f}$ is an *extension of a function* f if

$$\text{dom}(f) \subseteq \text{dom}(\tilde{f}) \quad \text{and} \quad f(x) = \tilde{f}(x) \quad \text{for all} \quad x \in \text{dom}(f).$$

Example 7
Let $f(x) = x \sin(\frac{1}{x})$ for $x \in (0, \frac{1}{\pi}]$. The function defined by

$$\tilde{f}(x) = \begin{cases} x\sin(\frac{1}{x}) & \text{for} \quad 0 < x \leq \frac{1}{\pi} \\ 0 & \text{for} \quad x = 0 \end{cases}$$

is an extension of f. Note that $\text{dom}(f) = (0, \frac{1}{\pi}]$ and $\text{dom}(\tilde{f}) = [0, \frac{1}{\pi}]$. In this case, $\tilde{f}$ is a continuous extension of f. See Figure 19.3 as well as Exercises 17.3(f) and 17.9(c).

Example 8
Let $g(x) = \sin(\frac{1}{x})$ for $x \in (0, \frac{1}{\pi}]$. The function g can be extended to a function $\tilde{g}$ with domain $[0, \frac{1}{\pi}]$ in many ways, but $\tilde{g}$ will not be continuous. See Figure 19.4.

The function f in Example 7 is uniformly continuous [since $\tilde{f}$ is], and f extends to a continuous function on the closed interval. The function g in Example 8 does not extend to a continuous function on the closed interval, and it turns out that g is not uniformly continuous. These examples illustrate the next theorem.

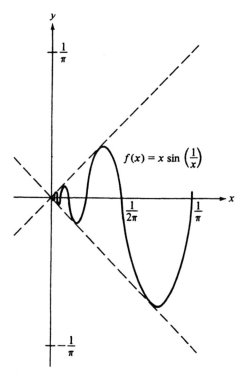

FIGURE 19.3

19.5 Theorem.
A real-valued function f on (a, b) is uniformly continuous on (a, b) if and only if it can be extended to a continuous function f̃ on [a, b].

Proof
First suppose that f can be extended to a continuous function $\tilde{f}$ on $[a, b]$. Then $\tilde{f}$ is uniformly continuous on $[a, b]$ by Theorem 19.2, so clearly f is uniformly continuous on (a, b).

Suppose now that f is uniformly continuous on (a, b). We need to define $\tilde{f}(a)$ and $\tilde{f}(b)$ so that the extended function will be continuous. It suffices for us to deal with $\tilde{f}(a)$. We make two claims:

$$\text{if } (s_n) \text{ is a sequence in } (a, b) \text{ converging} \atop \text{to } a, \text{ then } (f(s_n)) \text{ converges,} \tag{1}$$

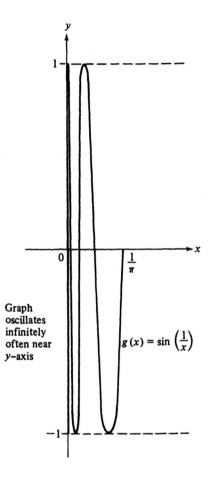

Graph
oscillates
infinitely
often near
y-axis

$g(x) = \sin\left(\frac{1}{x}\right)$

FIGURE 19.4

and

> if (s_n) and (t_n) are sequences in (a, b) converging
> to a, then $\lim f(s_n) = \lim f(t_n)$. $\hspace{1em}$ (2)

Momentarily accepting (1) and (2) as valid, we define

> $\tilde{f}(a) = \lim f(s_n)$ for any sequence
> (s_n) in (a, b) converging to a. $\hspace{1em}$ (3)

Assertion (1) guarantees that the limit exists, and assertion (2) guarantees that this definition is unambiguous. The continuity of $\tilde{f}$ at a
follows directly from (3); see Exercise 17.15.

To prove (1), note that (s_n) is a Cauchy sequence, so $(f(s_n))$ is also a Cauchy sequence by Theorem 19.4. Hence $(f(s_n))$ converges by Theorem 10.11. To prove (2) we create a third sequence (u_n) such that (s_n) and (t_n) are both subsequences of (u_n). In fact, we simply interleaf (s_n) and (t_n):

$$(u_n)_{n=1}^\infty = (s_1, t_1, s_2, t_2, s_3, t_3, s_4, t_4, s_5, t_5, \ldots).$$

It is evident that $\lim u_n = a$, so $\lim f(u_n)$ exists by (1). Theorem 11.2 shows that the subsequences $(f(s_n))$ and $(f(t_n))$ of $(f(u_n))$ both must converge to $\lim f(u_n)$, so $\lim f(s_n) = \lim f(t_n)$. ∎

Example 9
Let $h(x) = \frac{\sin x}{x}$ for $x \neq 0$. The function $\tilde{h}$ defined on $\mathbb{R}$ by

$$\tilde{h}(x) = \begin{cases} \frac{\sin x}{x} & \text{for} \quad x \neq 0 \\ 1 & \text{for} \quad x = 0 \end{cases}$$

is an extension of h. Clearly h and $\tilde{h}$ are continuous at all $x \neq 0$. It turns out that $\tilde{h}$ is continuous at $x = 0$ [see below], so h is uniformly continuous on $(a, 0)$ and $(0, b)$ for any $a < 0 < b$ by Theorem 19.5. In fact, $\tilde{h}$ is uniformly continuous on $\mathbb{R}$ [Exercise 19.11].

We cannot prove the continuity of $\tilde{h}$ at 0 in this book because we do not give a definition of $\sin x$. The continuity of $\tilde{h}$ at 0 reflects the fact that $\sin x$ is differentiable at 0 and that its derivative there is $\cos(0) = 1$, i.e.,

$$1 = \lim_{x \to 0} \frac{\sin x - \sin 0}{x - 0} = \lim_{x \to 0} \frac{\sin x}{x};$$

see Figure 19.5. The proof of this depends on how $\sin x$ is defined; see the brief discussion in 37.12. For a discussion of this limit and L'Hospital's rule, see Example 1 in §30.

Here is another useful criterion that implies uniform continuity.

19.6 Theorem.
Let f be a continuous function on an interval I [I may be bounded or unbounded]. Let I° be the interval obtained by removing from I any

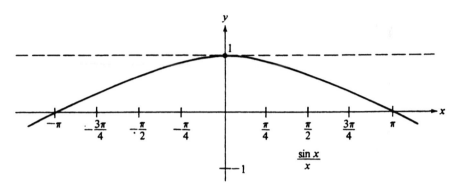

FIGURE 19.5

endpoints that happen to be in I. If f is differentiable on I° and if f' is bounded on I°, then f is uniformly continuous on I.

Proof
For this proof we need the Mean Value theorem, which can be found in most calculus texts or later in this book [Theorem 29.3].

Let M be a bound for f' on I so that $|f'(x)| \leq M$ for all x. Let $\epsilon > 0$ and let $\delta = \frac{\epsilon}{M}$. Consider $a, b \in I$ where $a < b$ and $|b - a| < \delta$. By the Mean Value theorem, there exists $x \in (a, b)$ such that $f'(x) = \frac{f(b)-f(a)}{b-a}$, so

$$|f(b) - f(a)| = |f'(x)| \cdot |b - a| \leq M|b - a| < M\delta = \epsilon.$$

This proves the uniform continuity of f on I. ∎

Example 10
Let $a > 0$ and consider $f(x) = \frac{1}{x^2}$. Since $f'(x) = -\frac{2}{x^3}$ we have $|f'(x)| \leq \frac{2}{a^3}$ on $[a, \infty)$. Hence f is uniformly continuous on $[a, \infty)$ by Theorem 19.6. For a direct proof of this fact, see Example 2.

Exercises

19.1. Which of the following continuous functions are uniformly continuous on the specified set? Justify your answers. Use any theorems you wish.

(a) $f(x) = x^{17} \sin x - e^x \cos 3x$ on $[0, \pi]$,

(b) $f(x) = x^3$ on $[0, 1]$,

(c) $f(x) = x^3$ on $(0, 1)$,

(d) $f(x) = x^3$ on $\mathbb{R}$,

(e) $f(x) = \frac{1}{x^3}$ on $(0, 1]$,

(f) $f(x) = \sin \frac{1}{x^2}$ on $(0, 1]$,

(g) $f(x) = x^2 \sin \frac{1}{x}$ on $(0, 1]$.

19.2. Prove that each of the following functions is uniformly continuous on the indicated set by directly verifying the ϵ-δ property in Definition 19.1.

(a) $f(x) = 3x + 11$ on $\mathbb{R}$,

(b) $f(x) = x^2$ on $[0, 3]$,

(c) $f(x) = \frac{1}{x}$ on $[\frac{1}{2}, \infty)$.

19.3. Repeat Exercise 19.2 for the following.

(a) $f(x) = \frac{x}{x+1}$ on $[0, 2]$,

(b) $f(x) = \frac{5x}{2x-1}$ on $[1, \infty)$.

19.4. (a) Prove that if f is uniformly continuous on a bounded set S, then f is a bounded function on S. *Hint*: Assume not. Use Theorems 11.5 and 19.4.

(b) Use (a) to give yet another proof that $\frac{1}{x^2}$ is not uniformly continuous on $(0, 1)$.

19.5. Which of the following continuous functions is uniformly continuous on the specified set? Justify your answers, using appropriate theorems or Exercise 19.4(a).

(a) $\tan x$ on $[0, \frac{\pi}{4}]$,

(b) $\tan x$ on $[0, \frac{\pi}{2})$,

(c) $\frac{1}{x} \sin^2 x$ on $(0, \pi]$,

(d) $\frac{1}{x-3}$ on $(0,3)$,

(e) $\frac{1}{x-3}$ on $(3,\infty)$,

(f) $\frac{1}{x-3}$ on $(4,\infty)$.

19.6. (a) Let $f(x) = \sqrt{x}$ for $x \geq 0$. Show that f' is unbounded on $(0,1]$ but that f is nevertheless uniformly continuous on $(0,1]$. Compare with Theorem 19.6.

(b) Show that f is uniformly continuous on $[1,\infty)$.

19.7. (a) Let f be a continuous function on $[0,\infty)$. Prove that if f is uniformly continuous on $[k,\infty)$ for some k, then f is uniformly continuous on $[0,\infty)$.

(b) Use (a) and Exercise 19.6(b) to prove that $\sqrt{x}$ is uniformly continuous on $[0,\infty)$.

19.8. (a) Use the Mean Value theorem to prove that

$$|\sin x - \sin y| \leq |x - y|$$

for all x, y in $\mathbb{R}$; see the proof of Theorem 19.6.

(b) Show that $\sin x$ is uniformly continuous on $\mathbb{R}$.

19.9. Let $f(x) = x \sin(\frac{1}{x})$ for $x \neq 0$ and $f(0) = 0$.

(a) Observe that f is continuous on $\mathbb{R}$; see Exercises 17.3(f) and 17.9(c).

(b) Why is f uniformly continuous on any bounded subset of $\mathbb{R}$?

(c) Is f uniformly continuous on $\mathbb{R}$?

19.10. Repeat Exercise 19.9 for the function g where $g(x) = x^2 \sin(\frac{1}{x})$ for $x \neq 0$ and $g(0) = 0$.

19.11. Accept the fact that the function $\bar{h}$ in Example 9 is continuous on $\mathbb{R}$; prove that it is uniformly continuous on $\mathbb{R}$.

§20 Limits of Functions

A function f is continuous at a point a provided the values $f(x)$ are near the value $f(a)$ for x near a [and $x \in \text{dom}(f)$]. See Definition 17.1

and Theorem 17.2. It would be reasonable to view $f(a)$ as the limit of the values $f(x)$, for x near a, and to write $\lim_{x \to a} f(x) = f(a)$. In this section we formalize this notion. This section is needed for our careful study of derivatives in Chapter 5, but it may be deferred until then.

We will be interested in ordinary limits, left-handed and right-handed limits and infinite limits. In order to handle these various concepts efficiently and also to emphasize their common features, we begin with a very general definition, which is not a standard definition.

20.1 Definition.
Let S be a subset of $\mathbb{R}$, let a be a real number or symbol ∞ or $-\infty$ that is the limit of some sequence in S, and let L be a real number or symbol $+\infty$ or $-\infty$. We write $\lim_{x \to a^S} f(x) = L$ if

$$f \text{ is a function defined on } S, \tag{1}$$

and

$$\begin{array}{l} \text{for every sequence } (x_n) \text{ in } S \text{ with limit } a, \\ \text{we have } \lim_{n \to \infty} f(x_n) = L. \end{array} \tag{2}$$

The expression "$\lim_{x \to a^S} f(x)$" is read "limit, as x tends to a along S, of $f(x)$."

20.2 Remarks.
(a) From Definition 17.1 we see that a function f is continuous at a in dom$(f) = S$ if and only if $\lim_{x \to a^S} f(x) = f(a)$.
(b) Observe that limits, when they exist, are unique. This follows from (2) of Definition 20.1, since limits of sequences are unique, a fact that is verified at the end of §7.

We now define the various standard limit concepts for functions.

20.3 Definition.
(a) For $a \in \mathbb{R}$ and a function f we write $\lim_{x \to a} f(x) = L$ provided $\lim_{x \to a^S} f(x) = L$ for some set $S = J \setminus \{a\}$ where J is an open interval containing a. $\lim_{x \to a} f(x)$ is called the [*two-sided*] *limit of f at a*. Note that f need not be defined at a and, even if f is defined at a, the value $f(a)$ need not equal $\lim_{x \to a} f(x)$. In

fact, $f(a) = \lim_{x \to a} f(x)$ if and only if f is defined on an open interval containing a and f is continuous at a.

(b) For $a \in \mathbb{R}$ and a function f we write $\lim_{x \to a^+} f(x) = L$ provided $\lim_{x \to a^s} f(x) = L$ for some open interval $S = (a, b)$. $\lim_{x \to a^+} f(x)$ is the *right-hand limit of f at a*. Again f need not be defined at a.

(c) For $a \in \mathbb{R}$ and a function f we write $\lim_{x \to a^-} f(x) = L$ provided $\lim_{x \to a^s} f(x) = L$ for some open interval $S = (c, a)$. $\lim_{x \to a^-} f(x)$ is the *left-hand limit of f at a*.

(d) For a function f we write $\lim_{x \to \infty} f(x) = L$ provided that $\lim_{x \to \infty^s} f(x) = L$ for some interval $S = (c, \infty)$. Likewise, we write $\lim_{x \to -\infty} f(x) = L$ provided $\lim_{x \to -\infty^s} f(x) = L$ for some interval $S = (-\infty, b)$.

The limits defined above are unique; i.e., they do not depend on the exact choice of the set S [Exercise 20.19].

Example 1

We have $\lim_{x \to 4} x^3 = 64$ and $\lim_{x \to 2} \frac{1}{x} = \frac{1}{2}$ because the functions x^3 and $\frac{1}{x}$ are continuous at 4 and 2, respectively. It is easy to show that $\lim_{x \to 0^+} \frac{1}{x} = +\infty$ and that $\lim_{x \to 0^-} \frac{1}{x} = -\infty$; see Exercise 20.14. It follows that $\lim_{x \to 0} \frac{1}{x}$ does *not* exist; see Theorem 20.10.

Example 2

Consider $\lim_{x \to 2} \frac{x^2 - 4}{x - 2}$. This is not like Example 1, because the function under the limit is not even defined at $x = 2$. However, we can rewrite the function as

$$\frac{x^2 - 4}{x - 2} = \frac{(x - 2)(x + 2)}{x - 2} = x + 2 \quad \text{for} \quad x \neq 2.$$

Now it is clear that $\lim_{x \to 2} \frac{x^2 - 4}{x - 2} = \lim_{x \to 2} (x + 2) = 4$. We should emphasize that the functions $\frac{x^2 - 4}{x - 2}$ and $x + 2$ are *not* identical. The domain of $f(x) = \frac{x^2 - 4}{x - 2}$ is $(-\infty, 2) \cup (2, \infty)$ while the domain of $\tilde{f}(x) = x + 2$ is $\mathbb{R}$, so that $\tilde{f}$ is an extension of f. This seems like nitpicking and this example may appear silly, but the function f, not $\tilde{f}$, arises naturally in computing the derivative of $g(x) = x^2$ at $x = 2$. Indeed, using the definition of derivative we have

$$g'(2) = \lim_{x \to 2} \frac{g(x) - g(2)}{x - 2} = \lim_{x \to 2} \frac{x^2 - 4}{x - 2},$$

so our modest computation above shows that $g'(2) = 4$. Of course, this is obvious from the formula $g'(x) = 2x$, but we are preparing the foundations of limits and derivatives, so we are beginning with simple examples.

Example 3

Consider $\lim_{x \to 1} \frac{\sqrt{x}-1}{x-1}$. We employ a trick that should be familiar by now; we multiply the numerator and denominator by $\sqrt{x} + 1$ and obtain

$$\frac{\sqrt{x} - 1}{x - 1} = \frac{x - 1}{(x - 1)(\sqrt{x} + 1)} = \frac{1}{\sqrt{x} + 1} \quad \text{for } x \neq 1.$$

Hence we have $\lim_{x \to 1} \frac{\sqrt{x}-1}{x-1} = \lim_{x \to 1} \frac{1}{\sqrt{x}+1} = \frac{1}{2}$. We have just laboriously verified that if $h(x) = \sqrt{x}$, then $h'(1) = \frac{1}{2}$.

Example 4

Let $f(x) = \frac{1}{(x-2)^3}$ for $x \neq 2$. Then $\lim_{x \to \infty} f(x) = \lim_{x \to -\infty} f(x) = 0$, $\lim_{x \to 2+} f(x) = +\infty$ and $\lim_{x \to 2-} f(x) = -\infty$.

To verify $\lim_{x \to \infty} f(x) = 0$, we consider a sequence (x_n) such that $\lim_{n \to \infty} x_n = +\infty$ and show that $\lim_{n \to \infty} f(x_n) = 0$. This will show that $\lim_{x \to \infty^s} f(x) = 0$ for $S = (2, \infty)$, for example. Exercise 9.11 and Theorem 9.9 show that $\lim_{n \to \infty} (x_n - 2)^3 = +\infty$, and then Theorem 9.10 shows that

$$\lim_{n \to \infty} f(x_n) = \lim_{n \to \infty} (x_n - 2)^{-3} = 0. \tag{1}$$

Here is a direct proof of (1). Consider $\epsilon > 0$. For large n, we need $|x_n - 2|^{-3} < \epsilon$ or $\epsilon^{-1} < |x_n-2|^3$ or $\epsilon^{-1/3} < |x_n-2|$. The last inequality holds if $x_n > \epsilon^{-1/3} + 2$. Since $\lim_{n \to \infty} x_n = +\infty$, there exists N so that

$$n > N \quad \text{implies} \quad x_n > \epsilon^{-1/3} + 2.$$

Reversing the algebraic steps above, we find

$$n > N \quad \text{implies} \quad |x_n - 2|^{-3} < \epsilon.$$

This establishes (1).

Similar arguments prove $\lim_{x \to -\infty} f(x) = 0$ and $\lim_{x \to 2^+} f(x) = +\infty$. To prove $\lim_{x \to 2^-} f(x) = -\infty$, consider a sequence (x_n) such that $x_n < 2$ for all n and $\lim_{n \to \infty} x_n = 2$. Then $2 - x_n > 0$ for all n and $\lim_{n \to \infty}(2 - x_n) = 0$. Hence $\lim_{n \to \infty}(2 - x_n)^3 = 0$ by Theorem 9.4, and Theorem 9.10 implies that $\lim_{n \to \infty}(2 - x_n)^{-3} = +\infty$. It follows [Exercise 9.10(b)] that

$$\lim_{n \to \infty} f(x_n) = \lim_{n \to \infty}(x_n - 2)^{-3} = -\infty. \tag{2}$$

This proves that $\lim_{x \to 2^s} f(x) = -\infty$ for $S = (-\infty, 2)$, so that $\lim_{x \to 2^-} f(x) = -\infty$. Of course, a direct proof of (2) also can be given. The limits discussed above are confirmed in Figure 20.1.

We will discuss the various limits defined in Definition 20.3 further at the end of this section. First we prove some limit theorems in considerable generality.

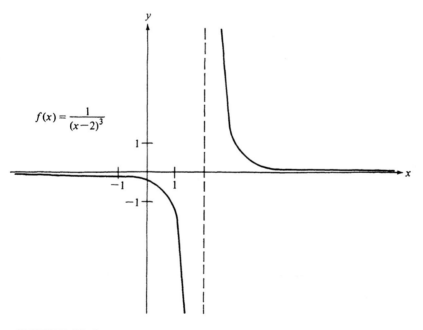

FIGURE 20.1

20.4 Theorem.
Let f_1 and f_2 be functions for which the limits $L_1 = \lim_{x \to a^s} f_1(x)$ and $L_2 = \lim_{x \to a^s} f_2(x)$ exist and are finite. Then
(i) $\lim_{x \to a^s} (f_1 + f_2)(x)$ exists and equals $L_1 + L_2$;
(ii) $\lim_{x \to a^s} (f_1 f_2)(x)$ exists and equals $L_1 L_2$;
(iii) $\lim_{x \to a^s} (f_1/f_2)(x)$ exists and equals L_1/L_2 provided $L_2 \neq 0$ and $f_2(x) \neq 0$ for $x \in S$.

Proof
The hypotheses imply that both f_1 and f_2 are defined on S and that a is the limit of some sequence in S. Clearly the functions $f_1 + f_2$ and $f_1 f_2$ are defined on S and so is f_1/f_2 if $f_2(x) \neq 0$ for $x \in S$.

Consider a sequence (x_n) in S with limit a. By hypotheses we have $L_1 = \lim_{n \to \infty} f_1(x_n)$ and $L_2 = \lim_{n \to \infty} f_2(x_n)$. Theorems 9.3 and 9.4 now show that

$$\lim_{n \to \infty} (f_1 + f_2)(x_n) = \lim_{n \to \infty} f_1(x_n) + \lim_{n \to \infty} f_2(x_n) = L_1 + L_2$$

and

$$\lim_{n \to \infty} (f_1 f_2)(x_n) = \left[\lim_{n \to \infty} f_1(x_n) \right] \cdot \left[\lim_{n \to \infty} f_2(x_n) \right] = L_1 L_2.$$

Thus (2) in Definition 20.1 holds for $f_1 + f_2$ and $f_1 f_2$, so that (i) and (ii) hold. Likewise (iii) follows by an application of Theorem 9.6. ∎

Some of the infinite variations of Theorem 20.4 appear in Exercise 20.20. The next theorem is less general than might have been expected; Example 7 shows why.

20.5 Theorem.
Let f be a function for which the limit $L = \lim_{x \to a^s} f(x)$ exists and is finite. If g is a function defined on $\{f(x) : x \in S\} \cup \{L\}$ that is continuous at L, then $\lim_{x \to a^s} g \circ f(x)$ exists and equals $g(L)$.

Proof
Note that $g \circ f$ is defined on S by our assumptions. Consider a sequence (x_n) in S with limit a. Then we have $L = \lim_{n \to \infty} f(x_n)$. Since g is continuous at L, it follows that

$$g(L) = \lim_{n \to \infty} g(f(x_n)) = \lim_{n \to \infty} g \circ f(x_n).$$

Hence $\lim_{x \to a^s} g \circ f(x) = g(L)$. ∎

Example 5

If f is a function for which the limit $L = \lim_{x \to a} f(x)$ exists and is finite, then we have $\lim_{x \to a} |f(x)| = |L|$. This follows immediately from Theorem 20.5 with $g(x) = |x|$. Similarly, we have $\lim_{x \to a} e^{f(x)} = e^L$ provided we accept the fact that $g(x) = e^x$ is continuous on $\mathbb{R}$.

Example 6

If f is a function for which $\lim_{x \to 0^+} f(x) = 0$ and $\lim_{x \to \infty} f(x) = \frac{\pi}{2}$, then we have $\lim_{x \to 0^+} e^{f(x)} = e^0 = 1$, $\lim_{x \to \infty} e^{f(x)} = e^{\frac{\pi}{2}}$, $\lim_{x \to 0^+} \sin(f(x)) = \sin(0) = 0$ and $\lim_{x \to \infty} \sin(f(x)) = \sin \frac{\pi}{2} = 1$.

Example 7

We give an example to show that continuity of g is needed in Theorem 20.5. Explicitly, we give examples of functions f and g such that $\lim_{x \to 0} f(x) = 1$, $\lim_{x \to 1} g(x) = 4$ and yet $\lim_{x \to 0} g \circ f(x)$ does not exist. One would expect this limit to exist and to equal 4, but in the example $f(x)$ will *equal* 1 for arbitrarily small x while $g(1) \neq 4$. The functions f and g are defined by $f(x) = 1 + x \sin \frac{\pi}{x}$ for $x \neq 0$, $g(x) = 4$ for $x \neq 1$, and $g(1) = -4$. Clearly $\lim_{x \to 0} f(x) = 1$ and $\lim_{x \to 1} g(x) = 4$. Let $x_n = \frac{2}{n}$ for $n \in \mathbb{N}$. Then $f(x_n) = 1 + \frac{2}{n} \sin(\frac{n\pi}{2})$; hence $f(x_n) = 1$ for even n and $f(x_n) \neq 1$ for odd n. Therefore $g \circ f(x_n) = -4$ for even n and $g \circ f(x_n) = 4$ for odd n. Since $\lim_{n \to \infty} x_n = 0$, we conclude that $\lim_{x \to 0} g \circ f(x)$ does not exist.

As in Theorem 17.2, the limits defined in Definitions 20.1 and 20.3 can be recast to avoid sequences. First we state and prove a typical result of this sort. Then, after Corollary 20.8, we give a general scheme without proof.

20.6 Theorem.
Let f be a function defined on a subset S of $\mathbb{R}$, let a be a real number that is the limit of some sequence in S, and let L be a real number. Then $\lim_{x \to a^s} f(x) = L$ if and only if

$$\text{for each } \epsilon > 0 \text{ there exists } \delta > 0 \text{ such that} \quad (1)$$
$$x \in S \text{ and } |x - a| < \delta \text{ imply } |f(x) - L| < \epsilon.$$

Proof

We imitate our proof of Theorem 17.2. Suppose that (1) holds, and consider a sequence (x_n) in S such that $\lim_{n\to\infty} x_n = a$. To show $\lim_{n\to\infty} f(x_n) = L$, consider $\epsilon > 0$. By (1) there exists $\delta > 0$ such that

$$x \in S \quad \text{and} \quad |x - a| < \delta \quad \text{imply} \quad |f(x) - L| < \epsilon.$$

Since $\lim_{n\to\infty} x_n = a$, there exists a number N such that $n > N$ implies $|x_n - a| < \delta$. Since $x_n \in S$ for all n, we conclude that

$$n > N \quad \text{implies} \quad |f(x_n) - L| < \epsilon.$$

Thus $\lim_{n\to\infty} f(x_n) = L$.

Now assume that $\lim_{x\to a^s} f(x) = L$, but that (1) fails. Then for some $\epsilon > 0$ the implication

$$\text{``} x \in S \quad \text{and} \quad |x - a| < \delta \quad \text{imply} \quad |f(x) - L| < \epsilon\text{''}$$

fails for each $\delta > 0$. Then for each $n \in \mathbb{N}$ there exists x_n in S where $|x_n - a| < \frac{1}{n}$ while $|f(x_n) - L| \geq \epsilon$. Hence (x_n) is a sequence in S with limit a for which $\lim_{n\to\infty} f(x_n) = L$ fails. Consequently $\lim_{x\to a^s} f(x) = L$ must also fail to hold. ∎

20.7 Corollary.

Let f be a function defined on $J \setminus \{a\}$ for some open interval J containing a, and let L be a real number. Then $\lim_{x\to a} f(x) = L$ if and only if

$$\begin{aligned}&\text{for each } \epsilon > 0 \text{ there exists } \delta > 0 \text{ such that}\\&0 < |x - a| < \delta \text{ implies } |f(x) - L| < \epsilon.\end{aligned} \tag{1}$$

20.8 Corollary.

Let f be a function defined on some interval (a, b), and let L be a real number. Then $\lim_{x\to a^+} f(x) = L$ if and only if

$$\begin{aligned}&\text{for each } \epsilon > 0 \text{ there exists } \delta > 0 \text{ such that}\\&a < x < a + \delta \text{ implies } |f(x) - L| < \epsilon.\end{aligned} \tag{1}$$

20.9 Discussion.

We now consider $\lim_{x\to s} f(x) = L$ where L can be finite, $+\infty$ or $-\infty$ and s is a symbol a, a^+, a^-, ∞ or $-\infty$ [here $a \in \mathbb{R}$]. Note that we have fifteen [$= 3 \cdot 5$] different sorts of limits here. It turns out that $\lim_{x \to s} f(x) = L$ if and only if

$$\text{for each} \underline{\qquad} \text{ there exists } \underline{\qquad} \text{ such that} \quad \underline{\qquad} \text{ implies } \underline{\qquad}. \tag{1}$$

For finite limits L, the first and last blanks are filled in by "$\epsilon > 0$" and "$|f(x) - L| < \epsilon$." For $L = +\infty$, the first and last blanks are filled in by "$M > 0$" and "$f(x) > M$," while for $L = -\infty$ they are filled in by "$M < 0$" and "$f(x) < M$." When we consider $\lim_{x \to a} f(x)$, then f is defined on $J \setminus \{a\}$ for some open interval J containing a, and the second and third blanks are filled in by "$\delta > 0$" and "$0 < |x - a| < \delta$." For $\lim_{x \to a^+} f(x)$ we require f to be defined on an interval (a, b) and the second and third blanks are filled in by "$\delta > 0$" and "$a < x < a + \delta$." For $\lim_{x \to a^-} f(x)$ we require f to be defined on an interval (c, a) and the second and third blanks are filled in by "$\delta > 0$" and "$a - \delta < x < a$." For $\lim_{x \to \infty} f(x)$ we require f to be defined on an interval (c, ∞) and the second and third blanks are filled in by "$\alpha < \infty$ and "$\alpha < x$." A similar remark applies to $\lim_{x \to -\infty} f(x)$.

The assertions above with L finite and s equal to a or a^+ are contained in Corollaries 20.7 and 20.8.

20.10 Theorem.
Let f be a function defined on $J \setminus \{a\}$ for some open interval J containing a. Then $\lim_{x \to a} f(x)$ exists if and only if the limits $\lim_{x \to a^+} f(x)$ and $\lim_{x \to a^-} f(x)$ both exist and are equal, in which case all three limits are equal.

Proof
Suppose that $\lim_{x \to a} f(x) = L$ and that L is finite. Then (1) in Corollary 20.7 holds, so (1) in Corollary 20.8 obviously holds. Thus we have $\lim_{x \to a^+} f(x) = L$; similarly $\lim_{x \to a^-} f(x) = L$.

Now suppose that $\lim_{x \to a^+} f(x) = \lim_{x \to a^-} f(x) = L$ where L is finite. Consider $\epsilon > 0$; we apply Corollary 20.8 and its analogue for a^- to obtain $\delta_1 > 0$ and $\delta_2 > 0$ such that

$$a < x < a + \delta_1 \quad \text{implies} \quad |f(x) - L| < \epsilon$$

and

$$a - \delta_2 < x < a \quad \text{implies} \quad |f(x) - L| < \epsilon.$$

If $\delta = \min\{\delta_1, \delta_2\}$, then

$$0 < |x - a| < \delta \quad \text{implies} \quad |f(x) - L| < \epsilon,$$

so $\lim_{x \to a} f(x) = L$ by Corollary 20.7.

Similar arguments apply if the limits L are infinite. For example, suppose that $\lim_{x \to a} f(x) = +\infty$ and consider $M > 0$. There exists $\delta > 0$ such that

$$0 < |x - a| < \delta \quad \text{implies} \quad f(x) > M. \tag{1}$$

Then clearly

$$a < x < a + \delta \quad \text{implies} \quad f(x) > M \tag{2}$$

and

$$a - \delta < x < a \quad \text{implies} \quad f(x) > M, \tag{3}$$

so that $\lim_{x \to a^+} f(x) = \lim_{x \to a^-} f(x) = +\infty$.

As a last example, suppose that $\lim_{x \to a^+} f(x) = \lim_{x \to a^-} f(x) = +\infty$. For each $M > 0$ there exists $\delta_1 > 0$ so that (2) holds, and there exists $\delta_2 > 0$ so that (3) holds. Then (1) holds with $\delta = \min\{\delta_1, \delta_2\}$. We conclude that $\lim_{x \to a} f(x) = +\infty$. ∎

20.11 Remark.
Note that $\lim_{x \to -\infty} f(x)$ is very similar to the right-hand limits $\lim_{x \to a^+} f(x)$. For example, if L is finite, then $\lim_{x \to a^+} f(x) = L$ if and only if

$$\text{for each } \epsilon > 0 \text{ there exists } \alpha > a \text{ such that} \atop 0 < x < \alpha \text{ implies } |f(x) - L| < \epsilon, \tag{1}$$

since $\alpha > a$ if and only if $\alpha = a + \delta$ for some $\delta > 0$; see Corollary 20.8. If we set $a = -\infty$ in (1), we obtain the condition 20.9(1) equivalent to $\lim_{x \to -\infty} f(x) = L$.

In the same way, the limits $\lim_{x \to \infty} f(x)$ and $\lim_{x \to a^-} f(x)$ will equal L [L finite] if and only if

$$\text{for each } \epsilon > 0 \text{ there exists } \alpha < a \text{ such that} \atop \alpha < x < a \text{ implies } |f(x) - L| < \epsilon. \tag{2}$$

Obvious changes are needed if L is infinite.

Exercises

20.1. Sketch the function $f(x) = \frac{x}{|x|}$. Determine, by inspection, the limits $\lim_{x \to \infty} f(x)$, $\lim_{x \to 0^+} f(x)$, $\lim_{x \to 0^-} f(x)$, $\lim_{x \to -\infty} f(x)$ and $\lim_{x \to 0} f(x)$ when they exist. Also indicate when they do not exist.

20.2. Repeat Exercise 20.1 for $f(x) = \frac{x^3}{|x|}$.

20.3. Repeat Exercise 20.1 for $f(x) = \frac{\sin x}{x}$. See Example 9 of §19.

20.4. Repeat Exercise 20.1 for $f(x) = x \sin \frac{1}{x}$.

20.5. Prove the limit assertions in Exercise 20.1.

20.6. Prove the limit assertions in Exercise 20.2.

20.7. Prove the limit assertions in Exercise 20.3.

20.8. Prove the limit assertions in Exercise 20.4.

20.9. Repeat Exercise 20.1 for $f(x) = \frac{1-x^2}{x}$.

20.10. Prove the limit assertions in Exercise 20.9.

20.11. Find the following limits.

   (a) $\lim_{x \to a} \frac{x^2 - a^2}{x-a}$ (b) $\lim_{x \to b} \frac{\sqrt{x} - \sqrt{b}}{x-b}, \; b > 0$

   (c) $\lim_{x \to a} \frac{x^3 - a^3}{x-a}$

   Hint for (c): $x^3 - a^3 = (x - a)(x^2 + ax + a^2)$.

20.12. (a) Sketch the function $f(x) = (x - 1)^{-1}(x - 2)^{-2}$.

   (b) Determine $\lim_{x \to 2^+} f(x)$, $\lim_{x \to 2^-} f(x)$, $\lim_{x \to 1^+} f(x)$ and $\lim_{x \to 1^-} f(x)$.

   (c) Determine $\lim_{x \to 2} f(x)$ and $\lim_{x \to 1} f(x)$ if they exist.

20.13. Prove that if $\lim_{x \to a} f(x) = 3$ and $\lim_{x \to a} g(x) = 2$, then

   (a) $\lim_{x \to a} [3f(x) + g(x)^2] = 13$,

   (b) $\lim_{x \to a} \frac{1}{g(x)} = \frac{1}{2}$,

   (c) $\lim_{x \to a} \sqrt{3f(x) + 8g(x)} = 5$.

20.14. Prove that $\lim_{x \to 0^+} \frac{1}{x} = +\infty$ and $\lim_{x \to 0^-} \frac{1}{x} = -\infty$.

20.15. Prove $\lim_{x \to -\infty} f(x) = 0$ and $\lim_{x \to 2^+} f(x) = +\infty$ for the function f in Example 4.

20.16. Suppose that the limits $L_1 = \lim_{x \to a^+} f_1(x)$ and $L_2 = \lim_{x \to a^+} f_2(x)$ exist.

(a) Show that if $f_1(x) \leq f_2(x)$ for all x in some interval (a, b), then $L_1 \leq L_2$.

(b) Suppose that, in fact, $f_1(x) < f_2(x)$ for all x in some interval (a, b). Can you conclude that $L_1 < L_2$?

20.17. Show that if $\lim_{x \to a^+} f_1(x) = \lim_{x \to a^+} f_3(x) = L$ and if $f_1(x) \leq f_2(x) \leq f_3(x)$ for all x in some interval (a, b), then $\lim_{x \to a^+} f_2(x) = L$. *Warning*: This is not immediate from Exercise 20.16(a), because we are not assuming that $\lim_{x \to a^+} f_2(x)$ exists; this must be proved.

20.18. Let $f(x) = \frac{\sqrt{1+3x^2}-1}{x^2}$ for $x \neq 0$. Show that $\lim_{x \to 0} f(x)$ exists and determine its value. Justify all claims.

20.19. The limits defined in Definition 20.3 do not depend on the choice of the set S. As an example, consider $a < b_1 < b_2$ and suppose that f is defined on (a, b_2). Show that if the limit $\lim_{x \to a^s} f(x)$ exists for either $S = (a, b_1)$ or $S = (a, b_2)$, then the limit exists for the other choice of S and these limits are identical. Their common value is what we write as $\lim_{x \to a^+} f(x)$.

20.20. Let f_1 and f_2 be functions such that $\lim_{x \to a^s} f_1(x) = +\infty$ and such that the limit $L_2 = \lim_{x \to a^s} f_2(x)$ exists.

(a) Prove that $\lim_{x \to a^s} (f_1 + f_2)(x) = +\infty$ if $L_2 \neq -\infty$. *Hint*: Use Exercise 9.11.

(b) Prove that $\lim_{x \to a^s} (f_1 f_2)(x) = +\infty$ if $0 < L_2 \leq +\infty$. *Hint*: Use Theorem 9.9.

(c) Prove that $\lim_{x \to a^s} (f_1 f_2)(x) = -\infty$ if $-\infty \leq L_2 < 0$.

(d) What can you say about $\lim_{x \to a^s} (f_1 f_2)(x)$ if $L_2 = 0$?

§21   * More on Metric Spaces: Continuity

In this section and the next section we continue the introduction to metric space ideas initiated in §13. More thorough treatments appear in [25], [33] and [36]. In particular, for this brief introduction we avoid the technical and somewhat confusing matter of relative topologies that is not, and should not be, avoided in the more thorough treatments.

We are interested in functions between metric spaces (S, d) and (S^*, d^*). We will write "$f: S \to S^*$" to signify that $\text{dom}(f) = S$ and that f takes values in S^*, i.e., $f(x) \in S^*$ for all $s \in S$.

21.1 Definition.
Consider metric spaces (S, d) and (S^*, d^*). A function $f: S \to S^*$ is *continuous at* s_0 in S if

> for each $\epsilon > 0$ there exists $\delta > 0$ such that
> $d(s, s_0) < \delta$ implies $d^*(f(s), f(s_0)) < \epsilon$. (1)

We say that f is *continuous on a subset* E of S if f is continuous at each point of E. The function f is *uniformly continuous on a subset* E of S if

> for each $\epsilon > 0$ there exists $\delta > 0$ such that
> $s, t \in E$ and $d(s, t) < \delta$ imply $d^*(f(s), f(t)) < \epsilon$. (2)

Example 1
Let $S = S^* = \mathbb{R}$ and $d = d^* = \text{dist}$ where, as usual, $\text{dist}(a, b) = |a-b|$. The definition of continuity given above is equivalent to that in §17 in view of Theorem 17.2. The definition of uniform continuity is equivalent to that in Definition 19.1.

Example 2
In several variable calculus, real-valued functions with domain $\mathbb{R}^2$ or $\mathbb{R}^3$, or even $\mathbb{R}^k$, are extensively studied. This corresponds to the case $S = \mathbb{R}^k$,

$$d(x, y) = \left[\sum_{j=1}^{k} (x_j - y_j)^2 \right]^{1/2},$$

$S^* = \mathbb{R}$ and $d^* = \text{dist}$. We will not develop the theory, but generally speaking, functions that look continuous will be. Some examples on $\mathbb{R}^2$ are $f(x_1, x_2) = x_1^2 + x_2^2$, $f(x_1, x_2) = x_1 x_2 \sqrt{x_1^2 + x_2^2 + 1}$, $f(x_1, x_2) = \cos(x_1 - x_2^5)$. Some examples on $\mathbb{R}^3$ are $g(x_1, x_2, x_3) = x_1^2 + x_2^2 + x_3^2$, $g(x_1, x_2, x_3) = x_1 x_2 + x_1 x_3 + x_2 x_3$, $g(x_1, x_2, x_3) = e^{x_1 + x_2} \log(x_3^2 + 2)$.

Example 3
Functions with domain $\mathbb{R}$ and values in $\mathbb{R}^2$ or $\mathbb{R}^3$, or generally $\mathbb{R}^k$, are also studied in several variable calculus. This corresponds to the

case $S = \mathbb{R}$, $d = $ dist, $S^* = \mathbb{R}^k$ and

$$d^*(x,y) = \left[\sum_{j=1}^{k}(x_j - y_j)^2\right]^{1/2}.$$

The images of such functions are what nonmathematicians often would call a "curve" or "path." In order to distinguish a function from its image, we will adhere to the following terminology. Suppose that $\gamma:\mathbb{R} \to \mathbb{R}^k$ is continuous. Then we will call γ a *path*; its image $\gamma(\mathbb{R})$ in $\mathbb{R}^k$ will be called a *curve*. We will also use this terminology if γ is defined and continuous on some subinterval of $\mathbb{R}$, such as $[a, b]$; see Exercise 21.7.

As an example, consider γ where $\gamma(t) = (\cos t, \sin t)$. This function maps $\mathbb{R}$ onto the circle in $\mathbb{R}^2$ about $(0, 0)$ with radius 1. More generally $\gamma_0(t) = (a\cos t, b\sin t)$ maps $\mathbb{R}$ onto the ellipse with equation $\frac{x^2}{a^2} + \frac{y^2}{b^2} = 1$; see Figure 21.1.

The graph of an ordinary continuous function $f:\mathbb{R} \to \mathbb{R}$ looks like a curve, and it is! It is the curve for the path $\gamma(t) = (t, f(t))$.

Curves in $\mathbb{R}^3$ can be quite exotic. For example, the curve for the path $h(t) = (\cos t, \sin t, \frac{t}{4})$ is a *helix*. See Figure 21.2.

We did not prove that any of the paths above are continuous, because we can easily prove the following general fact.

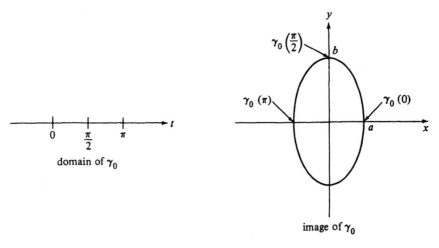

FIGURE 21.1

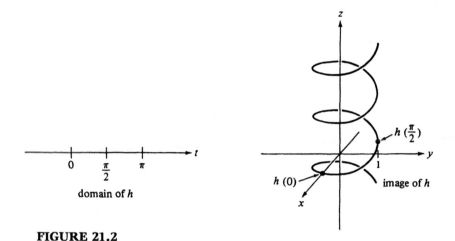

FIGURE 21.2

21.2 Proposition.

If $f_1, f_2, \ldots, f_k$ are continuous real-valued functions on $\mathbb{R}$, then

$$\gamma(t) = (f_1(t), f_2(t), \ldots, f_k(t))$$

defines a path in $\mathbb{R}^k$.

Proof

We need to show that γ is continuous. Recall formula (1) in the proof of Lemma 13.3 and Exercise 13.2:

$$d^*(\mathbf{x}, \mathbf{y}) \leq \sqrt{k} \max\{|x_j - y_j| : j = 1, 2, \ldots, k\}. \tag{1}$$

Consider $t_0 \in \mathbb{R}$ and $\epsilon > 0$. For each $j = 1, 2, \ldots, k$, there exists $\delta_j > 0$ such that

$$|t - t_0| < \delta_j \quad \text{implies} \quad |f_j(t) - f_j(t_0)| < \frac{\epsilon}{\sqrt{k}}.$$

For $\delta = \min\{\delta_1, \delta_2, \ldots, \delta_k\}$ and $|t - t_0| < \delta$, we have

$$\max\{|f_j(t) - f_j(t_0)| : j = 1, 2, \ldots, k\} < \frac{\epsilon}{\sqrt{k}},$$

so by (1) we have $d^*(\gamma(t), \gamma(t_0)) < \epsilon$. Thus γ is continuous at t_0. ∎

The next theorem shows that continuity is a topological property; see Discussion 13.7.

21.3 Theorem.
Consider metric spaces (S, d) and (S^*, d^*). A function $f: S \to S^*$ is continuous on S if and only if

$$f^{-1}(U) \text{ is an open subset of } S \text{ for every open subset } U \text{ of } S^*. \tag{1}$$

Recall that $f^{-1}(U) = \{s \in S : f(s) \in U\}$.

Proof
Suppose that f is continuous on S. Let U be an open subset of S^*, and consider $s_0 \in f^{-1}(U)$. We need to show that s_0 is interior to $f^{-1}(U)$. Since $f(s_0) \in U$ and U is open, we have

$$\{s^* \in S^* : d^*(s^*, f(s_0)) < \epsilon\} \subseteq U \tag{2}$$

for some $\epsilon > 0$. Since f is continuous at s_0, there exists $\delta > 0$ such that

$$d(s, s_0) < \delta \quad \text{implies} \quad d^*(f(s), f(s_0)) < \epsilon. \tag{3}$$

From (2) and (3) we conclude that $d(s, s_0) < \delta$ implies $f(s) \in U$, hence $s \in f^{-1}(U)$. That is,

$$\{s \in S : d(s, s_0) < \delta\} \subseteq f^{-1}(U),$$

so that s_0 is interior to $f^{-1}(U)$.

Conversely, suppose (1) holds, and consider $s_0 \in S$ and $\epsilon > 0$. Then $U = \{s^* \in S^* : d^*(s^*, f(s_0)) < \epsilon\}$ is open in S^*, so $f^{-1}(U)$ is open in S. Since $s_0 \in f^{-1}(U)$, for some $\delta > 0$ we have

$$\{s \in S : d(s, s_0) < \delta\} \subseteq f^{-1}(U).$$

It follows that

$$d(s, s_0) < \delta \quad \text{implies} \quad d^*(f(s), f(s_0)) < \epsilon.$$

Thus f is continuous at s_0. ∎

Continuity at a point is also a topological property; see Exercise 21.2. Uniform continuity is a topological property, too, but if we made this precise we would be led to a special class of topologies given by so-called "uniformities."

We will show that continuous functions preserve two important topological properties: compactness and connectedness, which will be defined in the next section. The next theorem and corollary illustrate the power of compactness.

21.4 Theorem.
Consider metric spaces (S, d), (S^*, d^*) and a continuous function $f: S \to S^*$. Let E be a compact subset of S. Then
    (i) $f(E)$ *is a compact subset of* S^*, *and*
    (ii) f *is uniformly continuous on* E.

Proof
To prove (i), let $\mathcal{U}$ be an open cover of $f(E)$. For each $U \in \mathcal{U}$, $f^{-1}(U)$ is open in S. Moreover, $\{f^{-1}(U) : U \in \mathcal{U}\}$ is a cover of E. Hence there exist $U_1, U_2, \ldots, U_m$ in $\mathcal{U}$ such that

$$E \subseteq f^{-1}(U_1) \cup f^{-1}(U_2) \cup \cdots \cup f^{-1}(U_m).$$

Then

$$f(E) \subseteq U_1 \cup U_2 \cup \cdots \cup U_m,$$

so $\{U_1, U_2, \ldots, U_m\}$ is the desired finite subcover of $\mathcal{U}$ for $f(E)$. This proves (i).

    To establish (ii), let $\epsilon > 0$. For each $s \in E$ there exists $\delta_s > 0$ [this δ depends on s] such that

$$d(s, t) < \delta_s \quad \text{implies} \quad d^*(f(s), f(t)) < \frac{\epsilon}{2}. \tag{1}$$

For each $s \in E$, let $V_s = \{t \in S : d(s, t) < \frac{1}{2}\delta_s\}$. Then the family $\mathcal{V} = \{V_s : s \in E\}$ is an open cover of E. By compactness, there exist finitely many points $s_1, s_2, \ldots, s_n$ in E such that

$$E \subseteq V_{s_1} \cup V_{s_2} \cup \cdots \cup V_{s_n}.$$

Let $\delta = \frac{1}{2} \min\{\delta_{s_1}, \delta_{s_2}, \ldots, \delta_{s_n}\}$. We complete the proof by showing

$$s, t \in E \quad \text{and} \quad d(s, t) < \delta \quad \text{imply} \quad d^*(f(s), f(t)) < \epsilon. \tag{2}$$

For some k in $\{1, 2, \ldots, n\}$ we have $s \in V_{s_k}$, i.e., $d(s, s_k) < \frac{1}{2}\delta_{s_k}$. Also we have

$$d(t, s_k) \le d(t, s) + d(s, s_k) < \delta + \frac{1}{2}\delta_{s_k} \le \delta_{s_k}.$$

Therefore applying (1) twice we have

$$d^*(f(s), f(s_k)) < \frac{\epsilon}{2} \quad \text{and} \quad d^*(f(t), f(s_k)) < \frac{\epsilon}{2}.$$

Hence $d^*(f(s), f(t)) < \epsilon$ as desired. ∎

Assertion (ii) in Theorem 21.4 generalizes Theorem 19.2. The next corollary should be compared with Theorem 18.1.

21.5 Corollary.
Let f be a continuous real-valued function on a metric space (S, d). If E is a compact subset of S, then
   (i) *f is bounded on E,*
   (ii) *f assumes its maximum and minimum on E.*

Proof
Since $f(E)$ is compact in $\mathbb{R}$, the set $f(E)$ must be bounded by Theorem 13.12. This implies (i).

Since $f(E)$ is compact, it contains $\sup f(E)$ by Exercise 13.13. Thus there exists $s_0 \in E$ so that $f(s_0) = \sup f(E)$. This tells us that f assumes it maximum value on E at the point s_0. Similarly, f assumes its minimum on E. ∎

Example 4
All the functions f in Example 2 are bounded on any compact subset of $\mathbb{R}^2$, i.e., on any closed and bounded set in $\mathbb{R}^2$. Likewise, all the functions g in Example 2 are bounded on each closed and bounded set in $\mathbb{R}^3$.

Example 5
Let γ be any path in $\mathbb{R}^k$; see Example 3. For $-\infty < a < b < \infty$, the image $\gamma([a, b])$ is closed and bounded in $\mathbb{R}^k$ by Theorem 21.4. Note that Corollary 21.5 does not apply in this case, since S^* is $\mathbb{R}^k$, not $\mathbb{R}$. Theorem 21.4 also tells us that γ is uniformly continuous on $[a, b]$. Thus if $\epsilon > 0$, there exists $\delta > 0$ such that

$$s, t \in [a, b] \quad \text{and} \quad |s - t| < \delta \quad \text{imply} \quad d(\gamma(s), \gamma(t)) < \epsilon.$$

This fact is useful in several variable calculus, where one integrates along paths γ; compare Discussion 19.3.

Exercises

21.1. Show that if the functions $f_1, f_2, \ldots, f_k$ in Proposition 21.2 are uniformly continuous, then so is γ.

21.2. Consider $f : S \rightarrow S^*$ where (S, d) and (S^*, d^*) are metric spaces. Show that f is continuous at $s_0 \in S$ if and only if

> for every open set U in S^* containing $f(s_0)$, there is an open set V in S containing s_0 such that $f(V) \subseteq U$.

21.3. Let (S, d) be a metric space and choose $s_0 \in S$. Show that $f(s) = d(s, s_0)$ defines a uniformly continuous real-valued function f on S.

21.4. Consider $f : S \rightarrow \mathbb{R}$ where (S, d) is a metric space. Show that the following are equivalent:
  (i) f is continuous;
  (ii) $f^{-1}((a, b))$ is open in S for all $a < b$;
  (iii) $f^{-1}((a, b))$ is open in S for all rational $a < b$.

21.5. Let E be a noncompact subset of $\mathbb{R}^k$.

  (a) Show that there is an unbounded continuous real-valued function on E. *Hint*: Either E is unbounded or else its closure E^- contains $x_0 \notin E$. In the latter case, use $\frac{1}{g}$ where $g(x) = d(x, x_0)$.

  (b) Show that there is a bounded continuous real-valued function on E that does not assume its maximum on E.

21.6. For metric spaces $(S_1, d_1), (S_2, d_2), (S_3, d_3)$, prove that if $f : S_1 \rightarrow S_2$ and $g : S_2 \rightarrow S_3$ are continuous, then $g \circ f$ is continuous from S_1 into S_3. *Hint*: It is somewhat easier to use Theorem 21.3 than to use the definition.

21.7. **(a)** Observe that if $E \subseteq S$ where (S, d) is a metric space, then (E, d) is also a metric space. In particular, if $E \subseteq \mathbb{R}$, then $d(a, b) = |a - b|$ for $a, b \in E$ defines a metric on E.

  (b) For $\gamma : [a, b] \rightarrow \mathbb{R}^k$, give the definition of continuity of γ.

21.8. Let (S, d) and (S^*, d^*) be metric spaces. Show that if $f : S \rightarrow S^*$ is uniformly continuous, and if (s_n) is a Cauchy sequence in S, then $(f(s_n))$ is a Cauchy sequence in S^*.

21.9. We say a function f *maps* a set E onto a set F provided $f(E) = F$.

(a) Show that there is a continuous function mapping the unit square

$$\{(x_1, x_2) \in \mathbb{R}^2 : 0 \leq x_1 \leq 1, 0 \leq x_2 \leq 1\}$$

onto $[0, 1]$.

(b) Do you think there is a continuous function mapping $[0, 1]$ onto the unit square?

21.10. Show that there exist continuous functions

(a) mapping $(0, 1)$ onto $[0, 1]$,

(b) mapping $(0, 1)$ onto $\mathbb{R}$,

(c) mapping $[0, 1] \cup [2, 3]$ onto $[0, 1]$.

21.11. Show that there do not exist continuous functions

(a) mapping $[0, 1]$ onto $(0, 1)$,

(b) mapping $[0, 1]$ onto $\mathbb{R}$.

§22   * More on Metric Spaces: Connectedness

Consider a subset E of $\mathbb{R}$ that is not an interval. As noted in the proof of Corollary 18.3, the property

$$y_1, y_2 \in E \quad \text{and} \quad y_1 < y < y_2 \quad \text{imply} \quad y \in E$$

must fail. So there exist y_1, y_2, y in $\mathbb{R}$ such that

$$y_1 < y < y_2, \qquad y_1, y_2 \in E, \qquad y \notin E. \tag{*}$$

The set E is not "connected" because y separates E into two pieces. Put another way, if we set $U_1 = (-\infty, y)$ and $U_2 = (y, \infty)$, then we obtain disjoint open sets such that

$$E \subseteq U_1 \cup U_2, \qquad E \cap U_1 \neq \emptyset, \qquad E \cap U_2 \neq \emptyset.$$

The last observation can be promoted to a useful general definition.

22.1 Definition.

Let E be a subset of a metric space (S, d). The set E is *disconnected* if there are disjoint open subsets U_1 and U_2 in S such that

$$E \subseteq U_1 \cup U_2, \tag{1}$$

$$E \cap U_1 \neq \varnothing \quad \text{and} \quad E \cap U_2 \neq \varnothing. \tag{2}$$

A set E is *connected* if it is not disconnected.

Example 1

As noted before the definition, sets in $\mathbb{R}$ that are not intervals are disconnected. Conversely, intervals in $\mathbb{R}$ are connected. To prove this from the definition, consider an interval I and assume open sets U_1 and U_2 exist as described in Definition 22.1. Select $a_1 \in I \cap U_1$ and $a_2 \in I \cap U_2$. We may suppose that $a_1 < a_2$. Let

$$b = \sup[a_1, a_2) \cap U_1.$$

Clearly $a_1 < b \leq a_2$. Since $b \in I$, we must have $b \in U_1$ or $b \in U_2$ and not both. Hence for some $\epsilon > 0$, we have either

$$(b - \epsilon, b + \epsilon) \subseteq U_1 \tag{1}$$

or

$$(b - \epsilon, b + \epsilon) \subseteq U_2. \tag{2}$$

In case (1), we have $a_1 < b < a_2$ and $(b, a_2) \cap U_1 \neq \varnothing$, so that b cannot be an upper bound of $[a_1, a_2) \cap U_1$ much less a least upper bound. In case (2), we have $U_1 \cap (b - \epsilon, b] = \varnothing$. If b is an upper bound for $[a_1, a_2) \cap U_1$, then so is $b - \epsilon$ in which case b cannot be the *least* upper bound for this set. Both cases lead to a contradiction, so I must be connected.

22.2 Theorem.

Consider metric spaces (S, d), (S^*, d^*), and let $f: S \rightarrow S^*$ be continuous. If E is a connected subset of S, then $f(E)$ is a connected subset of S^*.

Proof

Assume $f(E)$ is not connected in S^*. Then there exist disjoint open sets V_1 and V_2 in S^* such that

$$f(E) \subseteq V_1 \cup V_2, \tag{1}$$

$$f(E) \cap V_1 \neq \emptyset \quad \text{and} \quad f(E) \cap V_2 \neq \emptyset. \tag{2}$$

Let $U_1 = f^{-1}(V_1)$ and $U_2 = f^{-1}(V_2)$. Then U_1 and U_2 are disjoint open sets in S, $E \subseteq U_1 \cup U_2$, $E \cap U_1 \neq \emptyset$ and $E \cap U_2 \neq \emptyset$. ∎

The next corollary generalizes Theorem 18.2 and its corollary.

22.3 Corollary.
Let f be a continuous real-valued function on a metric space (S, d). If E is a connected subset of S, then $f(E)$ is an interval in $\mathbb{R}$. In particular, f has the intermediate value property.

Example 2
Curves are connected. That is, if γ is a path in $\mathbb{R}^k$ as described in Example 3 of §21 and I is a subinterval of $\mathbb{R}$, then the image $\gamma(I)$ is connected in $\mathbb{R}^k$.

22.4 Definition.
A subset E of a metric space (S, d) is said to be *path-connected* if, for each pair s, t of points in E, there exists a continuous function $\gamma: [a, b] \to E$ such that $\gamma(a) = s$ and $\gamma(b) = t$. We call γ a *path*.

22.5 Theorem.
If E in (S, d) is path-connected, then E is connected. [The failure of the converse is illustrated in Exercise 22.4.]

Proof
Assume E is disconnected by disjoint open sets U_1 and U_2:

$$E \subseteq U_1 \cup U_2, \tag{1}$$
$$E \cap U_1 \neq \emptyset \quad \text{and} \quad E \cap U_2 \neq \emptyset. \tag{2}$$

Select $s \in E \cap U_1$ and $t \in E \cap U_2$. Let $\gamma: [a, b] \to E$ be a path where $\gamma(a) = s$ and $\gamma(b) = t$. Let $F = \gamma([a, b])$. Then (1) and (2) hold with F in place of E. Thus F is disconnected, but F must be connected by Theorem 22.2. ∎

Figure 22.1 gives a path-connected set and a disconnected set in $\mathbb{R}^2$.

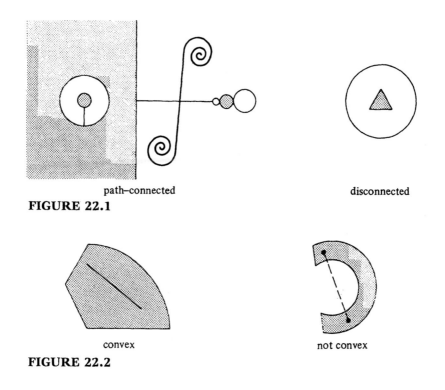

path–connected   disconnected

FIGURE 22.1

convex   not convex

FIGURE 22.2

Example 3

Many familiar sets in $\mathbb{R}^k$ such as the open ball $\{x : d(x, 0) < r\}$, the closed ball $\{x : d(x, 0) \le r\}$ and the cube

$$\{x : \max\{|x_j| : j = 1, 2, \ldots, k\} \le 1\}$$

are convex. A subset E of $\mathbb{R}^k$ is *convex* if

$$x, y \in E \quad \text{and} \quad 0 < t < 1 \quad \text{imply} \quad tx + (1 - t)y \in E,$$

i.e., whenever E contains two points it contains the line segment connecting them. See Figure 22.2. Convex sets E in $\mathbb{R}^k$ are always path-connected. This is because $\gamma(t) = tx + (1 - t)y$ defines a path $\gamma : [0, 1] \to E$ such that $\gamma(0) = y$ and $\gamma(1) = x$. For more details, see any book on several variable calculus.

We end this section with a discussion of some very different metric spaces. The points in these spaces are actually functions themselves.

22.6 Definition.
Let S be a subset of $\mathbb{R}$. Let $C(S)$ be the set of all bounded continuous real-valued functions on S and, for $f, g \in C(S)$, let

$$d(f, g) = \sup\{|f(x) - g(x)| : x \in S\}.$$

With this definition, $C(S)$ becomes a metric space [Exercise 22.6]. Now note that a sequence (f_n) in this metric space converges to a point [function!] f provided $\lim_{n \to \infty} d(f_n, f) = 0$, that is

$$\lim_{n \to \infty}[\sup\{|f_n(x) - f(x)| : x \in S\}] = 0. \qquad (*)$$

Put another way, for each $\epsilon > 0$ there exists a number N such that

$$|f_n(x) - f(x)| < \epsilon \quad \text{for all} \quad x \in S \quad \text{and} \quad n > N.$$

We will study this important concept in the next chapter, but without using metric space terminology. See Definition 24.2 and Remark 24.4 where $(*)$ is called *uniform convergence*.

A sequence (f_n) in $C(S)$ is a Cauchy sequence with respect to our metric exactly when it is *uniformly Cauchy* as defined in Definition 25.3. In our metric space terminology, Theorem 25.4 simply asserts that $C(S)$ is a *complete* metric space.

Exercises

22.1. Show that there do not exist continuous functions

   (a) mapping $[0, 1]$ onto $[0, 1] \cup [2, 3]$,

   (b) mapping $(0, 1)$ onto $\mathbb{Q}$.

22.2. Show that $\{(x_1, x_2) \in \mathbb{R}^2 : x_1^2 + x_2^2 = 1\}$ is a connected subset of $\mathbb{R}^2$.

22.3. Prove that if E is a connected subset of a metric space (S, d), then its closure E^- is also connected.

22.4. Consider the following subset of $\mathbb{R}^2$:

$$E = \left\{ \left(x, \sin\frac{1}{x}\right) : x \in (0, 1] \right\};$$

E is simply the graph of $f(x) = \sin\frac{1}{x}$ along the interval $(0, 1]$.

   (a) Sketch E and determine its closure E^-.

(b) Show that E^- is connected.

(c) Show that E^- is not path-connected.

22.5. Let E and F be connected sets in some metric space.

(a) Prove that if $E \cap F \neq \emptyset$, then $E \cup F$ is connected.

(b) Give an example to show that $E \cap F$ need not be connected. Incidentally, the empty set *is* connected.

22.6. (a) Show that $C(S)$ given in Definition 22.6 is a metric space.

(b) Why did we require the functions in $C(S)$ to be bounded when no such requirement appears in Definition 24.2?

22.7. Show that the metric space B in Exercise 13.3 can be regarded as $C(\mathbb{N})$.

22.8. Consider $C(S)$ for a subset S of $\mathbb{R}$. For a fixed s_0 in S, define $F(f) = f(s_0)$. Show that F is a uniformly continuous real-valued function on the metric space $C(S)$.

22.9. Consider $f, g \in C(S)$ where $S \subseteq \mathbb{R}$. Let $F(t) = tf + (1 - t)g$. Show that F is a uniformly continuous function from $\mathbb{R}$ into $C(S)$.

22.10. Let f be a uniformly continuous function in $C(\mathbb{R})$. For each $x \in \mathbb{R}$, let f_x be the function defined by $f_x(y) = f(x + y)$. Let $F(x) = f_x$; show that F is uniformly continuous from $\mathbb{R}$ into $C(\mathbb{R})$.

22.11. Consider $C(S)$ where $S \subseteq \mathbb{R}$, and let $\mathcal{E}$ consist of all f in $C(S)$ such that $\sup\{|f(x)| : x \in S\} \leq 1$.

(a) Show that $\mathcal{E}$ is closed in $C(S)$.

(b) Show that $C(S)$ is connected.

(c) Show that $\mathcal{E}$ is connected.

22.12. Consider a subset $\mathcal{E}$ of $C(S)$, $S \subseteq \mathbb{R}$. A function f_0 in $\mathcal{E}$ is *interior* to $\mathcal{E}$ if there exists a finite subset F of S and an $\epsilon > 0$ such that

$$\{f \in C(S) : |f(x) - f_0(x)| < \epsilon \text{ for } x \in F\} \subseteq \mathcal{E}.$$

The set $\mathcal{E}$ is *open* if every function in $\mathcal{E}$ is interior to $\mathcal{E}$.

(a) Reread Discussion 13.7.

(b) Show that the family of open sets defined above forms a topology for $C(S)$. *Remarks.* This topology is different from the one given by the metric in Definition 22.6. In fact, this topology does not come from any metric at all! It is called the *topology of pointwise convergence* and can be used to study the convergence in Definition 24.1 just as the metric in Definition 22.6 can be used to study the convergence in Definition 24.2.

22.13. Show that a function $f: \mathbb{R} \to \mathbb{R}$ is continuous if and only if its graph $G = \{(x, f(x)) : x \in \mathbb{R}\}$ is connected and closed in $\mathbb{R}^2$. See C.E. Burgess's article, Continuous Functions and Connected Graphs, *American Mathematical Monthly*, vol. **97** (1990), pp. 337-339.

4

CHAPTER

Sequences and Series of Functions

In this chapter we develop some of the basic properties of power series. In doing so, we will introduce uniform convergence and illustrate its importance. In §26 we prove that power series can be differentiated and integrated term-by-term.

§23 Power Series

Given a sequence $(a_n)_{n=0}^{\infty}$ of real numbers, the series $\sum_{n=0}^{\infty} a_n x^n$ is called a *power series*. Observe the variable x. Thus the power series is a function of x provided it converges for some or all x. Of course, it converges for $x = 0$; note the convention $0^0 = 1$. Whether it converges for other values of x depends on the choice of *coefficients* (a_n). It turns out that, given any sequence (a_n), one of the following holds for its power series:

- **(a)** the power series converges for all $x \in \mathbb{R}$;
- **(b)** the power series converges only for $x = 0$;
- **(c)** the power series converges for all x in some bounded interval centered at 0; the interval may be open, half-open or closed.

These remarks are consequences of the following important theorem.

23.1 Theorem.
For the power series $\sum a_n x^n$, let

$$\beta = \limsup |a_n|^{1/n} \quad and \quad R = \frac{1}{\beta}.$$

[If $\beta = 0$ we set $R = +\infty$, and if $\beta = +\infty$ we set $R = 0$.] Then
   (i) *the power series converges for $|x| < R$;*
   (ii) *the power series diverges for $|x| > R$.*

R is called the *radius of convergence* for the power series. Note that (i) is a vacuous statement if $R = 0$ and that (ii) is a vacuous statement if $R = +\infty$. Note also that (a) above corresponds to the case $R = +\infty$, (b) above corresponds to the case $R = 0$, and (c) above corresponds to the case $0 < R < +\infty$.

Proof of Theorem 23.1
The proof follows quite easily from the Root Test 14.9. Here are the details. We want to apply the Root Test to the series $\sum a_n x^n$. So for each $x \in \mathbb{R}$ let α_x be the number or symbol defined in 14.9 for the series $\sum a_n x^n$. Since the nth term of the series is $a_n x^n$, we have

$$\alpha_x = \limsup |a_n x^n|^{1/n} = \limsup |x||a_n|^{1/n} = |x| \cdot \limsup |a_n|^{1/n} = \beta|x|.$$

The third equality is justified by Exercise 12.6(a). Now we consider cases.
   Case 1. Suppose $0 < R < +\infty$. In this case $\alpha_x = \beta|x| = \frac{|x|}{R}$. If $|x| < R$ then $\alpha_x < 1$, so the series converges by the Root Test. Likewise, if $|x| > R$, then $\alpha_x > 1$ and the series diverges.
   Case 2. Suppose $R = +\infty$. Then $\beta = 0$ and $\alpha_x = 0$ no matter what x is. Hence the power series converges for all x by the Root Test.
   Case 3. Suppose $R = 0$. Then $\beta = +\infty$ and $\alpha_x = +\infty$ for $x \neq 0$. Thus by the Root Test the series diverges for $x \neq 0$. ∎

Recall that if $\lim |\frac{a_{n+1}}{a_n}|$ exists, then this limit equals β of the last theorem by Corollary 12.3. This limit is often easier to calculate than $\limsup |a_n|^{1/n}$; see the examples below.

Example 1

Consider $\sum_{n=0}^{\infty} \frac{1}{n!}x^n$. If $a_n = \frac{1}{n!}$, then $\frac{a_{n+1}}{a_n} = \frac{1}{n+1}$, so $\lim |\frac{a_{n+1}}{a_n}| = 0$. Therefore $\beta = 0$, $R = +\infty$ and this series has radius of convergence $+\infty$. That is, it converges for all x in $\mathbb{R}$. In fact, it converges to e^x for all x, but that is another story; see Example 1 in §31 and also §37.

Example 2

Consider $\sum_{n=0}^{\infty} x^n$. Then $\beta = 1$ and $R = 1$. Note that this series does not converge for $x = 1$ or $x = -1$, so the interval of convergence is exactly $(-1, 1)$. [By *interval of convergence* we mean the set of x for which the power series converges.] The series converges to $\frac{1}{1-x}$ by formula (2) of Example 1 in §14.

Example 3

Consider $\sum_{n=1}^{\infty} \frac{1}{n}x^n$. Since $\lim \frac{\frac{1}{n+1}}{\frac{1}{n}} = 1$, we again have $\beta = 1$ and $R = 1$. This series diverges for $x = 1$ [see Example 1 of §15], but it converges for $x = -1$ by the Alternating Series theorem 15.3. Hence the interval of convergence is exactly $[-1, 1)$.

Example 4

Consider $\sum_{n=1}^{\infty} \frac{1}{n^2}x^n$. Once again $\beta = 1$ and $R = 1$. This series converges at both $x = 1$ and $x = -1$, so its interval of convergence is exactly $[-1, 1]$.

Example 5

The series $\sum_{n=0}^{\infty} n!x^n$ has radius of convergence $R = 0$ because we have $\lim |\frac{(n+1)!}{n!}| = +\infty$. It diverges for every $x \neq 0$.

Examples 1–5 illustrate all the possibilities discussed in (a)–(c) prior to Theorem 23.1.

Example 6

Consider $\sum_{n=0}^{\infty} 2^{-n}x^{3n}$. This is deceptive, and it is tempting to calculate $\beta = \lim \sup (2^{-n})^{1/n} = \frac{1}{2}$ and conclude $R = 2$. *This is wrong* because 2^{-n} is the coefficient of x^{3n} not x^n, and the calculation of β must involve the coefficients a_n of x^n. We must handle this series more carefully. The series can be written $\sum_{n=0}^{\infty} a_n x^n$ where $a_{3k} = 2^{-k}$

and $a_n = 0$ if n is not a multiple of 3. We calculate β by using the subsequence of all nonzero terms, i.e., the subsequence given by $\sigma(k) = 3k$. This yields

$$\beta = \limsup |a_n|^{1/n} = \lim_{k\to\infty} |a_{3k}|^{1/3k} = \lim_{k\to\infty} (2^{-k})^{1/3k} = 2^{-1/3}.$$

Therefore the radius of convergence is $R = \frac{1}{\beta} = 2^{1/3}$.

One may consider more general power series of the form

$$\sum_{n=0}^{\infty} a_n(x - x_0)^n, \qquad (*)$$

where x_0 is a fixed real number, but they reduce to series of the form $\sum_{n=0}^{\infty} a_n y^n$ by the change of variable $y = x - x_0$. The interval of convergence for the series $(*)$ will be an interval centered at x_0.

Example 7
Consider the series

$$\sum_{n=1}^{\infty} \frac{(-1)^{n+1}}{n}(x - 1)^n. \qquad (1)$$

The radius of convergence for the series $\sum_{n=1}^{\infty} \frac{(-1)^{n+1}}{n} y^n$ is $R = 1$, so the interval of convergence for the series (1) is the interval $(0, 2)$ plus perhaps an endpoint or two. Direct substitution shows that the series (1) converges at $x = 2$ [it's an alternating series] and diverges to $-\infty$ at $x = 0$. So the exact interval of convergence is $(0, 2]$. It turns out that the series (1) represents the function $\log_e x$ on $(0, 2]$. See Examples 1 and 2 in §26.

One of our major goals is to understand the function given by a power series:

$$f(x) = \sum_{k=0}^{\infty} a_k x^k \quad \text{for} \quad |x| < R.$$

We are interested in questions like: Is f continuous? Is f differentiable? If so, can one differentiate f term-by-term:

$$f'(x) = \sum_{k=1}^{\infty} k a_k x^{k-1} ?$$

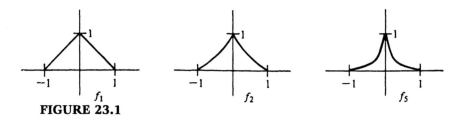

FIGURE 23.1

Can one integrate f term-by-term?

Returning to the question of continuity, what reason is there to believe that f must be continuous? Its partial sums $f_n = \sum_{k=0}^{n} a_k x^k$ are continuous, since they are polynomials. Moreover, we have $\lim_{n\to\infty} f_n(x) = f(x)$ for $|x| < R$. Therefore f would be continuous *if* a result like the following were true: If (f_n) is a sequence of continuous functions on (a, b) and if $\lim_{n\to\infty} f_n(x) = f(x)$ for all $x \in (a, b)$, then f is continuous on (a, b). However, this fine sounding result is false!

Example 8

Let $f_n(x) = (1 - |x|)^n$ for $x \in (-1, 1)$; see Figure 23.1. Let $f(x) = 0$ for $x \neq 0$ and let $f(0) = 1$. Then we have $\lim_{n\to\infty} f_n(x) = f(x)$ for all $x \in (-1, 1)$, since $\lim_{n\to\infty} a^n = 0$ if $|a| < 1$. Each f_n is a continuous function, but the limit function f is clearly discontinuous at $x = 0$.

This example, as well as Exercises 23.7–23.9, may be discouraging, but it turns out that power series do converge to continuous functions. This is because

$$\lim_{n\to\infty} \sum_{k=0}^{n} a_k x^k \quad \textit{converges uniformly to} \quad \sum_{k=0}^{\infty} a_k x^k$$

on sets $[-R_1, R_1]$ such that $R_1 < R$. The definition of uniform convergences is given in the next section, and the next two sections will be devoted to this important notion. We return to power series in §26.

Exercises

23.1. For each of the following power series, find the radius of convergence and determine the exact interval of convergence.

(a) $\sum n^2 x^n$ (b) $\sum (\frac{x}{n})^n$

(c) $\sum (\frac{2^n}{n^2}) x^n$ (d) $\sum (\frac{n^3}{3^n}) x^n$

(e) $\sum (\frac{2^n}{n!}) x^n$ (f) $\sum (\frac{1}{(n+1)^2 2^n}) x^n$

(g) $\sum (\frac{3^n}{n \cdot 4^n}) x^n$ (h) $\sum (\frac{(-1)^n}{n^2 \cdot 4^n}) x^n$

23.2. Repeat Exercise 23.1 for the following:

(a) $\sum \sqrt{n} x^n$ (b) $\sum \frac{1}{n \sqrt{n}} x^n$

(c) $\sum x^{n!}$ (d) $\sum \frac{3^n}{\sqrt{n}} x^{2n+1}$

23.3. Find the exact interval of convergence for the series in Example 6.

23.4. For $n = 0, 1, 2, 3, \ldots$, let $a_n = [\frac{4+2(-1)^n}{5}]^n$.

(a) Find $\limsup (a_n)^{1/n}$, $\liminf (a_n)^{1/n}$, $\limsup |\frac{a_{n+1}}{a_n}|$ and $\liminf |\frac{a_{n+1}}{a_n}|$.

(b) Do the series $\sum a_n$ and $\sum (-1)^n a_n$ converge? Explain briefly.

(c) Now consider the power series $\sum a_n x^n$ with the coefficients a_n as above. Find the radius of convergence and determine the exact interval of convergence for the series.

23.5. Consider a power series $\sum a_n x^n$ with radius of convergence R.

(a) Prove that if all the coefficients a_n are integers and if infinitely many of them are nonzero, then $R \leq 1$.

(b) Prove that if $\limsup |a_n| > 0$, then $R \leq 1$.

23.6. (a) Suppose that $\sum a_n x^n$ has finite radius of convergence R and that $a_n \geq 0$ for all n. Show that if the series converges at R, then it also converges at $-R$.

(b) Give an example of a power series whose interval of convergence is exactly $(-1, 1]$.

The next three exercises are designed to show that the notion of convergence of functions discussed prior to Example 8 has many defects.

23.7. For each $n \in \mathbb{N}$, let $f_n(x) = (\cos x)^n$. Each f_n is a continuous function. Nevertheless, show that

(a) $\lim f_n(x) = 0$ unless x is a multiple of π,

(b) $\lim f_n(x) = 1$ if x is an even multiple of π,

(c) $\lim f_n(x)$ does not exist if x is an odd multiple of π.

23.8. For each $n \in \mathbb{N}$, let $f_n(x) = \frac{1}{n} \sin nx$. Each f_n is a differentiable function. Show that

(a) $\lim f_n(x) = 0$ for all $x \in \mathbb{R}$,

(b) but $\lim f_n'(x)$ need not exist [at $x = \pi$ for instance].

23.9. Let $f_n(x) = nx^n$ for $x \in [0, 1]$ and $n \in \mathbb{N}$. Show that

(a) $\lim f_n(x) = 0$ for $x \in [0, 1)$. *Hint:* Use Exercise 9.12.

(b) However, $\lim_{n \to \infty} \int_0^1 f_n(x)\, dx = 1$.

§24 Uniform Convergence

We first formalize the notion of convergence discussed prior to Example 8 in the preceding section.

24.1 Definition.
Let (f_n) be a sequence of real-valued functions defined on a set $S \subseteq \mathbb{R}$. The sequence (f_n) *converges pointwise* [i.e., at each point] to a function f defined on S if

$$\lim_{n \to \infty} f_n(x) = f(x) \quad \text{for all} \quad x \in S.$$

We often write $\lim f_n = f$ *pointwise* [on S] or $f_n \to f$ *pointwise* [on S].

Example 1
All the functions f obtained in the last section as a limit of a sequence of functions were pointwise limits. See Example 8 of §23 and Exercises 23.7–23.9. In Exercise 23.8 we have $f_n \to 0$ pointwise on $\mathbb{R}$, and in Exercise 23.9 we have $f_n \to 0$ pointwise on $[0, 1)$.

Example 2
Let $f_n(x) = x^n$ for $x \in [0, 1]$. Then $f_n \to f$ pointwise on $[0, 1]$ where $f(x) = 0$ for $x \in [0, 1)$ and $f(1) = 1$.

Now observe that $f_n \to f$ pointwise on S means exactly the following:

> for each $\epsilon > 0$ and $x \in S$ there exists N such that
> $|f_n(x) - f(x)| < \epsilon$ for $n > N$.

Note that the value of N depends on both $\epsilon > 0$ and x in S. If for each $\epsilon > 0$ we could find N so that

$$|f_n(x) - f(x)| < \epsilon \quad \text{for } all \quad x \in S \quad \text{and} \quad n > N,$$

then the values $f_n(x)$ would be "uniformly" close to the values $f(x)$. Here N would depend on ϵ but not on x. This concept is extremely useful.

24.2 Definition.
Let (f_n) be a sequence of real-valued functions defined on a set $S \subseteq \mathbb{R}$. The sequence (f_n) *converges uniformly on S* to a function f defined on S if

> for each $\epsilon > 0$ there exists a number N such that
> $|f_n(x) - f(x)| < \epsilon$ for all $x \in S$ and all $n > N$.

We write $\lim f_n = f$ *uniformly on S* or $f_n \to f$ *uniformly on S*.

Note that if $f_n \to f$ uniformly on S and if $\epsilon > 0$, then there exists N such that $f(x) - \epsilon < f_n(x) < f(x) + \epsilon$ for *all* $x \in S$ and $n > N$. In other words, for $n > N$ the graph of f_n lies in the strip between the graphs of $f - \epsilon$ and $f + \epsilon$. In Figure 24.1 the graphs of f_n for $n > N$ would all lie between the dotted lines.

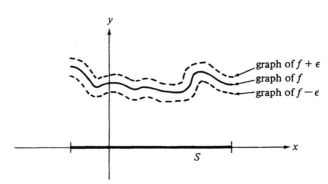

FIGURE 24.1

We return to our earlier examples.

Example 3
Let $f_n(x) = (1 - |x|)^n$ for $x \in (-1, 1)$. Also, let $f(x) = 0$ for $x \neq 0$ and $f(0) = 1$. As noted in Example 8 of §23, $f_n \to f$ pointwise on $(-1, 1)$. It turns out that the sequence (f_n) does not converge uniformly to f on $(-1, 1)$ in view of the next theorem. This can also be shown directly, as follows. Assume that $f_n \to f$ uniformly on $(-1, 1)$. Then [with $\epsilon = \frac{1}{2}$ in mind] we see that there exists N in $\mathbb{N}$ so that $|f(x) - f_n(x)| < \frac{1}{2}$ for *all* $x \in (-1, 1)$ and $n > N$. Hence

$$x \in (0, 1) \quad \text{and} \quad n > N \quad \text{imply} \quad |(1 - x)^n| < \frac{1}{2}.$$

In particular,

$$x \in (0, 1) \quad \text{implies} \quad (1 - x)^{N+1} < \frac{1}{2}.$$

However, this fails for sufficiently small x; for example, if we set $x = 1 - 2^{-1/(N+1)}$, then $1 - x = 2^{-1/(N+1)}$ and $(1 - x)^{N+1} = 2^{-1} = \frac{1}{2}$. This contradiction shows that (f_n) does not converge uniformly to f on $(-1, 1)$ as had been assumed.

Example 4
Let $f_n(x) = \frac{1}{n} \sin nx$ for $x \in \mathbb{R}$. Then $f_n \to 0$ pointwise on $\mathbb{R}$ as shown in Exercise 23.8. In fact, $f_n \to 0$ uniformly on $\mathbb{R}$. To see this, let $\epsilon > 0$ and let $N = \frac{1}{\epsilon}$. Then for $n > N$ and *all* $x \in \mathbb{R}$ we have

$$|f_n(x) - 0| = \left| \frac{1}{n} \sin nx \right| \leq \frac{1}{n} < \frac{1}{N} = \epsilon.$$

Example 5
Let $f_n(x) = nx^n$ for $x \in [0, 1)$. Since $\lim_{n \to \infty} f_n(1) = \lim_{n \to \infty} n = +\infty$, we have dropped the number 1 from the domain under consideration. Then $f_n \to 0$ pointwise on $[0, 1)$, as shown in Exercise 23.9. We show that the convergence is *not* uniform. If it were, there would exist N in $\mathbb{N}$ such that

$$|nx^n - 0| < 1 \quad \text{for } all \quad x \in [0, 1) \quad \text{and} \quad n > N.$$

In particular, we would have $(N + 1)x^{N+1} < 1$ for all $x \in [0, 1)$. But this fails for x sufficiently close to 1. Consider, for example, the reciprocal x of $(N + 1)^{1/(N+1)}$.

Example 6
As in Example 2, let $f_n(x) = x^n$ for $x \in [0, 1]$, $f(x) = 0$ for $x \in [0, 1)$ and $f(1) = 1$. Then $f_n \to f$ pointwise on $[0, 1]$, but (f_n) does not converge uniformly to f on $[0, 1]$, as can be seen directly or by applying the next theorem.

24.3 Theorem.
The uniform limit of continuous functions is continuous. More precisely, let (f_n) be a sequence of functions on a set $S \subseteq \mathbb{R}$, suppose that $f_n \to f$ uniformly on S, and suppose that $S = \text{dom}(f)$. If each f_n is continuous at x_0 in S, then f is continuous at x_0. [So if each f_n is continuous on S, then f is continuous on S.]

Proof
This involves the famous "$\frac{\epsilon}{3}$ argument." The critical inequality is

$$|f(x) - f(x_0)| \leq |f(x) - f_n(x)| + |f_n(x) - f_n(x_0)| + |f_n(x_0) - f(x_0)|. \quad (1)$$

If n is large enough, the first and third terms on the right side of (1) will be small, since $f_n \to f$ uniformly. *Once such n is selected*, the continuity of f_n implies that the middle term will be small provided x is close to x_0.

For the formal proof, let $\epsilon > 0$. There exists N in $\mathbb{N}$ such that

$$n > N \quad \text{implies} \quad |f_n(x) - f(x)| < \frac{\epsilon}{3} \quad \text{for all} \quad x \in S.$$

In particular,

$$|f_{N+1}(x) - f(x)| < \frac{\epsilon}{3} \quad \text{for all} \quad x \in S. \quad (2)$$

Since f_{N+1} is continuous at x_0 there is a $\delta > 0$ such that

$$x \in S \quad \text{and} \quad |x - x_0| < \delta \quad \text{imply} \quad |f_{N+1}(x) - f_{N+1}(x_0)| < \frac{\epsilon}{3}; \quad (3)$$

see Theorem 17.2. Now we apply (1) with $n = N + 1$, (2) twice [once for x and once for x_0] and (3) to conclude

$$x \in S \quad \text{and} \quad |x - x_0| < \delta \quad \text{imply} \quad |f(x) - f(x_0)| < 3 \cdot \frac{\epsilon}{3} = \epsilon.$$

This proves that f is continuous at x_0. ∎

One might think that this theorem would be useless in practice, since it should be easier to show that a single function f is continuous than to show that a sequence (f_n) consists of continuous functions and that the sequence converges to f uniformly. This would no doubt be true if f were given by a simple formula. But consider, for example,

$$f(x) = \sum_{n=1}^{\infty} \frac{1}{n^2} x^n \quad \text{for} \quad x \in [-1, 1]$$

or

$$J_0(x) = \sum_{n=0}^{\infty} \frac{(-1)^n (\frac{1}{2}x)^{2n}}{(n!)^2} \quad \text{for} \quad x \in \mathbb{R}.$$

The partial sums are clearly continuous, but neither f nor J_0 is given by a simple formula. Moreover, many functions that arise in mathematics and elsewhere, such as the Bessel function J_0, are defined by power series. It would be very useful to know when and where power series converge uniformly; an answer is given in §26.

24.4 Remark.
Uniform convergence can be reformulated as follows. *A sequence (f_n) of functions on a set $S \subseteq \mathbb{R}$ converges uniformly to a function f on S if and only if*

$$\lim_{n \to \infty} [\sup\{|f(x) - f_n(x)| : x \in S\}] = 0. \tag{1}$$

We leave the straightforward proof to Exercise 24.12.

According to (1) we can decide whether a sequence (f_n) converges uniformly to f by calculating $\sup\{|f(x) - f_n(x)| : x \in S\}$ for each n. If $f - f_n$ is differentiable, we may use calculus to find these suprema.

Example 7
Let $f_n(x) = \frac{x}{1+nx^2}$ for $x \in \mathbb{R}$. Clearly we have $\lim_{n \to \infty} f_n(0) = 0$. If $x \neq 0$, then $\lim_{n \to \infty}(1 + nx^2) = +\infty$, so $\lim_{n \to \infty} f_n(x) = 0$. Thus $f_n \to 0$ pointwise on $\mathbb{R}$. To find the maximum and minimum of f_n, we calculate $f_n'(x)$ and set it equal to 0. This leads to $(1+nx^2) \cdot 1 - x(2nx) = 0$ or $1 - nx^2 = 0$. Thus $f_n'(x) = 0$ if and only if $x = \frac{\pm 1}{\sqrt{n}}$. Further analysis or a sketch of f_n leads one to conclude that f_n takes its maximum at

$\frac{1}{\sqrt{n}}$ and its minimum at $-\frac{1}{\sqrt{n}}$. Since $f_n(\pm\frac{1}{\sqrt{n}}) = \pm\frac{1}{2\sqrt{n}}$, we conclude that

$$\lim_{n\to\infty}[\sup\{|f_n(x)| : x \in S\}] = \lim_{n\to\infty}\frac{1}{2\sqrt{n}} = 0.$$

Therefore $f_n \to 0$ uniformly on $\mathbb{R}$ by Remark 24.4.

Example 8

Let $f_n(x) = n^2x^n(1-x)$ for $x \in [0, 1]$. Then we have $\lim_{n\to\infty}f_n(1) = 0$. For $x \in [0, 1)$ we have $\lim_{n\to\infty}n^2x^n = 0$ by applying Exercise 9.12 [since

$$\frac{(n+1)^2x^{n+1}}{n^2x^n} = \left(\frac{n+1}{n}\right)^2 x \to x],$$

and hence $\lim_{n\to\infty}f_n(x) = 0$. Thus $f_n \to 0$ pointwise on $[0, 1]$. Again, to find the maximum and minimum of f_n we set its derivative equal to 0. We obtain $x^n(-1) + (1-x)nx^{n-1} = 0$ or $x^{n-1}[n - (n+1)x] = 0$. Since f_n takes the value 0 at both endpoints of the interval $[0, 1]$, it follows that f_n takes it maximum at $\frac{n}{n+1}$. We have

$$f_n\left(\frac{n}{n+1}\right) = n^2 \left(\frac{n}{n+1}\right)^n \left(1 - \frac{n}{n+1}\right) = \frac{n^2}{n+1}\left(\frac{n}{n+1}\right)^n. \quad (1)$$

The reciprocal of $(\frac{n}{n+1})^n$ is $(1+\frac{1}{n})^n$, the nth term of a sequence which has limit e. This was mentioned, but not proved, in Example 3 of §7; a proof is given in Theorem 37.11. Therefore we have $\lim(\frac{n}{n+1})^n = \frac{1}{e}$. Since $\lim[\frac{n^2}{n+1}] = +\infty$, we conclude from (1) that $\lim f_n(\frac{n}{n+1}) = +\infty$; see Exercise 12.9(a). In particular, (f_n) does *not* converge uniformly to 0.

Exercises

24.1. Let $f_n(x) = \frac{1+2\cos^2 nx}{\sqrt{n}}$. Prove carefully that (f_n) converges uniformly to 0 on $\mathbb{R}$.

24.2. For $x \in [0, \infty)$, let $f_n(x) = \frac{x}{n}$.

    (a) Find $f(x) = \lim f_n(x)$.

    (b) Determine whether $f_n \to f$ uniformly on $[0, 1]$.

(c) Determine whether $f_n \to f$ uniformly on $[0, \infty)$.

24.3. Repeat Exercise 24.2 for $f_n(x) = \frac{1}{1+x^n}$.

24.4. Repeat Exercise 24.2 for $f_n(x) = \frac{x^n}{1+x^n}$.

24.5. Repeat Exercise 24.2 for $f_n(x) = \frac{x^n}{n+x^n}$.

24.6. Let $f_n(x) = (x - \frac{1}{n})^2$ for $x \in [0, 1]$.

    (a) Does the sequence (f_n) converge pointwise on the set $[0, 1]$? If so, give the limit function.

    (b) Does (f_n) converge uniformly on $[0, 1]$? Prove your assertion.

24.7. Repeat Exercise 24.6 for $f_n(x) = x - x^n$.

24.8. Repeat Exercise 24.6 for $f_n(x) = \sum_{k=0}^{n} x^k$.

24.9. Consider $f_n(x) = nx^n(1 - x)$ for $x \in [0, 1]$.

    (a) Find $f(x) = \lim f_n(x)$.

    (b) Does $f_n \to f$ uniformly on $[0, 1]$? Justify.

    (c) Does $\int_0^1 f_n(x)\,dx$ converge to $\int_0^1 f(x)\,dx$? Justify.

24.10. **(a)** Prove that if $f_n \to f$ uniformly on a set S, and if $g_n \to g$ uniformly on S, then $f_n + g_n \to f + g$ uniformly on S.

    (b) Do you believe that the analogue of (a) holds for products? If so, see the next exercise.

24.11. Let $f_n(x) = x$ and $g_n(x) = \frac{1}{n}$ for all $x \in \mathbb{R}$. Let $f(x) = x$ and $g(x) = 0$ for $x \in \mathbb{R}$.

    (a) Observe that $f_n \to f$ uniformly on $\mathbb{R}$ [obvious!] and that $g_n \to g$ uniformly on $\mathbb{R}$ [almost obvious].

    (b) Observe that the sequence $(f_n g_n)$ does not converge uniformly to fg on $\mathbb{R}$. Compare Exercise 24.2.

24.12. Prove the assertion in Remark 24.4.

24.13. Prove that if (f_n) is a sequence of uniformly continuous functions on an interval (a, b), and if $f_n \to f$ uniformly on (a, b), then f is also uniformly continuous on (a, b). *Hint:* Try an $\frac{\varepsilon}{3}$ argument as in the proof of Theorem 24.3.

24.14. Let $f_n(x) = \frac{nx}{1+n^2x^2}$.

    (a) Show that $f_n \to 0$ pointwise on $\mathbb{R}$.

(b) Does $f_n \to 0$ uniformly on $[0, 1]$? Justify.

(c) Does $f_n \to 0$ uniformly on $[1, \infty)$? Justify.

24.15. Let $f_n(x) = \frac{nx}{1+nx}$ for $x \in [0, \infty)$.

(a) Find $f(x) = \lim f_n(x)$.

(b) Does $f_n \to f$ uniformly on $[0, 1]$? Justify.

(c) Does $f_n \to f$ uniformly on $[1, \infty)$? Justify.

24.16. Repeat Exercise 24.15 for $f_n(x) = \frac{nx}{1+nx^2}$.

24.17. Let (f_n) be a sequence of continuous functions on $[a, b]$ that converges uniformly to f on $[a, b]$. Show that if (x_n) is a sequence in $[a, b]$ and if $x_n \to x$, then $\lim_{n \to \infty} f_n(x_n) = f(x)$.

§25 More on Uniform Convergence

Our next theorem shows that one can interchange integrals and *uniform* limits. The adjective "uniform" here is important; compare Exercise 23.9.

25.1 Discussion.
To prove Theorem 25.2 below we merely use some basic facts about integration which should be familiar [or believable] even if your calculus is rusty. Specifically, we use:

(a) *If g and h are integrable on $[a, b]$ and if $g(x) \leq h(x)$ for all $x \in [a, b]$, then $\int_a^b g(x)\, dx \leq \int_a^b h(x)\, dx$.* See Theorem 33.4.
  We also use the following corollary:

(b) *If g is integrable on $[a, b]$, then*

$$\left| \int_a^b g(x)\, dx \right| \leq \int_a^b |g(x)|\, dx.$$

Continuous functions on closed intervals are integrable, as noted in Discussion 19.3 and proved in Theorem 33.2.

25.2 Theorem.
Let (f_n) be a sequence of continuous functions on $[a, b]$, and suppose that $f_n \to f$ uniformly on $[a, b]$. Then

$$\lim_{n \to \infty} \int_a^b f_n(x)\, dx = \int_a^b f(x)\, dx. \tag{1}$$

Proof
By Theorem 24.3 f is continuous, so the functions $f_n - f$ are all integrable on $[a, b]$. Let $\epsilon > 0$. Since $f_n \to f$ uniformly on $[a, b]$, there exists a number N such that $|f_n(x) - f(x)| < \frac{\epsilon}{b-a}$ for all $x \in [a, b]$ and all $n > N$. Consequently $n > N$ implies

$$\left| \int_a^b f_n(x)\, dx - \int_a^b f(x)\, dx \right| = \left| \int_a^b [f_n(x) - f(x)]\, dx \right|$$

$$\leq \int_a^b |f_n(x) - f(x)|\, dx \leq \int_a^b \frac{\epsilon}{b - a}\, dx = \epsilon.$$

The first $\leq$ follows from 25.1(b) applied to $g = f_n - f$ and the second $\leq$ follows from 25.1(a) applied to $g = |f_n - f|$ and $h = \frac{\epsilon}{b-a}$; h happens to be a constant function, but this does no harm.

The last paragraph shows that given $\epsilon > 0$, there exists N such that $| \int_a^b f_n(x)\, dx - \int_a^b f(x)\, dx | \leq \epsilon$ for $n > N$. Therefore (1) holds. ∎

Recall one of the advantages of the notion of Cauchy sequence: A sequence (s_n) of real numbers can be shown to converge *without knowing its limit* by simply verifying that it is a Cauchy sequence. Clearly a similar result for sequences of functions would be valuable, since it is likely that we will not know the limit function in advance. What we need is the idea of "uniformly Cauchy."

25.3 Definition.
A sequence (f_n) of functions defined on a set $S \subseteq \mathbb{R}$ is *uniformly Cauchy* on S if

> for each $\epsilon > 0$ there exists a number N such that
> $|f_n(x) - f_m(x)| < \epsilon$ for all $x \in S$ and all $m, n > N$.

Compare this definition with that of a Cauchy sequence of real numbers [Definition 10.8] and that of uniform convergence [Definition 24.2]. It is an easy exercise to show that uniformly convergent sequences of functions are uniformly Cauchy; see Exercise 25.4. The interesting and useful result is the converse, just as in the case of sequences of real numbers.

25.4 Theorem.
Let (f_n) be a sequence of functions defined and uniformly Cauchy on a set $S \subseteq \mathbb{R}$. Then there exists a function f on S such that $f_n \to f$ uniformly on S.

Proof
First we have to "find" f. We begin by showing

$$\text{for each } x_0 \in S \text{ the sequence } (f_n(x_0)) \text{ is a Cauchy} \atop \text{sequence of real numbers.} \tag{1}$$

For each $\epsilon > 0$, there exists N such that

$$|f_n(x) - f_m(x)| < \epsilon \quad \text{for} \quad x \in S \quad \text{and} \quad m, n > N.$$

In particular, we have

$$|f_n(x_0) - f_m(x_0)| < \epsilon \quad \text{for} \quad m, n > N.$$

This shows that $(f_n(x_0))$ is a Cauchy sequence, so (1) holds.

Now for each $x_0 \in S$, assertion (1) implies that $\lim_{n \to \infty} f_n(x_0)$ must exist; this is proved in Theorem 10.11 which in the end depends on the Completeness Axiom 4.4. Hence we *define* $f(x_0) = \lim_{n \to \infty} f_n(x_0)$. This defines a function f on S such that $f_n \to f$ *pointwise* on S.

Now that we have "found" f, we need to prove that $f_n \to f$ uniformly on S. Let $\epsilon > 0$. There is a number N such that

$$|f_n(x) - f_m(x)| < \frac{\epsilon}{2} \quad \text{for all} \quad x \in S \quad \text{and all} \quad m, n > N. \tag{2}$$

Consider $m > N$ and $x \in S$. Assertion (2) tells us that $f_n(x)$ lies in the open interval $(f_m(x) - \frac{\epsilon}{2}, f_m(x) + \frac{\epsilon}{2})$ for all $n > N$. Therefore, as noted in Exercise 8.9, the limit $f(x) = \lim_{n \to \infty} f_n(x)$ must lie in the closed interval $[f_m(x) - \frac{\epsilon}{2}, f_m(x) + \frac{\epsilon}{2}]$. In other words,

$$|f(x) - f_m(x)| \le \frac{\epsilon}{2} \quad \text{for all} \quad x \in S \quad \text{and} \quad m > N.$$

Then of course

$$|f(x) - f_m(x)| < \epsilon \quad \text{for all} \quad x \in S \quad \text{and} \quad m > N.$$

This shows that $f_m \to f$ uniformly on S, as desired. ■

Theorem 25.4 is especially useful for "series of functions." Let us recall what $\sum_{k=1}^{\infty} a_k$ signifies, where the a_k's are real numbers. This signifies $\lim_{n\to\infty} \sum_{k=1}^{n} a_k$ provided this limit exists [as a real number, $+\infty$ or $-\infty$]. Otherwise the symbol $\sum_{k=1}^{\infty} a_k$ has no meaning. Thus the infinite series is the limit of the sequence of partial sums $\sum_{k=1}^{n} a_k$. Similar remarks apply to series of functions. A *series of functions* is an expression $\sum_{k=0}^{\infty} g_k$ or $\sum_{k=0}^{\infty} g_k(x)$ which makes sense provided the sequence of partial sums $\sum_{k=0}^{n} g_k$ converges, or diverges to $+\infty$ or $-\infty$ pointwise. If the sequence of partial sums converges uniformly on a set S to $\sum_{k=0}^{\infty} g_k$, then we say that the *series is uniformly convergent on S*.

Example 1
Any power series is a series of functions, since $\sum_{k=0}^{\infty} a_k x^k$ has the form $\sum_{k=0}^{\infty} g_k$ where $g_k(x) = a_k x^k$.

Example 2
$\sum_{k=0}^{\infty} \frac{x^k}{1+x^k}$ is a series of functions, but is not a power series, at least not in its present form. This is a series $\sum_{k=0}^{\infty} g_k$ where $g_0(x) = \frac{1}{2}$ for all x, $g_1(x) = \frac{x}{1+x}$ for all x, $g_2(x) = \frac{x^2}{1+x^2}$ for all x, etc.

Example 3
Let g be the function drawn in Figure 25.1, and let $g_n(x) = g(4^n x)$ for all $x \in \mathbb{R}$. Then $\sum_{n=0}^{\infty} (\frac{3}{4})^n g_n(x)$ is a series of functions. The limit

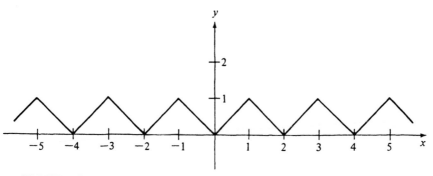

function f is continuous on $\mathbb{R}$, but has the amazing property that it is not differentiable at any point! The proof of the nondifferentiability of f is somewhat delicate; see 7.18 of [36].

Theorems for sequences of functions translate easily into theorems for series of functions. Here is an example.

25.5 Theorem.
Consider a series $\sum_{k=0}^{\infty} g_k$ of functions on a set $S \subseteq \mathbb{R}$. Suppose that each g_k is continuous on S and that the series converges uniformly on S. Then the series $\sum_{k=0}^{\infty} g_k$ represents a continuous function on S.

Proof
Each partial sum $f_n = \sum_{k=1}^{n} g_k$ is continuous and the sequence (f_n) converges uniformly on S. Hence the limit function is continuous by Theorem 24.3. ∎

Recall the Cauchy criterion for series $\sum a_k$ given in Definition 14.3:

> for each $\epsilon > 0$ there exists a number N such that
> $n \geq m > N$ implies $|\sum_{k=m}^{n} a_k| < \epsilon$.

The analogue for series of functions is also useful. The sequence of partial sums of a series $\sum_{k=0}^{\infty} g_k$ of functions is uniformly Cauchy on a set S if and only if the series satisfies the *Cauchy criterion [uniformly on S]*:

> for each $\epsilon > 0$ there exists a number N such that
> $n \geq m > N$ implies $|\sum_{k=m}^{n} g_k(x)| < \epsilon$ for all $x \in S$.

25.6 Theorem.
If a series $\sum_{k=0}^{\infty} g_k$ of functions satisfies the Cauchy criterion uniformly on a set S, then the series converges uniformly on S.

Proof
Let $f_n = \sum_{k=0}^{n} g_k$. The sequence (f_n) of partial sums is uniformly Cauchy on S, so (f_n) converges uniformly on S by Theorem 25.4. ∎

Here is a useful corollary.

25.7 Weierstrass M-test.

Let (M_k) be a sequence of nonnegative real numbers where $\sum M_k < \infty$. If $|g_k(x)| \leq M_k$ for all x in a set S, then $\sum g_k$ converges uniformly on S.

Proof

To verify the Cauchy criterion on S, let $\epsilon > 0$. Since the series $\sum M_k$ converges, it satisfies the Cauchy criterion 14.3. So there exists a number N such that

$$n \geq m > N \quad \text{implies} \quad \left| \sum_{k=m}^{n} M_k \right| < \epsilon.$$

Hence if $n \geq m > N$ and $x \in S$, then

$$\left| \sum_{k=m}^{n} g_k(x) \right| \leq \sum_{k=m}^{n} |g_k(x)| \leq \sum_{k=m}^{n} M_k < \epsilon.$$

Thus the series $\sum g_k$ satisfies the Cauchy criterion uniformly on S and Theorem 25.6 shows that it converges uniformly on S. ∎

Example 4

Show that $\sum_{n=1}^{\infty} 2^{-n} x^n$ represents a continuous function f on $(-2, 2)$, but that the convergence is not uniform.

Solution

This is a power series with radius of convergence 2. Clearly the series does not converge at $x = 2$ or at $x = -2$, so its interval of convergence is $(-2, 2)$.

Consider $0 < a < 2$ and note that $\sum_{n=1}^{\infty} 2^{-n} a^n = \sum_{n=1}^{\infty} (\frac{a}{2})^n$ converges. Since $|2^{-n} x^n| \leq 2^{-n} a^n = (\frac{a}{2})^n$ for $x \in [-a, a]$, the Weierstrass M-test 25.7 shows that the series $\sum_{n=1}^{\infty} 2^{-n} x^n$ converges uniformly to a function on $[-a, a]$. By Theorem 25.5 the limit function f is continuous at each point of the set $[-a, a]$. Since a can be any number less than 2, we conclude that f represents a continuous function on $(-2, 2)$.

Since we have $\sup\{|2^{-n} x^n| : x \in (-2, 2)\} = 1$ for all n, the convergence of the series cannot be uniform on $(-2, 2)$ in view of the next example. □

Example 5

Show that if the series $\sum g_n$ converges uniformly on a set S, then

$$\lim_{n\to\infty} [\sup\{|g_n(x)| : x \in S\}] = 0. \tag{1}$$

Solution

Let $\epsilon > 0$. Since the series $\sum g_n$ satisfies the Cauchy criterion, there exists N such that

$$n \geq m > N \quad \text{implies} \quad \left|\sum_{k=m}^{n} g_k(x)\right| < \epsilon \quad \text{for all} \quad x \in S.$$

In particular,

$$n > N \quad \text{implies} \quad |g_n(x)| < \epsilon \quad \text{for all} \quad x \in S.$$

Therefore

$$n > N \quad \text{implies} \quad \sup\{|g_n(x)| : x \in S\} \leq \epsilon.$$

This establishes (1). $\qquad\qquad\qquad\qquad\qquad\qquad\qquad\qquad\qquad\square$

Exercises

25.1. Derive 25.1(b) from 25.1(a). *Hint*: Apply (a) twice, once to g and $|g|$ and once to $-|g|$ and g.

25.2. Let $f_n(x) = \frac{x^n}{n}$. Show that (f_n) is uniformly convergent on $[-1, 1]$ and specify the limit function.

25.3. Let $f_n(x) = \frac{n+\cos x}{2n+\sin^2 x}$ for all real numbers x.

   (a) Show that (f_n) converges uniformly on $\mathbb{R}$. *Hint*: First decide what the limit function is; then show (f_n) converges uniformly to it.

   (b) Calculate $\lim_{n\to\infty} \int_2^7 f_n(x)\, dx$. *Hint*: Don't integrate f_n.

25.4. Let (f_n) be a sequence of functions on a set $S \subseteq \mathbb{R}$, and suppose that $f_n \to f$ uniformly on S. Prove that (f_n) is uniformly Cauchy on S. *Hint*: Use the proof of Lemma 10.9 as a model, but be careful.

25.5. Let (f_n) be a sequence of bounded functions on a set S, and suppose that $f_n \to f$ uniformly on S. Prove that f is a bounded function on S.

25.6. **(a)** Show that if $\sum |a_k| < \infty$, then $\sum a_k x^k$ converges uniformly on $[-1, 1]$ to a continuous function.

(b) Does $\sum_{n=1}^{\infty} \frac{1}{n^2} x^n$ represent a continuous function on $[-1, 1]$?

25.7. Show that $\sum_{n=1}^{\infty} \frac{1}{n^2} \cos nx$ converges uniformly on $\mathbb{R}$ to a continuous function.

25.8. Show that $\sum_{n=1}^{\infty} \frac{x^n}{n^2 2^n}$ has radius of convergence 2 and that the series converges uniformly to a continuous function on $[-2, 2]$.

25.9. **(a)** Let $0 < a < 1$. Show that the series $\sum_{n=0}^{\infty} x^n$ converges uniformly on $[-a, a]$ to $\frac{1}{1-x}$.

(b) Does the series $\sum_{n=0}^{\infty} x^n$ converge uniformly on $(-1, 1)$ to $\frac{1}{1-x}$? Explain.

25.10. **(a)** Show that $\sum \frac{x^n}{1+x^n}$ converges for $x \in [0, 1)$.

(b) Show that the series converges uniformly on $[0, a]$ for each $a, 0 < a < 1$.

(c) Does the series converge uniformly on $[0, 1)$? Explain.

25.11. **(a)** Sketch the functions g_0, g_1, g_2 and g_3 in Example 3.

(b) Prove that the function f in Example 3 is continuous.

25.12. Suppose that $\sum_{k=1}^{\infty} g_k$ is a series of continuous functions g_k on $[a, b]$ that converges uniformly to g on $[a, b]$. Prove that

$$\int_a^b g(x)\,dx = \sum_{k=1}^{\infty} \int_a^b g_k(x)\,dx.$$

25.13. Suppose that $\sum_{k=1}^{\infty} g_k$ and $\sum_{k=1}^{\infty} h_k$ converge uniformly on a set S. Show that $\sum_{k=1}^{\infty} (g_k + h_k)$ converges uniformly on S.

25.14. Prove that if $\sum g_k$ converges uniformly on a set S and if h is a bounded function on S, then $\sum h g_k$ converges uniformly on S.

25.15. Let (f_n) be a sequence of continuous functions on $[a, b]$. Suppose that, for each $x \in [a, b]$, $(f_n(x))$ is a nonincreasing sequence of real numbers.

(a) Prove that if $f_n \to 0$ pointwise on $[a, b]$, then $f_n \to 0$ uniformly on $[a, b]$. *Hint:* If not, there exists $\epsilon > 0$ and a sequence (x_n) in $[a, b]$ such that $f_n(x_n) \geq \epsilon$ for all n. Obtain a contradiction.

(b) Prove that if $f_n \to f$ pointwise on $[a, b]$ and if f is continuous on $[a, b]$, then $f_n \to f$ uniformly on $[a, b]$. This is Dini's theorem.

§26 Differentiation and Integration of Power Series

The following result was mentioned in §23 after Example 8.

26.1 Theorem.
Let $\sum_{n=0}^{\infty} a_n x^n$ be a power series with radius of convergence $R > 0$ [possibly $R = +\infty$]. If $0 < R_1 < R$, then the power series converges uniformly on $[-R_1, R_1]$ to a continuous function.

Proof
Consider $0 < R_1 < R$. A glance at Theorem 23.1 shows that the series $\sum a_n x^n$ and $\sum |a_n| x^n$ have the same radius of convergence, since β and R are defined in terms of $|a_n|$. Since $|R_1| < R$, we must have $\sum |a_n| R_1^n < \infty$. Clearly we have $|a_n x^n| \leq |a_n| R_1^n$ for all x in $[-R_1, R_1]$, so the series $\sum a_n x^n$ converges uniformly on $[-R_1, R_1]$ by the Weierstrass M-test 25.7. The limit function is continuous at each point of $[-R_1, R_1]$ by Theorem 25.5. ∎

26.2 Corollary.
The power series $\sum a_n x^n$ converges to a continuous function on the open interval $(-R, R)$.

Proof
If $x_0 \in (-R, R)$, then $x_0 \in (-R_1, R_1)$ for some $R_1 < R$. The theorem shows that the limit of the series is continuous at x_0. ∎

We emphasize that a power series need *not* converge uniformly on its interval of convergence though it might; see Example 4 of §25 and Exercise 25.8.

We are going to differentiate and integrate power series term-by-term, so clearly it would be useful to know where the new series converge. The next lemma tells us.

26.3 Lemma.

If the power series $\sum_{n=0}^{\infty} a_n x^n$ has radius of convergence R, then the power series

$$\sum_{n=1}^{\infty} n a_n x^{n-1} \quad and \quad \sum_{n=0}^{\infty} \frac{a_n}{n+1} x^{n+1}$$

also have radius of convergence R.

Proof

First observe that the series $\sum n a_n x^{n-1}$ and $\sum n a_n x^n$ have the same radius of convergence; since the second series is x times the first series, they converge for exactly the same values of x. Likewise $\sum \frac{a_n}{n+1} x^{n+1}$ and $\sum \frac{a_n}{n+1} x^n$ have the same radius of convergence.

Next recall that $R = \frac{1}{\beta}$ where $\beta = \limsup |a_n|^{1/n}$. For the series $\sum n a_n x^n$, we consider $\limsup (n|a_n|)^{1/n} = \limsup n^{1/n} |a_n|^{1/n}$. By 9.7(c) we have $\lim n^{1/n} = 1$, so $\limsup (n|a_n|)^{1/n} = \beta$ by Theorem 12.1. Hence the series $\sum n a_n x^n$ has radius of convergence R.

For the series $\sum \frac{a_n}{n+1} x^n$, we consider $\limsup (\frac{|a_n|}{n+1})^{1/n}$. It is easy to show that $\lim (n+1)^{1/n} = 1$; therefore $\lim (\frac{1}{n+1})^{1/n} = 1$. Hence by Theorem 12.1 we have $\limsup (\frac{|a_n|}{n+1})^{1/n} = \beta$, so the series $\sum \frac{a_n}{n+1} x^n$ has radius of convergence R. ∎

26.4 Theorem.

Suppose that $f(x) = \sum_{n=0}^{\infty} a_n x^n$ has radius of convergence R > 0. Then

$$\int_0^x f(t)\, dt = \sum_{n=0}^{\infty} \frac{a_n}{n+1} x^{n+1} \quad for \quad |x| < R. \tag{1}$$

Proof

We fix x and assume $x < 0$; the case $x > 0$ is similar [Exercise 26.1]. On the interval $[x, 0]$, the sequence of partial sums $\sum_{k=0}^{n} a_k t^k$ converges uniformly to $f(t)$ by Theorem 26.1. Consequently, by Theorem 25.2 we have

$$\int_x^0 f(t)\, dt = \lim_{n \to \infty} \int_x^0 \left(\sum_{k=0}^{n} a_k t^k \right) dt = \lim_{n \to \infty} \sum_{k=0}^{n} a_k \int_x^0 t^k\, dt$$

$$= \lim_{n \to \infty} \sum_{k=0}^{n} a_k \left[\frac{0^{k+1} - x^{k+1}}{k+1} \right] = - \sum_{k=0}^{\infty} \frac{a_k}{k+1} x^{k+1}. \quad (2)$$

The second equality is valid because we can interchange integrals and *finite* sums; this is a basic property of integrals [Theorem 33.3]. Since $\int_0^x f(t) \, dt = - \int_x^0 f(t) \, dt$, equation (2) implies equation (1). ∎

The theorem just proved shows that a power series can be integrated term-by-term inside its interval of convergence. Term-by-term differentiation is also legal.

26.5 Theorem.
Let $f(x) = \sum_{n=0}^{\infty} a_n x^n$ have radius of convergence $R > 0$. Then f is differentiable on $(-R, R)$ and

$$f'(x) = \sum_{n=1}^{\infty} n a_n x^{n-1} \quad \text{for} \quad |x| < R. \quad (1)$$

The proof of Theorem 26.4 was a straightforward application of Theorem 25.2, but the direct analogue of Theorem 25.2 for derivatives is not true [see Exercise 23.8 and Example 4 of §24]. So we give a devious indirect proof of the theorem.

Proof
We begin with the series $g(x) = \sum_{n=1}^{\infty} n a_n x^{n-1}$ and observe that this series converges for $|x| < R$ by Lemma 26.3. Theorem 26.4 shows that we can *integrate* g term-by-term:

$$\int_0^x g(t) \, dt = \sum_{n=1}^{\infty} a_n x^n = f(x) - a_0 \quad \text{for} \quad |x| < R.$$

Thus if $0 < R_1 < R$, then

$$f(x) = \int_{-R_1}^{x} g(t) \, dt + k \quad \text{for} \quad |x| < R_1,$$

where k is a constant; in fact, $k = a_0 - \int_{-R_1}^{0} g(t) \, dt$. Since g is continuous, one of the versions of the Fundamental Theorem of Calculus [Theorem 34.3] shows that f is differentiable and that $f'(x) = g(x)$.

Thus

$$f'(x) = g(x) = \sum_{n=1}^{\infty} n a_n x^{n-1} \quad \text{for} \quad |x| < R.$$

■

Example 1
Recall that

$$\sum_{n=0}^{\infty} x^n = \frac{1}{1-x} \quad \text{for} \quad |x| < 1. \tag{1}$$

Differentiating term-by-term, we obtain

$$\sum_{n=1}^{\infty} n x^{n-1} = \frac{1}{(1-x)^2} \quad \text{for} \quad |x| < 1.$$

Integrating (1) term-by-term, we get

$$\sum_{n=0}^{\infty} \frac{1}{n+1} x^{n+1} = \int_0^x \frac{1}{1-t} \, dt = -\log(1-x)$$

or

$$\log(1-x) = -\sum_{n=1}^{\infty} \frac{1}{n} x^n \quad \text{for} \quad |x| < 1. \tag{2}$$

Replacing x by $-x$, we find

$$\log(1+x) = x - \frac{x^2}{2} + \frac{x^3}{3} - \frac{x^4}{4} + \cdots \quad \text{for} \quad |x| < 1. \tag{3}$$

It turns out that this equality is also valid for $x = 1$ [see Example 2], so we have the interesting identity

$$\log_e 2 = 1 - \frac{1}{2} + \frac{1}{3} - \frac{1}{4} + \frac{1}{5} - \frac{1}{6} + \cdots. \tag{4}$$

In equation (2) set $x = \frac{m-1}{m}$. Then

$$\sum_{n=1}^{\infty} \frac{1}{n} \left(\frac{m-1}{m} \right)^n = -\log \left(1 - \frac{m-1}{m} \right) = -\log \left(\frac{1}{m} \right) = \log m.$$

Hence we have

$$\sum_{n=1}^{\infty} \frac{1}{n} \geq \sum_{n=1}^{\infty} \frac{1}{n} \left(\frac{m-1}{m} \right)^n = \log m \quad \text{for all} \quad m.$$

Here is yet another proof that $\sum_{n=1}^{\infty} \frac{1}{n} = +\infty$.

To establish (4) we need a relatively difficult theorem about convergence of a power series at the endpoints of its interval of convergence.

26.6 Abel's Theorem.

Let $f(x) = \sum_{n=0}^{\infty} a_n x^n$ be a power series with finite positive radius of convergence R. If the series converges at $x = R$, then f is continuous at $x = R$. If the series converges at $x = -R$, then f is continuous at $x = -R$.

Example 2

As promised, we return to (3) in Example 1:

$$\log(1 + x) = x - \frac{x^2}{2} + \frac{x^3}{3} - \frac{x^4}{4} + \cdots \quad \text{for} \quad |x| < 1.$$

For $x = 1$ the series converges by the Alternating Series Theorem 15.3. Thus the series represents a function f on $(-1, 1]$ that is continuous at $x = 1$ by Abel's theorem. The function $\log(1 + x)$ is also continuous at $x = 1$, so the functions agree at $x = 1$. [In detail, if (x_n) is a sequence in $(-1, 1)$ converging to 1, then $f(1) = \lim_{n \to \infty} f(x_n) = \lim_{n \to \infty} \log(1 + x_n) = \log 2$.] Therefore we have

$$\log_e 2 = 1 - \frac{1}{2} + \frac{1}{3} - \frac{1}{4} + \frac{1}{5} - \frac{1}{6} + \cdots.$$

Another proof of this identity is given in Example 2 of §31.

Example 3

Recall that $\sum_{n=0}^{\infty} x^n = \frac{1}{1-x}$ for $|x| < 1$. Note that at $x = -1$ the function $\frac{1}{1-x}$ is continuous and takes the value $\frac{1}{2}$. However, the series does *not* converge for $x = -1$, so Abel's theorem does not apply.

Proof of Abel's Theorem

The heart of the proof is in Case 1.

   Case 1. Suppose $f(x) = \sum_{n=0}^{\infty} a_n x^n$ has radius of convergence 1 and that the series converges at $x = 1$. We need to prove that f is continuous at $x = 1$.

Let $f_n(x) = \sum_{k=0}^{n} a_k x^k$ and $s_n = \sum_{k=0}^{n} a_k = f_n(1)$ for $n = 0, 1, 2, \ldots$, and let $s = \sum_{k=0}^{\infty} a_k = f(1)$ so that $\lim s_n = s$. For $0 < x < 1$ we have

$$
\begin{aligned}
f_n(x) &= \sum_{k=0}^{n} a_k x^k = s_0 + \sum_{k=1}^{n} (s_k - s_{k-1}) x^k \\
&= s_0 + \sum_{k=1}^{n} s_k x^k - x \sum_{k=1}^{n} s_{k-1} x^{k-1} \\
&= s_0 + \sum_{k=1}^{n} s_k x^k - x \sum_{k=0}^{n-1} s_k x^k \\
&= s_0 + s_n x^n + \sum_{k=1}^{n-1} s_k (1-x) x^k - x \cdot s_0 \\
&= \sum_{k=0}^{n-1} s_k (1-x) x^k + s_n x^n.
\end{aligned}
$$

We now take limits as n tends to ∞. We have $\lim_{n \to \infty} f_n(x) = f(x)$ and $\lim_{n \to \infty} s_n x^n = \lim s_n \cdot \lim x^n = s \cdot 0 = 0$. Therefore we conclude

$$
f(x) = \sum_{n=0}^{\infty} s_n (1-x) x^n \quad \text{for} \quad 0 < x < 1.
$$

Since $\sum_{n=0}^{\infty} x^n = \frac{1}{1-x}$ we also have

$$
f(1) = s = \sum_{n=0}^{\infty} s(1-x) x^n.
$$

Hence we have

$$
f(1) - f(x) = \sum_{n=0}^{\infty} (s - s_n)(1-x) x^n. \tag{1}
$$

Now let $\epsilon > 0$. Since $\lim s_n = s$ there exists N in $\mathbb{N}$ such that $n > N$ implies $|s - s_n| < \frac{\epsilon}{2}$. Let $g_N(x) = \sum_{n=0}^{N} |s - s_n|(1-x)x^n$. From (1) we obtain

$$
|f(1) - f(x)| \leq g_N(x) + \sum_{n=N+1}^{\infty} |s - s_n|(1-x)x^n
$$

$$
\leq g_N(x) + \sum_{n=N+1}^{\infty} \frac{\epsilon}{2}(1-x)x^n < g_N(x) + \frac{\epsilon}{2} \tag{2}
$$

for $0 < x < 1$. The function g_N is continuous and $g_N(1) = 0$. Hence there exists $\delta > 0$ such that

$$1 - \delta < x < 1 \quad \text{implies} \quad g_N(x) < \frac{\epsilon}{2}.$$

Then from (2) we see that

$$1 - \delta < x < 1 \quad \text{implies} \quad |f(1) - f(x)| < \frac{\epsilon}{2} + \frac{\epsilon}{2} = \epsilon.$$

This proves that f is continuous at $x = 1$. [We do not consider $x > 1$ because dom(f) $\subseteq [-1, 1]$.]

Case 2. Suppose $f(x) = \sum_{n=0}^{\infty} a_n x^n$ has radius of convergence R, $0 < R < \infty$, and that the series converges at $x = R$. Let $g(x) = f(Rx)$ and note that

$$g(x) = \sum_{n=0}^{\infty} a_n R^n x^n \quad \text{for} \quad |x| < 1.$$

This series has radius of convergence 1 and it converges at $x = 1$. By Case 1, g is continuous at $x = 1$. Since $f(x) = g(\frac{x}{R})$, it follows that f is continuous at $x = R$.

Case 3. Suppose $f(x) = \sum_{n=0}^{\infty} a_n x^n$ has radius of convergence R, $0 < R < \infty$, and that the series converges at $x = -R$. Let $h(x) = f(-x)$ and note that

$$h(x) = \sum_{n=0}^{\infty} (-1)^n a_n x^n \quad \text{for} \quad |x| < R.$$

The series for h converges at $x = R$, so h is continuous at $x = R$ by Case 2. It follows that $f(x) = h(-x)$ is continuous at $x = -R$. ∎

The point of view in our extremely brief introduction to power series has been: For a given power series $\sum a_n x^n$, what can one say about the function $f(x) = \sum a_n x^n$? This point of view was misleading. Often, in real life, one begins with a function f and seeks a power series that represents the function for some or all values of x. This is because power series, being limits of polynomials, are in some sense basic objects.

If we have $f(x) = \sum_{n=0}^{\infty} a_n x^n$ for $|x| < R$, then we can differentiate f term-by-term forever. At each step, we may calculate the kth derivative of f at 0, written $f^{(k)}(0)$. It is easy to show that $f^{(k)}(0) = k! a_k$

for $k \geq 0$. This tells us that if f can be represented by a power series, then that power series must be $\sum_{k=0}^{\infty} \frac{f^{(k)}(0)}{k!} x^k$. This is the *Taylor series* for f about 0. Frequently, but not always, the Taylor series will agree with f on the interval of convergence. This turns out to be true for many familiar functions. Thus the following relations can be proved:

$$e^x = \sum_{k=0}^{\infty} \frac{1}{k!} x^k, \quad \cos x = \sum_{k=0}^{\infty} \frac{(-1)^k}{(2k)!} x^{2k}, \quad \sin x = \sum_{k=0}^{\infty} \frac{(-1)^k}{(2k+1)!} x^{2k+1}$$

for all x in $\mathbb{R}$. A more detailed study of Taylor series is given in §31.

Exercises

26.1. Prove Theorem 26.4 for $x > 0$.

26.2. **(a)** Observe that $\sum_{n=1}^{\infty} n x^n = \frac{x}{(1-x)^2}$ for $|x| < 1$; see Example 1.

(b) Evaluate $\sum_{n=1}^{\infty} \frac{n}{2^n}$. Compare with Exercise 14.13(d).

(c) Evaluate $\sum_{n=1}^{\infty} \frac{n}{3^n}$ and $\sum_{n=1}^{\infty} \frac{(-1)^n n}{3^n}$.

26.3. **(a)** Use Exercise 26.2 to derive an explicit formula for $\sum_{n=1}^{\infty} n^2 x^n$.

(b) Evaluate $\sum_{n=1}^{\infty} \frac{n^2}{2^n}$ and $\sum_{n=1}^{\infty} \frac{n^2}{3^n}$.

26.4. **(a)** Observe that $e^{-x^2} = \sum_{n=0}^{\infty} \frac{(-1)^n}{n!} x^{2n}$ for $x \in \mathbb{R}$, since we have $e^x = \sum_{n=0}^{\infty} \frac{1}{n!} x^n$ for $x \in \mathbb{R}$.

(b) Express $F(x) = \int_0^x e^{-t^2} \, dt$ as a power series.

26.5. Let $f(x) = \sum_{n=0}^{\infty} \frac{1}{n!} x^n$ for $x \in \mathbb{R}$. Show that $f' = f$. Do *not* use the fact that $f(x) = e^x$; this is true but has not been established at this point in the text.

26.6. Let $s(x) = x - \frac{x^3}{3!} + \frac{x^5}{5!} - \cdots$ and $c(x) = 1 - \frac{x^2}{2!} + \frac{x^4}{4!} - \cdots$ for $x \in \mathbb{R}$.

(a) Prove that $s' = c$ and $c' = -s$.

(b) Prove that $(s^2 + c^2)' = 0$.

(c) Prove that $s^2 + c^2 = 1$.

Actually $s(x) = \sin x$ and $c(x) = \cos x$, but you do *not* need these facts.

26.7. Let $f(x) = |x|$ for $x \in \mathbb{R}$. Is there a power series $\sum a_n x^n$ such that $f(x) = \sum_{n=0}^{\infty} a_n x^n$ for all x? Discuss.

26.8. (a) Show that $\sum_{n=0}^{\infty} (-1)^n x^{2n} = \frac{1}{1+x^2}$ for $x \in (-1, 1)$. *Hint:* $\sum_{n=0}^{\infty} y^n = \frac{1}{1-y}$. Let $y = -x^2$.

(b) Show that $\arctan x = \sum_{n=0}^{\infty} \frac{(-1)^n}{2n+1} x^{2n+1}$ for $x \in (-1, 1)$.

(c) Show that the equality in (b) also holds for $x = 1$. Use this to find a nice formula for π.

(d) What happens at $x = -1$?

§27 * Weierstrass's Approximation Theorem

Suppose that a power series has radius of convergence greater than 1, and let f denote the function given by the power series. Theorem 26.1 tells us that the partial sums of the power series get uniformly close to f on $[-1, 1]$. In other words, f can be approximated uniformly on $[-1, 1]$ by polynomials. Weierstrass's approximation theorem is a generalization of this last observation, for it tells us that *any* continuous function on $[-1, 1]$ can be uniformly approximated by polynomials on $[-1, 1]$. This result is quite different because such a function need not be given by a power series; see Exercise 26.7. The approximation theorem is valid for any closed interval $[a, b]$ and can be deduced from the case $[0, 1]$; see Exercise 27.1.

We give the beautiful proof due to S. N. Bernstein. Bernstein was motivated by probabilistic considerations, but we will not use any probability here. One of the attractive features of Bernstein's proof is that the approximating polynomials will be given explicitly. There are more abstract proofs in which this is not the case. On the other hand, the abstract proofs lead to far-reaching and important generalizations. See the treatment in [23] or [36].

We need some preliminary facts about polynomials involving binomial coefficients.

27.1 Lemma.

For every $x \in \mathbb{R}$ and $n \geq 0$, we have

$$\sum_{k=0}^{n} \binom{n}{k} x^k (1-x)^{n-k} = 1.$$

Proof

This is just the binomial theorem [Exercise 1.12] applied to $a = x$ and $b = 1 - x$, since in this case $(a+b)^n = 1^n = 1$. ∎

27.2 Lemma.

For $x \in \mathbb{R}$ and $n \geq 0$, we have

$$\sum_{k=0}^{n} (nx - k)^2 \binom{n}{k} x^k (1-x)^{n-k} = nx(1-x) \leq \frac{n}{4}. \tag{1}$$

Proof

Since $k\binom{n}{k} = n\binom{n-1}{k-1}$ for $k \geq 1$, we have

$$\sum_{k=0}^{n} k \binom{n}{k} x^k (1-x)^{n-k} = n \sum_{k=1}^{n} \binom{n-1}{k-1} x^k (1-x)^{n-k}$$

$$= nx \sum_{j=0}^{n-1} \binom{n-1}{j} x^j (1-x)^{n-1-j} = nx. \tag{2}$$

Since $k(k-1)\binom{n}{k} = n(n-1)\binom{n-2}{k-2}$ for $k \geq 2$, we have

$$\sum_{k=0}^{n} k(k-1) \binom{n}{k} x^k (1-x)^{n-k} = n(n-1)x^2 \sum_{j=0}^{n-2} \binom{n-2}{j} x^j (1-x)^{n-2-j}$$

$$= n(n-1)x^2. \tag{3}$$

Adding the results in (2) and (3), we find

$$\sum_{k=0}^{n} k^2 \binom{n}{k} x^k (1-x)^{n-k} = n(n-1)x^2 + nx = n^2 x^2 + nx(1-x). \tag{4}$$

Since $(nx - k)^2 = n^2 x^2 - 2nx \cdot k + k^2$, we use Lemma 27.1, (2) and (4) to obtain

$$\sum_{k=0}^{n} (nx-k)^2 \binom{n}{k} x^k (1-x)^{n-k} = n^2 x^2 - 2nx(nx) + [n^2 x^2 + nx(1-x)]$$

$$= nx(1-x).$$

This establishes the equality in (1). The inequality in (1) simply reflects the inequality $x(1-x) \leq \frac{1}{4}$, which is equivalent to $4x^2 - 4x + 1 \geq 0$ or $(2x-1)^2 \geq 0$. ∎

27.3 Definition.
Let f be a function defined on $[0, 1]$. The polynomials $B_n f$ defined by

$$B_n f(x) = \sum_{k=0}^{n} f\left(\frac{k}{n}\right) \cdot \binom{n}{k} x^k (1-x)^{n-k}$$

are called *Bernstein polynomials* for the function f.

Here is Bernstein's version of the Weierstrass approximation theorem.

27.4 Theorem.
For every continuous function f on $[0, 1]$, we have

$$B_n f \to f \quad uniformly\ on \quad [0, 1].$$

Proof
We assume that f is not identically zero, and we let

$$M = \sup\{|f(x)| : x \in [0, 1]\}.$$

Consider $\epsilon > 0$. Since f is uniformly continuous by Theorem 19.2, there exists $\delta > 0$ such that

$$x, y \in [0, 1] \quad \text{and} \quad |x - y| < \delta \quad \text{imply} \quad |f(x) - f(y)| < \frac{\epsilon}{2}. \quad (1)$$

Let $N = \frac{M}{\epsilon \delta^2}$. This choice of N is unmotivated at this point, but we make it here to emphasize that it does not depend on the choice of x. We will show that

$$|B_n f(x) - f(x)| < \epsilon \quad \text{for all} \quad x \in [0, 1] \quad \text{and all} \quad n > N, \quad (2)$$

completing the proof of the theorem.

To prove (2), consider a fixed $x \in [0, 1]$ and $n > N$. In view of Lemma 27.1, we have

$$f(x) = \sum_{k=0}^{n} f(x) \binom{n}{k} x^k (1-x)^{n-k},$$

so

$$|B_n f(x) - f(x)| \leq \sum_{k=0}^{n} \left| f\left(\frac{k}{n}\right) - f(x) \right| \cdot \binom{n}{k}' x^k (1-x)^{n-k}. \qquad (3)$$

To estimate this sum, we divide the set $\{0, 1, 2, \ldots, n\}$ into two sets:

$$k \in A \text{ if } \left| \frac{k}{n} - x \right| < \delta \quad \text{while} \quad k \in B \text{ if } \left| \frac{k}{n} - x \right| \geq \delta.$$

For $k \in A$ we have $|f(\frac{k}{n}) - f(x)| < \frac{\epsilon}{2}$ by (1), so

$$\sum_{k \in A} \left| f\left(\frac{k}{n}\right) - f(x) \right| \cdot \binom{n}{k} x^k (1-x)^{n-k} \leq \sum_{k \in A} \frac{\epsilon}{2} \binom{n}{k} x^k (1-x)^{n-k} \leq \frac{\epsilon}{2}$$

$$(4)$$

using Lemma 27.1. For $k \in B$, we have $|\frac{k-nx}{n}| \geq \delta$ or $(k-nx)^2 \geq n^2\delta^2$, so

$$\sum_{k \in B} \left| f\left(\frac{k}{n}\right) - f(x) \right| \cdot \binom{n}{k} x^k (1-x)^{n-k} \leq 2M \sum_{k \in B} \binom{n}{k} x^k (1-x)^{n-k}$$

$$\leq \frac{2M}{n^2\delta^2} \sum_{k \in B} (k-nx)^2 \binom{n}{k} x^k (1-x)^{n-k}.$$

By Lemma 27.2, this is bounded by

$$\frac{2M}{n^2\delta^2} \cdot \frac{n}{4} = \frac{M}{2n\delta^2} < \frac{M}{2N\delta^2} = \frac{\epsilon}{2}.$$

This observation, (4) and (3) show that

$$|B_n f(x) - f(x)| < \epsilon.$$

That is, (2) holds. ∎

27.5 Weierstrass's Approximation Theorem.
Every continuous function on a closed interval $[a, b]$ can be uniformly approximated by polynomials on $[a, b]$.

In other words, given a continuous function f on $[a, b]$, there exists a sequence (p_n) of polynomials such that $p_n \to f$ uniformly on $[a, b]$.

Exercises

27.1. Prove Theorem 27.5 from Theorem 27.4. *Hint*: Let $\phi(x) = (b-a)x + a$ so that ϕ maps $[0, 1]$ onto $[a, b]$. If f is continuous on $[a, b]$, then $f \circ \phi$ is continuous on $[0, 1]$.

27.2. Show that if f is continuous on $\mathbb{R}$, then there exists a sequence (p_n) of polynomials such that $p_n \to f$ uniformly on each bounded subset of $\mathbb{R}$. *Hint*: Arrange for $|f(x) - p_n(x)| < \frac{1}{n}$ for $|x| \leq n$.

27.3. Show that there does not exist a sequence of polynomials converging uniformly on $\mathbb{R}$ to f if
  (a) $f(x) = \sin x$,   **(b)** $f(x) = e^x$.

27.4. Let f be a continuous function on $[a, b]$. Show that there exists a sequence (p_n) of polynomials such that $p_n \to f$ uniformly on $[a, b]$ and such that $p_n(a) = f(a)$ for all n.

27.5. Find the sequence $(B_n f)$ of Bernstein polynomials in case
  (a) $f(x) = x$,   **(b)** $f(x) = x^2$.

27.6. The Bernstein polynomials were defined for any function f on $[0, 1]$. Show that if $B_n f \to f$ uniformly on $[0, 1]$, then f is continuous on $[0, 1]$.

27.7. Let f be a bounded function on $[0, 1]$, say $|f(x)| \leq M$ for all $x \in [0, 1]$. Show that all the Bernstein polynomials $B_n f$ are bounded by M.

5

C H A P T E R

Differentiation

In this chapter we give a theoretical treatment of differentiation and related concepts, most or all of which will be familiar from the standard calculus course. Three of the most useful results are the Mean Value Theorem, which is treated in §29, L'Hospital's Rule, which is treated in §30, and Taylor's Theorem, which is given in §31.

§28 Basic Properties of the Derivative

The reader may wish to review the theory of limits treated in §20.

28.1 Definition.
Let f be a real-valued function defined on an open interval containing a point a. We say that f is *differentiable at a*, or that *f has a derivative at a*, if the limit

$$\lim_{x \to a} \frac{f(x) - f(a)}{x - a}$$

exists and is finite. We will write $f'(a)$ for the derivative of f at a:

$$f'(a) = \lim_{x \to a} \frac{f(x) - f(a)}{x - a} \tag{1}$$

whenever this limit exists and is finite.

Generally speaking, we will be interested in f' as a function in its own right. The domain of f' is the set of points at which f is differentiable; thus $\text{dom}(f') \subseteq \text{dom}(f)$.

Example 1
The derivative of the function $g(x) = x^2$ at $x = 2$ was calculated in Example 2 of §20:

$$g'(2) = \lim_{x \to 2} \frac{x^2 - 4}{x - 2} = \lim_{x \to 2}(x + 2) = 4.$$

We can calculate $g'(a)$ just as easily:

$$g'(a) = \lim_{x \to a} \frac{x^2 - a^2}{x - a} = \lim_{x \to a}(x + a) = 2a.$$

This computation is even valid for $a = 0$. We may write $g'(x) = 2x$ since the name of the variable a or x is immaterial. Thus the derivative of the function given by $g(x) = x^2$ is the function given by $g'(x) = 2x$, as every calculus student knows.

Example 2
The derivative of $h(x) = \sqrt{x}$ at $x = 1$ was calculated in Example 3 of §20: $h'(1) = \frac{1}{2}$. In fact, $h(x) = x^{1/2}$ for $x \geq 0$ and $h'(x) = \frac{1}{2}x^{-1/2}$ for $x > 0$; see Exercise 28.3.

Example 3
Let n be a positive integer, and let $f(x) = x^n$ for all $x \in \mathbb{R}$. We show that $f'(x) = nx^{n-1}$ for all $x \in \mathbb{R}$. Fix a in $\mathbb{R}$ and observe that

$$f(x) - f(a) = x^n - a^n = (x-a)(x^{n-1} + ax^{n-2} + a^2 x^{n-3} + \cdots + a^{n-2}x + a^{n-1}),$$

so

$$\frac{f(x) - f(a)}{x - a} = x^{n-1} + ax^{n-2} + a^2 x^{n-3} + \cdots + a^{n-2}x + a^{n-1}$$

for $x \neq a$. It follows that

$$f'(a) = \lim_{x \to a} \frac{f(x) - f(a)}{x - a}$$
$$= a^{n-1} + aa^{n-2} + a^2 a^{n-3} + \cdots + a^{n-2}a + a^{n-1} = na^{n-1};$$

we are using Theorem 20.4 and the fact that $\lim_{x \to a} x^k = a^k$ for $k \in \mathbb{N}$.

We first prove that differentiability at a point implies continuity at the point. This may seem obvious from all the pictures of familiar differentiable functions. However, Exercise 28.8 contains an example of a function that is differentiable at 0 and of course continuous at 0 [by the next theorem], but is discontinuous at all other points.

28.2 Theorem.
If f is differentiable at a point a, then f is continuous at a.

Proof
We are given $f'(a) = \lim_{x \to a} \frac{f(x)-f(a)}{x-a}$ and we need to prove that $\lim_{x \to a} f(x) = f(a)$. We have

$$f(x) = (x - a)\frac{f(x) - f(a)}{x - a} + f(a)$$

for $x \in \mathrm{dom}(f)$, $x \neq a$. Since $\lim_{x \to a}(x - a) = 0$ and $\lim_{x \to a} \frac{f(x)-f(a)}{x-a}$ exists and is finite, Theorem 20.4(ii) shows $\lim_{x \to a}(x-a) \cdot \frac{f(x)-f(a)}{x-a} = 0$. Therefore $\lim_{x \to a} f(x) = f(a)$, as desired. ∎

We next prove some results about sums, products, etc. of derivatives. Let us first recall why the product rule is *not* $(fg)' = f'g'$ [as many naive calculus students wish!] even though the product of limits does behave as expected:

$$\lim_{x \to a}(f_1 f_2)(x) = \left[\lim_{x \to a} f_1(x)\right] \cdot \left[\lim_{x \to a} f_2(x)\right]$$

provided the limits on the right side exist and are finite; see Theorem 20.4(ii). The difficulty is that the limit for the derivative of the product is not the product of the limits of the derivatives, i.e.,

$$\frac{f(x)g(x) - f(a)g(a)}{x - a} \neq \frac{f(x) - f(a)}{x - a} \cdot \frac{g(x) - g(a)}{x - a}.$$

The correct product rule is obtained by shrewdly writing the left hand side in terms of $\frac{f(x)-f(a)}{x-a}$ and $\frac{g(x)-g(a)}{x-a}$ as in the proof of 28.3(iii) below.

28.3 Theorem.
Let f and g be functions that are differentiable at the point a. Each of the functions cf [c a constant], $f + g$, fg and f/g is also differentiable at a, except f/g if $g(a) = 0$ since f/g is not defined at a in this case. The formulas are
   (i) $(cf)'(a) = c \cdot f'(a)$;
   (ii) $(f + g)'(a) = f'(a) + g'(a)$;
   (iii) *[product rule]* $(fg)'(a) = f(a)g'(a) + f'(a)g(a)$;
   (iv) *[quotient rule]* $(f/g)'(a) = [g(a)f'(a) - f(a)g'(a)]/g^2(a)$ *if* $g(a) \neq 0$.

Proof
(i) By definition of cf we have $(cf)(x) = c \cdot f(x)$ for all $x \in \operatorname{dom}(f)$; hence

$$(cf)'(a) = \lim_{x \to a} \frac{(cf)(x) - (cf)(a)}{x - a} = \lim_{x \to a} c \cdot \frac{f(x) - f(a)}{x - a} = c \cdot f'(a).$$

(ii) This follows from the identity

$$\frac{(f + g)(x) - (f + g)(a)}{x - a} = \frac{f(x) - f(a)}{x - a} + \frac{g(x) - g(a)}{x - a}$$

upon taking the limit as $x \to a$ and applying Theorem 20.4(i).
   (iii) Observe that

$$\frac{(fg)(x) - (fg)(a)}{x - a} = f(x)\frac{g(x) - g(a)}{x - a} + g(a)\frac{f(x) - f(a)}{x - a}$$

for $x \in \operatorname{dom}(fg)$, $x \neq a$. We take the limit as $x \to a$ and note that $\lim_{x \to a} f(x) = f(a)$ by Theorem 28.2. We obtain [again using Theorem 20.4]

$$(fg)'(a) = f(a)g'(a) + g(a)f'(a).$$

(iv) Since $g(a) \neq 0$ and g is continuous at a, there exists an open interval I containing a such that $g(x) \neq 0$ for $x \in I$. For $x \in I$ we can write

$$(f/g)(x) - (f/g)(a) = \frac{f(x)}{g(x)} - \frac{f(a)}{g(a)} = \frac{g(a)f(x) - f(a)g(x)}{g(x)g(a)}$$
$$= \frac{g(a)f(x) - g(a)f(a) + g(a)f(a) - f(a)g(x)}{g(x)g(a)},$$

so

$$\frac{(f/g)(x) - (f/g)(a)}{x - a}$$
$$= \left\{ g(a)\frac{f(x) - f(a)}{x - a} - f(a)\frac{g(x) - g(a)}{x - a} \right\} \frac{1}{g(x)g(a)}$$

for $x \in I$, $x \neq a$. Now we take the limit as $x \to a$ to obtain (iv); note that $\lim_{x \to a} \frac{1}{g(x)g(a)} = \frac{1}{g^2(a)}$. ∎

Example 4
Let m be a positive integer, and let $h(x) = x^{-m}$ for $x \neq 0$. Then $h(x) = f(x)/g(x)$ where $f(x) = 1$ and $g(x) = x^m$ for all x. By the quotient rule,

$$h'(a) = \frac{g(a)f'(a) - f(a)g'(a)}{g^2(a)} = \frac{a^m \cdot 0 - 1 \cdot ma^{m-1}}{a^{2m}}$$
$$= \frac{-m}{a^{m+1}} = -ma^{-m-1}$$

for $a \neq 0$. If we write n for $-m$, then we see that the derivative of x^n is nx^{n-1} for negative integers n as well as for positive integers. The result is also trivially valid for $n = 0$. For fractional exponents, see Exercise 29.15.

28.4 Theorem [Chain Rule].
If f is differentiable at a and g is differentiable at $f(a)$, then the composite function $g \circ f$ is differentiable at a and $(g \circ f)'(a) = g'(f(a)) \cdot f'(a)$.

Discussion. Here is a faulty "proof" which nevertheless contains the essence of a valid proof. We write

$$\frac{g \circ f(x) - g \circ f(a)}{x - a} = \frac{g(f(x)) - g(f(a))}{f(x) - f(a)} \cdot \frac{f(x) - f(a)}{x - a} \tag{1}$$

for $x \neq a$. Since $\lim_{x \to a} f(x) = f(a)$, we have

$$\lim_{x \to a} \frac{g(f(x)) - g(f(a))}{f(x) - f(a)} = \lim_{y \to f(a)} \frac{g(y) - g(f(a))}{y - f(a)} = g'(f(a)). \quad (2)$$

We also have $\lim_{x \to a} \frac{f(x) - f(a)}{x - a} = f'(a)$, so (1) shows that $(g \circ f)'(a) = g'(f(a)) \cdot f'(a)$.

This "proof" can be made rigorous provided $f(x) \neq f(a)$ for $x \neq a$. In this case, the only vague part of the "proof" is the first equality in (2) which is justified by Exercise 28.16 with $h(y) = \frac{g(y) - g(f(a))}{y - f(a)}$. If $f(x) = f(a)$ for some x's near a, the "proof" cannot be repaired using (2). In fact, Exercise 28.5 gives an example of differentiable functions f and g for which $\lim_{x \to 0} \frac{g(f(x)) - g(f(0))}{f(x) - f(0)}$ is meaningless. In the formal proof, we will avoid writing $f(x) - f(a)$ as a denominator and we will appeal to Theorem 20.5 instead of the awkward Exercise 28.16. For a recent enlightening proof, see the article by Stephen Kenton, *College Math. J.* **30** (1999), 216-218.

Proof

It is easy to check that $g \circ f$ is defined on some open interval containing a; see Exercise 28.13. Let

$$h(y) = \frac{g(y) - g(f(a))}{y - f(a)} \quad \text{for} \quad y \in \text{dom}(g) \quad \text{and} \quad y \neq f(a),$$

and let $h(f(a)) = g'(f(a))$. Since $\lim_{y \to f(a)} h(y) = h(f(a))$, the function h is continuous at $f(a)$. Note that $g(y) - g(f(a)) = h(y)[y - f(a)]$ for *all* $y \in \text{dom}(g)$, so

$$g \circ f(x) - g \circ f(a) = h(f(x))[f(x) - f(a)] \quad \text{for} \quad x \in \text{dom}(g \circ f).$$

Hence

$$\frac{g \circ f(x) - g \circ f(a)}{x - a} = h(f(x)) \frac{f(x) - f(a)}{x - a} \quad (3)$$

for $x \in \text{dom}(g \circ f)$, $x \neq a$. Since $\lim_{x \to a} f(x) = f(a)$ and the function h is continuous at $f(a)$, Theorem 20.5 shows that

$$\lim_{x \to a} h(f(x)) = h(f(a)) = g'(f(a)).$$

Of course, $\lim_{x \to a} \frac{f(x) - f(a)}{x - a} = f'(a)$, so taking the limit in (3) as $x \to a$ we obtain $(g \circ f)'(a) = g'(f(a)) \cdot f'(a)$. ∎

It is worth emphasizing that if f is differentiable on an interval I and if g is differentiable on $\{f(x) : x \in I\}$, then $(g \circ f)'$ is exactly the function $(g' \circ f) \cdot f'$ on I.

Example 5
Let $h(x) = \sin(x^3 + 7x)$ for $x \in \mathbb{R}$. The reader can undoubtedly verify that $h'(x) = (3x^2 + 7)\cos(x^3 + 7x)$ for $x \in \mathbb{R}$ using some automatic technique learned in calculus. Whatever the automatic technique, it is justified by the chain rule. In this case, $h = g \circ f$ where $f(x) = x^3 + 7x$ and $g(y) = \sin y$. Then $f'(x) = 3x^2 + 7$ and $g'(y) = \cos y$ so that

$$h'(x) = g'(f(x)) \cdot f'(x) = [\cos f(x)] \cdot f'(x) = [\cos(x^3 + 7x)] \cdot (3x^2 + 7).$$

We do not want the reader to unlearn the automatic technique, but the reader should be aware that the chain rule stands behind it.

Exercises

28.1. For each of the following functions defined on $\mathbb{R}$, give the set of points at which it is *not* differentiable. Sketches will be helpful.

(a) $e^{|x|}$ (b) $\sin |x|$
(c) $|\sin x|$ (d) $|x| + |x - 1|$
(e) $|x^2 - 1|$ (f) $|x^3 - 8|$

28.2. Use the *definition* of derivative to calculate the derivatives of the following functions at the indicated points.

(a) $f(x) = x^3$ at $x = 2$;

(b) $g(x) = x + 2$ at $x = a$;

(c) $f(x) = x^2 \cos x$ at $x = 0$;

(d) $r(x) = \frac{3x+4}{2x-1}$ at $x = 1$.

28.3. (a) Let $h(x) = \sqrt{x} = x^{1/2}$ for $x \geq 0$. Use the *definition* of derivative to prove that $h'(x) = \frac{1}{2}x^{-1/2}$ for $x > 0$.

(b) Let $f(x) = x^{1/3}$ for $x \in \mathbb{R}$ and use the definition of derivative to prove that $f'(x) = \frac{1}{3}x^{-2/3}$ for $x \neq 0$.

(c) Is the function f in part (b) differentiable at $x = 0$? Explain.

28.4. Let $f(x) = x^2 \sin \frac{1}{x}$ for $x \neq 0$ and $f(0) = 0$.

(a) Use Theorems 28.3 and 28.4 to show that f is differentiable at each $a \neq 0$ and calculate $f'(a)$. Use, without proof, the fact that $\sin x$ is differentiable and that $\cos x$ is its derivative.

(b) Use the definition to show that f is differentiable at $x = 0$ and that $f'(0) = 0$.

(c) Show that f' is not continuous at $x = 0$.

28.5. Let $f(x) = x^2 \sin \frac{1}{x}$ for $x \neq 0$, $f(0) = 0$, and $g(x) = x$ for $x \in \mathbb{R}$.

(a) Observe that f and g are differentiable on $\mathbb{R}$.

(b) Calculate $f(x)$ for $x = \frac{1}{\pi n}$, $n = \pm 1, \pm 2, \ldots$.

(c) Explain why $\lim_{x \to 0} \frac{g(f(x)) - g(f(0))}{f(x) - f(0)}$ is meaningless.

28.6. Let $f(x) = x \sin \frac{1}{x}$ for $x \neq 0$ and $f(0) = 0$.

(a) Observe that f is continuous at $x = 0$ by Exercise 17.9(c).

(b) Is f differentiable at $x = 0$? Justify your answer.

28.7. Let $f(x) = x^2$ for $x \geq 0$ and $f(x) = 0$ for $x < 0$.

(a) Sketch the graph of f.

(b) Show that f is differentiable at $x = 0$. *Hint*: You will have to use the definition of derivative.

(c) Calculate f' on $\mathbb{R}$ and sketch its graph.

(d) Is f' continuous on $\mathbb{R}$? differentiable on $\mathbb{R}$?

28.8. Let $f(x) = x^2$ for x rational and $f(x) = 0$ for x irrational.

(a) Prove that f is continuous at $x = 0$.

(b) Prove that f is discontinuous at all $x \neq 0$.

(c) Prove that f is differentiable at $x = 0$. *Warning*: You cannot simply claim $f'(x) = 2x$.

28.9. Let $h(x) = (x^4 + 13x)^7$.

(a) Calculate $h'(x)$.

(b) Show how the chain rule justifies your computation in part (a) by writing $h = g \circ f$ for suitable f and g.

28.10. Repeat Exercise 28.9 for the function $h(x) = [\cos x + e^x]^{12}$.

28.11. Suppose that f is differentiable at a, g is differentiable at $f(a)$, and h is differentiable at $g \circ f(a)$. State and prove the chain rule for $(h \circ g \circ f)'(a)$. *Hint:* Apply Theorem 28.4 twice.

28.12. (a) Differentiate the function whose value at x is $\cos(e^{x^5 - 3x})$.

   (b) Use Exercise 28.11 or Theorem 28.4 to justify your computation in part (a).

28.13. Show that if f is defined on an open interval containing a, if g is defined on an open interval containing $f(a)$, and if f is continuous at a, then $g \circ f$ is defined on an open interval containing a.

28.14. Suppose that f is differentiable at a. Prove

   (a) $\lim_{h \to 0} \frac{f(a+h) - f(a)}{h} = f'(a)$,   (b) $\lim_{h \to 0} \frac{f(a+h) - f(a-h)}{2h} = f'(a)$.

28.15. Prove Leibniz' rule

$$(fg)^{(n)}(a) = \sum_{k=0}^{n} \binom{n}{k} f^{(k)}(a) g^{(n-k)}(a)$$

provided both f and g have n derivatives at a. Here $h^{(j)}$ signifies the jth derivative of h so that $h^{(0)} = h$, $h^{(1)} = h'$, $h^{(2)} = h''$, etc. Also, $\binom{n}{k}$ is the binomial coefficient that appears in the binomial expansion; see Exercise 1.12. *Hint:* Use mathematical induction. For $n = 1$, apply Theorem 28.3(iii).

28.16. Let f be a function defined on an open interval I containing a. Let h be a function defined on an open interval J containing $f(a)$, except at $f(a)$, and suppose that $f(x) \in J$ and $f(x) \neq f(a)$ for all $x \in I \setminus \{a\}$. Then $h \circ f$ is defined on $I \setminus \{a\}$. Use Corollary 20.7 to prove that if $\lim_{x \to a} f(x) = f(a)$ and if $\lim_{y \to f(a)} h(y)$ exists and is finite, then $\lim_{x \to a} h \circ f(x) = \lim_{y \to f(a)} h(y)$.

§29 The Mean Value Theorem

Our first result justifies the following strategy in calculus: To find the maximum and minimum of a continuous function f on an interval $[a, b]$ it suffices to consider (a) the points x where $f'(x) = 0$; (b) the points where f is not differentiable; and (c) the endpoints a and b. These are the candidates for maxima and minima.

29.1 Theorem.

If f is defined on an open interval containing x_0, if f assumes its maximum or minimum at x_0, and if f is differentiable at x_0, then $f'(x_0) = 0$.

Proof

We suppose that f is defined on (a, b) where $a < x_0 < b$. Since either f or $-f$ assumes its maximum at x_0, we may assume that f assumes its maximum at x_0.

Assume first that $f'(x_0) > 0$. Since

$$f'(x_0) = \lim_{x \to x_0} \frac{f(x) - f(x_0)}{x - x_0},$$

there exists $\delta > 0$ such that $a < x_0 - \delta < x_0 + \delta < b$ and

$$0 < |x - x_0| < \delta \quad \text{implies} \quad \frac{f(x) - f(x_0)}{x - x_0} > 0; \qquad (1)$$

see Corollary 20.7. If we select x so that $x_0 < x < x_0 + \delta$, then (1) shows that $f(x) > f(x_0)$, contrary to the assumption that f assumes its maximum at x_0. Likewise, if $f'(x_0) < 0$, there exists $\delta > 0$ such that

$$0 < |x - x_0| < \delta \quad \text{implies} \quad \frac{f(x) - f(x_0)}{x - x_0} < 0. \qquad (2)$$

If we select x so that $x_0 - \delta < x < x_0$, then (2) implies $f(x) > f(x_0)$, again a contradiction. Thus we must have $f'(x_0) = 0$. ∎

Our next result is fairly obvious except for one subtle point: one must know or believe that a continuous function on a closed interval assumes its maximum and minimum. We proved this in Theorem 18.1 using the Bolzano-Weierstrass theorem.

29.2 Rolle's Theorem.

Let f be a continuous function on $[a, b]$ that is differentiable on (a, b) and satisfies $f(a) = f(b)$. There exists [at least one] x in (a, b) such that $f'(x) = 0$.

Proof

By Theorem 18.1, there exist $x_0, y_0 \in [a, b]$ such that $f(x_0) \leq f(x) \leq f(y_0)$ for all $x \in [a, b]$. If x_0 and y_0 are both endpoints of $[a, b]$, then f is

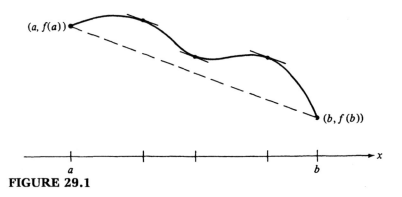

$(a, f(a))$

$(b, f(b))$

x

a

b

FIGURE 29.1

a constant function [since $f(a) = f(b)$] and $f'(x) = 0$ for *all* $x \in (a, b)$. Otherwise, f assumes either a maximum or a minimum at a point x in (a, b), in which case $f'(x) = 0$ by Theorem 29.1. ∎

The Mean Value Theorem tells us that a differentiable function on $[a, b]$ must somewhere have its derivative equal to the slope of the line connecting $(a, f(a))$ to $(b, f(b))$, namely $\frac{f(b)-f(a)}{b-a}$. See Figure 29.1.

29.3 Mean Value Theorem.
Let f be a continuous function on $[a, b]$ that is differentiable on (a, b). Then there exists [at least one] x in (a, b) such that

$$f'(x) = \frac{f(b) - f(a)}{b - a}. \tag{1}$$

Note that Rolle's Theorem is the special case of the Mean Value Theorem where $f(a) = f(b)$.

Proof
Let L be the function whose graph is the straight line connecting $(a, f(a))$ to $(b, f(b))$, i.e., the dotted line in Figure 29.1. Observe that $L(a) = f(a)$, $L(b) = f(b)$ and $L'(x) = \frac{f(b)-f(a)}{b-a}$ for all x. Let $g(x) = f(x) - L(x)$ for $x \in [a, b]$. Clearly g is continuous on $[a, b]$ and differentiable on (a, b). Also $g(a) = 0 = g(b)$, so $g'(x) = 0$ for some $x \in (a, b)$ by Rolle's Theorem 29.2. For this x, we have $f'(x) = L'(x) = \frac{f(b)-f(a)}{b-a}$. ∎

29.4 Corollary.
Let f be a differentiable function on (a, b) such that $f'(x) = 0$ for all $x \in (a, b)$. Then f is a constant function on (a, b).

Proof
If f is not constant on (a, b), then there exist x_1, x_2 such that

$$a < x_1 < x_2 < b \quad \text{and} \quad f(x_1) \neq f(x_2).$$

By the Mean Value Theorem, for some $x \in (x_1, x_2)$ we have $f'(x) = \frac{f(x_2)-f(x_1)}{x_2-x_1} \neq 0$, a contradiction. ■

29.5 Corollary.
Let f and g be differentiable functions on (a, b) such that $f' = g'$ on (a, b). Then there exists a constant c such that $f(x) = g(x) + c$ for all $x \in (a, b)$.

Proof
Apply Corollary 29.4 to the function $f - g$. ■

Corollary 29.5 is important for integral ·calculus because it guarantees that all anti-derivatives, alias indefinite integrals, for a function differ by a constant. Integral tables [and sophisticated calculators] contain formulas like

$$\int x^2 \cos x \, dx = 2x \cos x + (x^2 - 2) \sin x + C.$$

It is straightforward to show that the derivative of each function $2x \cos x + (x^2 - 2) \sin x + C$ is in fact $x^2 \cos x$. Corollary 29.5 shows that *these must be the only antiderivatives of* $x^2 \cos x$.

We need some terminology in order to give another useful corollary of the Mean Value Theorem.

29.6 Definition.
Let f be a real-valued function defined on an interval I. We say that f is *strictly increasing on I* if

$$x_1, x_2 \in I \quad \text{and} \quad x_1 < x_2 \quad \text{imply} \quad f(x_1) < f(x_2),$$

strictly decreasing on I if

$$x_1, x_2 \in I \quad \text{and} \quad x_1 < x_2 \quad \text{imply} \quad f(x_1) > f(x_2),$$

increasing on I if

$$x_1, x_2 \in I \quad \text{and} \quad x_1 < x_2 \quad \text{imply} \quad f(x_1) \leq f(x_2),$$

decreasing on I if

$$x_1, x_2 \in I \quad \text{and} \quad x_1 < x_2 \quad \text{imply} \quad f(x_1) \geq f(x_2).$$

Example 1
The functions e^x on $\mathbb{R}$ and $\sqrt{x}$ on $[0, \infty)$ are strictly increasing. The function $\cos x$ is strictly decreasing on $[0, \pi]$. The signum function and the postage-stamp function in Exercise 17.10 are increasing functions but not strictly increasing functions.

29.7 Corollary.
Let f be a differentiable function on an interval (a, b). Then
  (i) *f is strictly increasing if $f'(x) > 0$ for all $x \in (a, b)$;*
  (ii) *f is strictly decreasing if $f'(x) < 0$ for all $x \in (a, b)$;*
  (iii) *f is increasing if $f'(x) \geq 0$ for all $x \in (a, b)$;*
  (iv) *f is decreasing if $f'(x) \leq 0$ for all $x \in (a, b)$.*

Proof
(i) Consider x_1, x_2 where $a < x_1 < x_2 < b$. By the Mean Value Theorem, for some $x \in (x_1, x_2)$ we have

$$\frac{f(x_2) - f(x_1)}{x_2 - x_1} = f'(x) > 0.$$

Since $x_2 - x_1 > 0$, we see that $f(x_2) - f(x_1) > 0$ or $f(x_2) > f(x_1)$.
  The remaining cases are left to Exercise 29.8. ∎

Exercise 28.4 shows that the derivative f' of a differentiable function f need not be continuous. Nevertheless, like a continuous function, f' has the intermediate value property [see Theorem 18.2].

29.8 Intermediate Value Theorem for Derivatives.
Let f be a differentiable function on (a, b). Whenever $a < x_1 < x_2 < b$ and c lies between $f'(x_1)$ and $f'(x_2)$, there exists [at least one] x in (x_1, x_2) such that $f'(x) = c$.

Proof

We may assume that $f'(x_1) < c < f'(x_2)$. Let $g(x) = f(x) - cx$ for $x \in (a, b)$. Then we have $g'(x_1) < 0 < g'(x_2)$. Theorem 18.1 shows that g assumes its minimum on $[x_1, x_2]$ at some point $x_0 \in [x_1, x_2]$. Since

$$g'(x_1) = \lim_{y \to x_1} \frac{g(y) - g(x_1)}{y - x_1} < 0,$$

$g(y) - g(x_1)$ must be negative for y close to and larger than x_1. In particular, there exists $y_1 \in (x_1, x_2)$ such that $g(y_1) < g(x_1)$. Therefore g does not take its minimum at x_1, so we must have $x_0 \neq x_1$. Similarly, there exists $y_2 \in (x_1, x_2)$ such that $g(y_2) < g(x_2)$, so $x_0 \neq x_2$. We have shown that x_0 is in (x_1, x_2), so $g'(x_0) = 0$ by Theorem 29.1. Therefore $f'(x_0) = g'(x_0) + c = c$. ∎

We next show how to differentiate the inverse of a differentiable function. Let f be a one-to-one differentiable function on an open interval I. By Theorem 18.6, f is strictly increasing or strictly decreasing on I, and by Corollary 18.3 the image $f(I)$ is an interval J. The set J is the domain of f^{-1} and

$$f^{-1} \circ f(x) = x \quad \text{for} \quad x \in I; \qquad f \circ f^{-1}(y) = y \quad \text{for} \quad y \in J.$$

The formula for the derivative of f^{-1} is easy to obtain [or remember] from the Chain Rule: $x = f^{-1} \circ f(x)$, so

$$1 = (f^{-1})'(f(x)) \cdot f'(x) \quad \text{for all} \quad x \in I.$$

If $x_0 \in I$ and $y_0 = f(x_0)$, then we can write $1 = (f^{-1})'(y_0) \cdot f'(x_0)$ or

$$(f^{-1})'(y_0) = \frac{1}{f'(x_0)} \quad \text{where} \quad y_0 = f(x_0).$$

This is *not* a proof because the Chain Rule requires that the functions, f^{-1} and f in this case, be differentiable. We assumed that f is differentiable, but we must *prove* that f^{-1} is also differentiable. In addition, observe that $f'(x_0)$ might be 0 [consider $f(x) = x^3$ at $x_0 = 0$], so our final result will have to avoid this possibility.

29.9 Theorem.

Let f be a one-to-one continuous function on an open interval I, and let $J = f(I)$. If f is differentiable at $x_0 \in I$ and if $f'(x_0) \neq 0$, then f^{-1} is *differentiable at* $y_0 = f(x_0)$ *and*

$$(f^{-1})'(y_0) = \frac{1}{f'(x_0)}.$$

Proof

Note that J is also an open interval. We have $\lim_{x \to x_0} \frac{f(x)-f(x_0)}{x-x_0} = f'(x_0)$. Since $f'(x_0) \neq 0$ and since $f(x) \neq f(x_0)$ for $x \neq x_0$, we can write

$$\lim_{x \to x_0} \frac{x - x_0}{f(x) - f(x_0)} = \frac{1}{f'(x_0)}; \tag{1}$$

see Theorem 20.4(iii). Let $\epsilon > 0$. By (1) and Corollary 20.7, there exists $\delta > 0$ such that

$$0 < |x - x_0| < \delta \quad \text{implies} \quad \left| \frac{x - x_0}{f(x) - f(x_0)} - \frac{1}{f'(x_0)} \right| < \epsilon. \tag{2}$$

Let $g = f^{-1}$ and observe that g is continuous at y_0 by Theorems 18.6 and 18.4 [or Exercise 18.11]. Hence there exists $\eta > 0$ [lower case Greek eta] such that

$$0 < |y - y_0| < \eta \quad \text{implies} \quad |g(y) - g(y_0)| < \delta, \quad \text{i.e.,} \quad |g(y) - x_0| < \delta. \tag{3}$$

Combining (3) and (2) we obtain

$$0 < |y - y_0| < \eta \quad \text{implies} \quad \left| \frac{g(y) - x_0}{f(g(y)) - f(x_0)} - \frac{1}{f'(x_0)} \right| < \epsilon.$$

Since $\frac{g(y)-x_0}{f(g(y))-f(x_0)} = \frac{g(y)-g(y_0)}{y-y_0}$, this shows that

$$\lim_{y \to y_0} \frac{g(y) - g(y_0)}{y - y_0} = \frac{1}{f'(x_0)}.$$

Hence $g'(y_0)$ exists and equals $\frac{1}{f'(x_0)}$. ∎

Example 2

Let n be a positive integer, and let $g(y) = \sqrt[n]{y} = y^{1/n}$. If n is even, the domain of g is $[0, \infty)$ and, if n is odd, the domain is $\mathbb{R}$. In either case, g is strictly increasing and its inverse is $f(x) = x^n$; here $\text{dom}(f) = [0, \infty)$ if n is even. Consider $y_0 \in \text{dom}(g)$ where $y_0 \neq 0$, and write $y_0 = x_0^n$ where $x_0 \in \text{dom}(f)$. Since $f'(x_0) = nx_0^{n-1}$, Theorem 29.9

shows that

$$g'(y_0) = \frac{1}{nx_0^{n-1}} = \frac{1}{ny_0^{(n-1)/n}} = \frac{1}{n}y_0^{1/n-1}.$$

This shows that the function g is differentiable for $y \neq 0$ and that the rule for differentiating x^n holds for exponents of the form $1/n$; see also Exercise 29.15.

Theorem 29.9 applies to the various inverse functions encountered in calculus. We give one example.

Example 3
The function $f(x) = \sin x$ is one-to-one on $[-\frac{\pi}{2}, \frac{\pi}{2}]$, and it is traditional to use the inverse g of f restricted to this domain; g is usually denoted $\sin^{-1}$ or arcsin. Note that $\text{dom}(g) = [-1, 1]$. For $y_0 = \sin x_0$ in $(-1, 1)$ where $x_0 \in (-\frac{\pi}{2}, \frac{\pi}{2})$, Theorem 29.9 shows that $g'(y_0) = \frac{1}{\cos x_0}$. Since $1 = \sin^2 x_0 + \cos^2 x_0 = y_0^2 + \cos^2 x_0$ and $\cos x_0 > 0$, we may write

$$g'(y_0) = \frac{1}{\sqrt{1-y_0^2}} \quad \text{for} \quad y_0 \in (-1, 1).$$

Exercises

29.1. Determine whether the conclusion of the Mean Value Theorem holds for the following functions on the specified intervals. If the conclusion holds, give an example of a point x satisfying (1) of Theorem 29.3. If the conclusion fails, state which *hypotheses* of the Mean Value Theorem fail.
(a) x^2 on $[-1, 2]$,
(b) $\sin x$ on $[0, \pi]$,
(c) $|x|$ on $[-1, 2]$,
(d) $\frac{1}{x}$ on $[-1, 1]$,
(e) $\frac{1}{x}$ on $[1, 3]$,
(f) $\text{sgn}(x)$ on $[-2, 2]$.
The function sgn is defined in Exercise 17.10.

29.2. Prove that $|\cos x - \cos y| \leq |x - y|$ for all $x, y \in \mathbb{R}$.

29.3. Suppose that f is differentiable on $\mathbb{R}$ and that $f(0) = 0$, $f(1) = 1$ and $f(2) = 1$.

(a) Show that $f'(x) = \frac{1}{2}$ for some $x \in (0, 2)$.

(b) Show that $f'(x) = \frac{1}{7}$ for some $x \in (0, 2)$.

29.4. Let f and g be differentiable functions on an open interval I. Suppose that a, b in I satisfy $a < b$ and $f(a) = f(b) = 0$. Show that $f'(x) + f(x)g'(x) = 0$ for some $x \in (a, b)$. *Hint*: Consider $h(x) = f(x)e^{g(x)}$.

29.5. Let f be defined on $\mathbb{R}$, and suppose that $|f(x) - f(y)| \le (x - y)^2$ for all $x, y \in \mathbb{R}$. Prove that f is a constant function.

29.6. Give the equation of the straight line used in the proof of the Mean Value Theorem 29.3.

29.7. **(a)** Suppose that f is twice differentiable on an open interval I and that $f''(x) = 0$ for all $x \in I$. Show that f has the form $f(x) = ax + b$ for suitable constants a and b.

    (b) Suppose f is three times differentiable on an open interval I and that $f''' = 0$ on I. What form does f have? Prove your claim.

29.8. Prove (ii)–(iv) of Corollary 29.7.

29.9. Show that $ex \le e^x$ for all $x \in \mathbb{R}$.

29.10. Let $f(x) = x^2 \sin(\frac{1}{x}) + \frac{x}{2}$ for $x \neq 0$ and $f(0) = 0$.

    (a) Show that $f'(0) > 0$; see Exercise 28.4.

    (b) Show that f is not increasing on any open interval containing 0.

    (c) Compare this example with Corollary 29.7(i).

29.11. Show that $\sin x \le x$ for all $x \ge 0$. *Hint*: Show that $f(x) = x - \sin x$ is increasing on $[0, \infty)$.

29.12. **(a)** Show that $x < \tan x$ for all $x \in (0, \frac{\pi}{2})$.

    (b) Show that $\frac{x}{\sin x}$ is a strictly increasing function on $(0, \frac{\pi}{2})$.

    (c) Show that $x \le \frac{\pi}{2} \sin x$ for $x \in [0, \frac{\pi}{2}]$.

29.13. Prove that if f and g are differentiable on $\mathbb{R}$, if $f(0) = g(0)$ and if $f'(x) \le g'(x)$ for all $x \in \mathbb{R}$, then $f(x) \le g(x)$ for $x \ge 0$.

29.14. Suppose that f is differentiable on $\mathbb{R}$, that $1 \le f'(x) \le 2$ for $x \in \mathbb{R}$, and that $f(0) = 0$. Prove that $x \le f(x) \le 2x$ for all $x \ge 0$.

29.15. Let r be a nonzero rational number $\frac{m}{n}$ where n is a positive integer, m is any nonzero integer, and m and n have no common factors. Let $h(x) = x^r$ where $\mathrm{dom}(h) = [0, \infty)$ if n is even and $m > 0$, dom(h) = $(0, \infty)$ if n is even and $m < 0$, dom(h) = $\mathbb{R}$ if n is odd and $m > 0$, and dom(h) = $\mathbb{R} \setminus \{0\}$ if n is odd and $m < 0$. Show that $h'(x) = rx^{r-1}$ for $x \in$ dom(h), $x \neq 0$. *Hint*: Use Example 2.

29.16. Use Theorem 29.9 to obtain the derivative of the inverse $g =$ arctan of f where $f(x) = \tan x$ for $x \in (-\frac{\pi}{2}, \frac{\pi}{2})$.

29.17. Let f and g be differentiable on an open interval I and consider $a \in I$. Define h on I by the rules: $h(x) = f(x)$ for $x < a$, and $h(x) = g(x)$ for $x \geq a$. Prove that h is differentiable at a if and only if both $f(a) = g(a)$ and $f'(a) = g'(a)$ hold. *Suggestion*: Draw a picture to see what is going on.

29.18. Let f be differentiable on $\mathbb{R}$ with $a = \sup\{|f'(x)| : x \in \mathbb{R}\} < 1$. Select $s_0 \in \mathbb{R}$ and define $s_n = f(s_{n-1})$ for $n \geq 1$. Thus $s_1 = f(s_0)$, $s_2 = f(s_1)$, etc. Prove that (s_n) is a convergence sequence. *Hint*: To show (s_n) is Cauchy, first show that $|s_{n+1} - s_n| \leq a|s_n - s_{n-1}|$ for $n \geq 1$.

§30 * L'Hospital's Rule

In analysis one frequently encounters limits of the form

$$\lim_{x \to s} \frac{f(x)}{g(x)}$$

where s signifies a, a^+, a^-, ∞ or $-\infty$. See Definition 20.3 concerning such limits. The limit exists and is simply $\frac{\lim_{x \to s} f(x)}{\lim_{x \to s} g(x)}$ provided the limits $\lim_{x \to s} f(x)$ and $\lim_{x \to s} g(x)$ exist and are finite and provided $\lim_{x \to s} g(x) \neq 0$; see Theorem 20.4. If these limits lead to an indeterminate form such as $\frac{0}{0}$ or $\frac{\infty}{\infty}$, then L'Hospital's rule can often be used. Moreover, other indeterminate forms, such as $\infty - \infty$, 1^{∞}, ∞^0, 0^0 or $0 \cdot \infty$, can usually be reformulated so as to take the form $\frac{0}{0}$ or $\frac{\infty}{\infty}$; see Examples 5–9. Before we state and prove L'Hospital's rule, we will prove a generalized mean value theorem.

30.1 Generalized Mean Value Theorem.
Let f and g be continuous functions on $[a, b]$ that are differentiable on (a, b). Then there exists [at least one] x in (a, b) such that

$$f'(x)[g(b) - g(a)] = g'(x)[f(b) - f(a)]. \tag{1}$$

This result reduces to the standard Mean Value Theorem 29.3 when g is the function given by $g(x) = x$ for all x.

Proof
The trick is to look at the difference of the two quantities in (1) and hope that Rolle's Theorem will help. Thus we define

$$h(x) = f(x)[g(b) - g(a)] - g(x)[f(b) - f(a)];$$

it suffices to show that $h'(x) = 0$ for some $x \in (a, b)$. Note that

$$h(a) = f(a)[g(b) - g(a)] - g(a)[f(b) - f(a)] = f(a)g(b) - g(a)f(b)$$

and

$$h(b) = f(b)[g(b) - g(a)] - g(b)[f(b) - f(a)] = -f(b)g(a) + g(b)f(a) = h(a).$$

Clearly h is continuous on $[a, b]$ and differentiable on (a, b), so Rolle's Theorem 29.2 shows that $h'(x) = 0$ for at least one x in (a, b). ∎

Our proof of L'Hospital's rule below is somewhat wordy but is really quite straightforward. It is based on the elegant presentation in Rudin [36]. Many texts give more complicated proofs.

30.2 L'Hospital's Rule.
Let s signify a, a^+, a^-, ∞ or $-\infty$ where $a \in \mathbb{R}$, and suppose f and g are differentiable functions for which the following limit exists:

$$\lim_{x \to s} \frac{f'(x)}{g'(x)} = L. \tag{1}$$

If

$$\lim_{x \to s} f(x) = \lim_{x \to s} g(x) = 0 \tag{2}$$

or if

$$\lim_{x \to s} |g(x)| = +\infty, \tag{3}$$

then

$$\lim_{x \to s} \frac{f(x)}{g(x)} = L. \tag{4}$$

Note that the hypothesis (1) includes some implicit assumptions: f and g must be defined and differentiable "near" s and $g'(x)$ must be *nonzero* "near" s. For example, if $\lim_{x\to a+} \frac{f'(x)}{g'(x)}$ exists, then there must be an interval (a, b) on which f and g are differentiable and g' is nonzero. The requirement that g' be nonzero is crucial; see Exercise 30.7.

Proof

We first make some reductions. The case of $\lim_{x\to a}$ follows from the cases $\lim_{x\to a+}$ and $\lim_{x\to a-}$, since $\lim_{x\to a} h(x)$ exists if and only if the limits $\lim_{x\to a+} h(x)$ and $\lim_{x\to a-} h(x)$ exist and are equal; see Theorem 20.10. In fact, we restrict our attention to $\lim_{x\to a+}$ and $\lim_{x\to -\infty}$, since the other two cases are treated in an entirely analogous manner. Finally, we are able to handle these cases together in view of Remark 20.11.

We assume $a \in \mathbb{R}$ or $a = -\infty$. We will show that if $-\infty \le L < \infty$ and $L_1 > L$, then there exists $\alpha_1 > a$ such that

$$a < x < \alpha_1 \quad \text{implies} \quad \frac{f(x)}{g(x)} < L_1. \tag{5}$$

A similar argument [which we omit] shows that if $-\infty < L \le \infty$ and $L_2 < L$, then there exists $\alpha_2 > a$ such that

$$a < x < \alpha_2 \quad \text{implies} \quad \frac{f(x)}{g(x)} > L_2. \tag{6}$$

We now show how to complete the proof *using* (5) and (6); (5) will be proved in the next paragraph. If L is finite and $\epsilon > 0$, we can apply (5) to $L_1 = L + \epsilon$ and (6) to $L_2 = L - \epsilon$ to obtain $\alpha_1 > a$ and $\alpha_2 > a$ satisfying

$$a < x < \alpha_1 \quad \text{implies} \quad \frac{f(x)}{g(x)} < L + \epsilon,$$

$$a < x < \alpha_2 \quad \text{implies} \quad \frac{f(x)}{g(x)} > L - \epsilon.$$

Consequently if $\alpha = \min\{\alpha_1, \alpha_2\}$ then

$$a < x < \alpha \quad \text{implies} \quad \left| \frac{f(x)}{g(x)} - L \right| < \epsilon;$$

in view of Remark 20.11 this shows that $\lim_{x\to a^+} \frac{f(x)}{g(x)} = L$ [if $a = -\infty$, then $a^+ = -\infty$]. If $L = -\infty$, then (5) and the fact that L_1 is arbitrary show that $\lim_{x\to a^+} \frac{f(x)}{g(x)} = -\infty$. If $L = \infty$, then (6) and the fact that L_2 is arbitrary show that $\lim_{x\to a^+} \frac{f(x)}{g(x)} = \infty$.

It remains for us to consider $L_1 > L \geq -\infty$ and show that there exists $\alpha_1 > a$ satisfying (5). Let (a, b) be an interval on which f and g are differentiable and on which g' never vanishes. Theorem 29.8 shows that either g' is positive on (a, b) or else g' is negative on (a, b). The former case can be reduced to the latter case by replacing g by $-g$. So we assume $g'(x) < 0$ for $x \in (a, b)$, so that g is strictly decreasing on (a, b) by Corollary 29.7. Since g is one-to-one on (a, b), $g(x)$ can equal 0 for at most one x in (a, b). By choosing b smaller if necessary, we may assume that g never vanishes on (a, b). Now select K so that $L < K < L_1$. By (1) there exists $\alpha > a$ such that

$$a < x < \alpha \quad \text{implies} \quad \frac{f'(x)}{g'(x)} < K.$$

If $a < x < y < \alpha$, then Theorem 30.1 shows that

$$\frac{f(x) - f(y)}{g(x) - g(y)} = \frac{f'(z)}{g'(z)} \quad \text{for some} \quad z \in (x, y).$$

Therefore

$$a < x < y < \alpha \quad \text{implies} \quad \frac{f(x) - f(y)}{g(x) - g(y)} < K. \tag{7}$$

If hypothesis (2) holds, then

$$\lim_{x\to a^+} \frac{f(x) - f(y)}{g(x) - g(y)} = \frac{f(y)}{g(y)},$$

so (7) shows that

$$\frac{f(y)}{g(y)} \leq K < L_1 \quad \text{for} \quad a < y < \alpha;$$

hence (5) holds in this case. If hypothesis (3) holds, then $\lim_{x\to a^+} g(x) = +\infty$ since g is strictly decreasing on (a, b). Also $g(x) > 0$ for $x \in (a, b)$ since g never vanishes on (a, b). We multiply both sides of (7) by

$\frac{g(x)-g(y)}{g(x)}$, which is positive, to see that

$$a < x < y < \alpha \quad \text{implies} \quad \frac{f(x)-f(y)}{g(x)} < K \cdot \frac{g(x)-g(y)}{g(x)}$$

and hence

$$\frac{f(x)}{g(x)} < K \cdot \frac{g(x)-g(y)}{g(x)} + \frac{f(y)}{g(x)} = K + \frac{f(y)-Kg(y)}{g(x)}.$$

We regard y as fixed and observe that

$$\lim_{x \to a^+} \frac{f(y)-Kg(y)}{g(x)} = 0.$$

Hence there exists $\alpha_2 > a$ such that $\alpha_2 \leq y < \alpha$ and

$$a < x < \alpha_2 \quad \text{implies} \quad \frac{f(x)}{g(x)} < L_1.$$

Thus again (5) holds. ∎

Example 1
If we assume familiar properties of the trigonometric functions, then $\lim_{x \to 0} \frac{\sin x}{x}$ is easy to calculate by L'Hospital's rule:

$$\lim_{x \to 0} \frac{\sin x}{x} = \lim_{x \to 0} \frac{\cos x}{1} = \cos(0) = 1. \tag{1}$$

Note that $f(x) = \sin x$ and $g(x) = x$ satisfy the hypotheses in Theorem 30.2. This particular computation is really dishonest because the limit (1) is needed to *prove* that the derivative of $\sin x$ is $\cos x$. This fact reduces to the assertion that the derivative of $\sin x$ at 0 is 1, i.e., to the assertion

$$\lim_{x \to 0} \frac{\sin x - \sin 0}{x - 0} = \lim_{x \to 0} \frac{\sin x}{x} = 1.$$

Example 2
We calculate $\lim_{x \to 0} \frac{\cos x - 1}{x^2}$. L'Hospital's rule will apply provided the limit $\lim_{x \to 0} \frac{-\sin x}{2x}$ exists. But $\frac{-\sin x}{2x} = -\frac{1}{2}\frac{\sin x}{x}$ and this has limit $-\frac{1}{2}$ by Example 1. We conclude that

$$\lim_{x \to 0} \frac{\cos x - 1}{x^2} = -\frac{1}{2}.$$

Example 3

We show that $\lim_{x\to\infty} \frac{x^2}{e^{3x}} = 0$. As written we have an indeterminate of the form $\frac{\infty}{\infty}$. By L'Hospital's rule, this limit will exist provided $\lim_{x\to\infty} \frac{2x}{3e^{3x}}$ exists and by L'Hospital's rule again, this limit will exist provided $\lim_{x\to\infty} \frac{2}{9e^{3x}}$ exists. The last limit is 0, so we conclude that $\lim_{x\to\infty} \frac{x^2}{e^{3x}} = 0$.

Example 4

Consider $\lim_{x\to 0^+} \frac{\log x}{x}$ if it exists. By L'Hospital's rule, this appears to be

$$\lim_{x\to 0^+} \frac{1/x}{1} = +\infty$$

and yet this is *incorrect*. The difficulty is that we should have checked the hypotheses. Since $\lim_{x\to 0^+} \log x = -\infty$ and $\lim_{x\to 0^+} x = 0$, neither of the hypotheses (2) or (3) in Theorem 30.2 hold. To find the limit, we rewrite $\frac{\log x}{x}$ as $-\frac{\log(1/x)}{x}$. It is easy to show that $\lim_{x\to 0^+} \frac{\log(1/x)}{x}$ will agree with $\lim_{y\to\infty} y \log y$ provided the latter limit exists; see Exercise 30.4. It follows that $\lim_{x\to 0^+} \frac{\log x}{x} = - \lim_{y\to\infty} y \log y = -\infty$.

The next five examples illustrate how indeterminate limits of various forms can be modified so that L'Hospital's rule applies.

Example 5

Consider $\lim_{x\to 0^+} x \log x$. As written this limit is of the indeterminate form $0 \cdot (-\infty)$ since $\lim_{x\to 0^+} x = 0$ and $\lim_{x\to 0^+} \log x = -\infty$. By writing $x \log x$ as $\frac{\log x}{1/x}$ we obtain an indeterminate of the form $\frac{-\infty}{\infty}$, so we may apply L'Hospital's rule:

$$\lim_{x\to 0^+} x \log x = \lim_{x\to 0^+} \frac{\log x}{\frac{1}{x}} = \lim_{x\to 0^+} \frac{\frac{1}{x}}{-\frac{1}{x^2}} = - \lim_{x\to 0^+} x = 0.$$

We could also write $x \log x$ as $\frac{x}{1/\log x}$ to obtain an indeterminate of the form $\frac{0}{0}$. However, an attempt to apply L'Hospital's rule only makes the problem more complicated:

$$\lim_{x\to 0^+} x \log x = \lim_{x\to 0^+} \frac{x}{\frac{1}{\log x}} = \lim_{x\to 0^+} \frac{1}{\frac{-1}{x(\log x)^2}} = - \lim_{x\to 0^+} x(\log x)^2.$$

Example 6

The limit $\lim_{x \to 0^+} x^x$ is of the indeterminate form 0^0. We write x^x as $e^{x \log x}$ [remember $x = e^{\log x}$] and note that $\lim_{x \to 0^+} x \log x = 0$ by Example 5. Since $g(x) = e^x$ is continuous at 0, Theorem 20.5 shows that

$$\lim_{x \to 0^+} x^x = \lim_{x \to 0^+} e^{x \log x} = e^0 = 1.$$

Example 7

The limit $\lim_{x \to \infty} x^{1/x}$ is of the indeterminate form ∞^0. We write $x^{1/x}$ as $e^{(\log x)/x}$. By L'Hospital's rule

$$\lim_{x \to \infty} \frac{\log x}{x} = \lim_{x \to \infty} \frac{\frac{1}{x}}{1} = 0.$$

Theorem 20.5 now shows that $\lim_{x \to \infty} x^{1/x} = e^0 = 1$.

Example 8

The limit $\lim_{x \to \infty} (1 - \frac{1}{x})^x$ is indeterminate of the form 1^∞. Since

$$\left(1 - \frac{1}{x}\right)^x = e^{x \log(1 - 1/x)}$$

we evaluate

$$\lim_{x \to \infty} x \log \left(1 - \frac{1}{x}\right) = \lim_{x \to \infty} \frac{\log \left(1 - \frac{1}{x}\right)}{\frac{1}{x}} = \lim_{x \to \infty} \frac{\left(1 - \frac{1}{x}\right)^{-1} x^{-2}}{-x^{-2}}$$

$$= \lim_{x \to \infty} - \left(1 - \frac{1}{x}\right)^{-1} = -1.$$

So by Theorem 20.5 we have

$$\lim_{x \to \infty} \left(1 - \frac{1}{x}\right)^x = e^{-1},$$

as should have been expected since $\lim_{n \to \infty} (1 - \frac{1}{n})^n = e^{-1}$.

Example 9

Consider $\lim_{x \to 0} h(x)$ where

$$h(x) = \frac{1}{e^x - 1} - \frac{1}{x} = (e^x - 1)^{-1} - x^{-1} \quad \text{for} \quad x \neq 0.$$

Neither of the limits $\lim_{x \to 0}(e^x - 1)^{-1}$ or $\lim_{x \to 0} x^{-1}$ exists, so $\lim_{x \to 0} h(x)$ is not an indeterminate form as written. However, $\lim_{x \to 0^+} h(x)$ is indeterminate of the form $\infty - \infty$ and $\lim_{x \to 0^-} h(x)$ is indeterminate of the form $(-\infty) - (-\infty)$. By writing

$$h(x) = \frac{x - e^x + 1}{x(e^x - 1)}$$

the limit $\lim_{x \to 0} h(x)$ becomes an indeterminate of the form $\frac{0}{0}$. By L'Hospital's rule this should be

$$\lim_{x \to 0} \frac{1 - e^x}{xe^x + e^x - 1},$$

which is still indeterminate $\frac{0}{0}$. Note that $xe^x + e^x - 1 \neq 0$ for $x \neq 0$ so that the hypotheses of Theorem 30.2 hold. Applying L'Hospital's rule again, we obtain

$$\lim_{x \to 0} \frac{-e^x}{xe^x + 2e^x} = -\frac{1}{2}.$$

Note that we have $xe^x + 2e^x \neq 0$ for x in $(-2, \infty)$. We conclude that $\lim_{x \to 0} h(x) = -\frac{1}{2}$.

Exercises

30.1. Find the following limits if they exist.
(a) $\lim_{x \to 0} \frac{e^{2x} - \cos x}{x}$
(b) $\lim_{x \to 0} \frac{1 - \cos x}{x^2}$
(c) $\lim_{x \to \infty} \frac{x^3}{e^{2x}}$
(d) $\lim_{x \to 0} \frac{\sqrt{1+x} - \sqrt{1-x}}{x}$

30.2. Find the following limits if they exist.
(a) $\lim_{x \to 0} \frac{x^3}{\sin x - x}$
(b) $\lim_{x \to 0} \frac{\tan x - x}{x^3}$
(c) $\lim_{x \to 0} [\frac{1}{\sin x} - \frac{1}{x}]$
(d) $\lim_{x \to 0}(\cos x)^{1/x^2}$

30.3. Find the following limits if they exist.
(a) $\lim_{x \to \infty} \frac{x - \sin x}{x}$
(b) $\lim_{x \to \infty} x^{\sin(1/x)}$
(c) $\lim_{x \to 0^+} \frac{1 + \cos x}{e^x - 1}$
(d) $\lim_{x \to 0} \frac{1 - \cos 2x - 2x^2}{x^4}$

30.4. Let f be a function defined on some interval $(0, a)$, and define $g(y) = f(\frac{1}{y})$ for $y \in (a^{-1}, \infty)$; here we set $a^{-1} = 0$ if $a = \infty$. Show that $\lim_{x \to 0^+} f(x)$ exists if and only if $\lim_{y \to \infty} g(y)$ exists, in which case they are equal.

30.5. Find the limits

(a) $\lim_{x\to 0}(1+2x)^{1/x}$ (b) $\lim_{y\to\infty}(1+\frac{2}{y})^y$

(c) $\lim_{x\to\infty}(e^x+x)^{1/x}$

30.6. Let f be differentiable on some interval (c,∞) and suppose that $\lim_{x\to\infty}[f(x)+f'(x)]=L$, where L is finite. Prove that $\lim_{x\to\infty}f(x)=L$ and that $\lim_{x\to\infty}f'(x)=0$. *Hint:* $f(x)=\frac{f(x)e^x}{e^x}$.

30.7. This example is taken from [38] and is due to Otto Stolz, *Math. Annalen* **15** (1879), 556-559. The requirement in Theorem 30.2 that $g'(x)\neq 0$ for x "near" s is important. In a careless application of L'Hospital's rule in which the zeros of g' "cancel" the zeros of f', erroneous results can be obtained. For $x\in\mathbb{R}$, let

$$f(x)=x+\cos x\sin x \quad \text{and} \quad g(x)=e^{\sin x}(x+\cos x\sin x).$$

(a) Show that $\lim_{x\to\infty}f(x)=\lim_{x\to\infty}g(x)=+\infty$.

(b) Show $f'(x)=2(\cos x)^2$ and $g'(x)=e^{\sin x}\cos x[2\cos x+f(x)]$.

(c) Show that $\frac{f'(x)}{g'(x)}=\frac{2e^{-\sin x}\cos x}{2\cos x+f(x)}$ if $\cos x\neq 0$ and $x>3$.

(d) Show that $\lim_{x\to\infty}\frac{2e^{-\sin x}\cos x}{2\cos x+f(x)}=0$ and yet the limit $\lim_{x\to\infty}\frac{f(x)}{g(x)}$ does *not* exist.

§31 Taylor's Theorem

31.1 Discussion.

Consider a power series with radius of convergence $R>0$ [R may be $+\infty$]:

$$f(x)=\sum_{k=0}^{\infty}a_k x^k. \tag{1}$$

By Theorem 26.5 the function f is differentiable in the interval $|x|<R$ and

$$f'(x)=\sum_{k=1}^{\infty}ka_k x^{k-1}.$$

The same theorem shows that f' is differentiable for $|x| < R$ and

$$f''(x) = \sum_{k=2}^{\infty} k(k-1)a_k x^{k-2}.$$

Continuing in this way, we find that the nth derivative $f^{(n)}$ exists for $|x| < R$ and

$$f^{(n)}(x) = \sum_{k=n}^{\infty} k(k-1)\cdots(k-n+1)a_k x^{k-n}.$$

In particular,

$$f^{(n)}(0) = n(n-1)\cdots(n-n+1)a_n = n!a_n.$$

This relation even holds for $n = 0$ if we make the convention $f^{(0)} = f$ and recall the convention $0! = 1$. Since $f^{(k)}(0) = k!a_k$, the original power series (1) has the form

$$f(x) = \sum_{k=0}^{\infty} \frac{f^{(k)}(0)}{k!}x^k, \qquad |x| < R. \tag{2}$$

As suggested at the end of §26, we now begin with a function f and seek a power series for f. The last paragraph shows that f should possess derivatives of all orders at 0, i.e., $f'(0), f''(0), f'''(0), \ldots$ should all exist. For such f formula (2) might hold for some $R > 0$, in which case we have found a power series for f.

31.2 Definition.
Let f be a function defined on some open interval containing 0. If f possesses derivatives of all orders at 0, then the series

$$\sum_{k=0}^{\infty} \frac{f^{(k)}(0)}{k!}x^k \tag{1}$$

is called the *Taylor series for f about 0*. The *remainder $R_n(x)$* is defined by

$$R_n(x) = f(x) - \sum_{k=0}^{n-1} \frac{f^{(k)}(0)}{k!}x^k. \tag{2}$$

Of course the remainder R_n depends on f, so a more accurate notation would be something like $R_n(f;x)$. The remainder is important because, for any x,

$$f(x) = \sum_{k=0}^{\infty} \frac{f^{(k)}(0)}{k!} x^k \quad \text{if and only if} \quad \lim_{n \to \infty} R_n(x) = 0.$$

We will show in Example 3 that f need not be given by its Taylor series, i.e., that $\lim_{n \to \infty} R_n(x) = 0$ can fail. Since we want to know when f is given by its Taylor series, our various versions of Taylor's theorem all concern the nature of the remainder R_n.

31.3 Taylor's Theorem.

Let f be defined on (a, b) where $a < 0 < b$, and suppose the nth derivative $f^{(n)}$ exists on (a, b). Then for each nonzero x in (a, b) there is some y between 0 and x such that

$$R_n(x) = \frac{f^{(n)}(y)}{n!} x^n.$$

The proof we give is due to James Wolfe [41]; compare Exercise 31.6.

Proof

Fix $x \neq 0$. Let M be the unique solution of

$$f(x) = \sum_{k=0}^{n-1} \frac{f^{(k)}(0)}{k!} x^k + \frac{M x^n}{n!} \tag{1}$$

and observe that we need only show that

$$f^{(n)}(y) = M \quad \text{for some} \quad y \quad \text{between} \quad 0 \quad \text{and} \quad x. \tag{2}$$

[To see this, replace M by $f^{(n)}(y)$ in equation (1) and recall the definition of $R_n(x)$.] To prove (2), consider the difference

$$g(t) = \sum_{k=0}^{n-1} \frac{f^{(k)}(0)}{k!} t^k + \frac{M t^n}{n!} - f(t). \tag{3}$$

A direct calculation shows that $g(0) = 0$ and that $g^{(k)}(0) = 0$ for $k < n$. Also $g(x) = 0$ by the choice of M in (1). By Rolle's theorem 29.2 we have $g'(x_1) = 0$ for some x_1 between 0 and x. Since $g'(0) = 0$, a second application of Rolle's theorem shows that $g''(x_2) = 0$ some x_2 between 0 and x_1. Again, since $g''(0) = 0$ we have $g'''(x_3) = 0$ for some x_3 between 0 and x_2. This process continues until we obtain x_n between 0 and x_{n-1} such that $g^{(n)}(x_n) = 0$. From (3) it follows that $g^{(n)}(t) = M - f^{(n)}(t)$ for all $t \in (a, b)$, so (2) holds with $y = x_n$. ∎

31.4 Corollary.
Let f be defined on (a, b) where $a < 0 < b$. If all the derivatives $f^{(n)}$ exist on (a, b) and are bounded by a single constant C, then

$$\lim_{n \to \infty} R_n(x) = 0 \quad \text{for all} \quad x \in (a, b).$$

Proof
Consider x in (a, b). From Theorem 31.3 we see that

$$|R_n(x)| \leq \frac{C}{n!} |x|^n \quad \text{for all} \quad n.$$

Since $\lim_{n \to \infty} \frac{|x|^n}{n!} = 0$ by Exercise 9.15, we conclude that $\lim_{n \to \infty} R_n(x) = 0$. ∎

Example 1
We assume the familiar differentiation properties of e^x, $\sin x$, etc.

(a) Let $f(x) = e^x$ for $x \in \mathbb{R}$. Then $f^{(n)}(x) = e^x$ for all $n = 0, 1, 2, \ldots$, so $f^{(n)}(0) = 1$ for all n. The Taylor series for e^x about 0 is

$$\sum_{k=0}^{\infty} \frac{1}{k!} x^k.$$

For any bounded interval $(-M, M)$ in $\mathbb{R}$ all the derivatives of f are bounded [by e^M, in fact], so Corollary 31.4 shows that

$$e^x = \sum_{k=0}^{\infty} \frac{1}{k!} x^k \quad \text{for all} \quad x \in \mathbb{R}.$$

(b) If $f(x) = \sin x$ for $x \in \mathbb{R}$, then

$$f^{(n)}(x) = \begin{cases} \cos x & n = 1, 5, 9, \ldots \\ -\sin x & n = 2, 6, 10, \ldots \\ -\cos x & n = 3, 7, 11, \ldots \\ \sin x & n = 0, 4, 8, 12, \ldots; \end{cases}$$

thus

$$f^{(n)}(0) = \begin{cases} 1 & n = 1, 5, 9, \ldots \\ -1 & n = 3, 7, 11, \ldots \\ 0 & \text{otherwise.} \end{cases}$$

Hence the Taylor series for $\sin x$ is

$$\sum_{k=0}^{\infty} \frac{(-1)^k}{(2k+1)!} x^{2k+1}.$$

The derivatives of f are all bounded by 1, so

$$\sin x = \sum_{k=0}^{\infty} \frac{(-1)^k}{(2k+1)!} x^{2k+1} \quad \text{for all} \quad x \in \mathbb{R}.$$

Example 2
In Example 2 of §26 we used Abel's theorem to prove

$$\log_e 2 = 1 - \frac{1}{2} + \frac{1}{3} - \frac{1}{4} + \frac{1}{5} - \frac{1}{6} + \frac{1}{7} - \cdots. \tag{1}$$

Here is another proof, based on Taylor's theorem. Consider the function $f(x) = \log(1+x)$ for $x \in (-1, \infty)$. Differentiating, we find

$$f'(x) = (1+x)^{-1}, \qquad f''(x) = -(1+x)^{-2}, \qquad f'''(x) = 2(1+x)^{-3},$$

etc. A simple induction argument shows that

$$f^{(n)}(x) = (-1)^{n+1}(n-1)!(1+x)^{-n}. \tag{2}$$

In particular, $f^{(n)}(0) = (-1)^{n+1}(n-1)!$, so the Taylor series for f about 0 is

$$\sum_{k=1}^{\infty} \frac{(-1)^{k+1}}{k} x^k = x - \frac{x^2}{2} + \frac{x^3}{3} - \frac{x^4}{4} + \frac{x^5}{5} - \cdots.$$

We also could have obtained this Taylor series using Example 1 in §26, but we need formula (2) anyway. We now apply Theorem 31.3 with $a = -1$, $b = +\infty$, and $x = 1$. Thus for each n there exists $y_n \in (0, 1)$ such that $R_n(1) = \frac{f^{(n)}(y_n)}{n!}$. Equation (2) shows that

$$R_n(1) = \frac{(-1)^{n+1}(n-1)!}{(1+y_n)^n n!};$$

hence

$$|R_n(1)| = \frac{1}{(1+y_n)^n n} < \frac{1}{n} \quad \text{for all} \quad n.$$

Therefore $\lim_{n\to\infty} R_n(1) = 0$, so (1) holds.

The next version of Taylor's theorem gives the remainder in integral form. The proof uses results from integration theory that should be familiar from calculus; they also appear in the next chapter.

31.5 Taylor's Theorem.
Let f be defined on (a, b) where $a < 0 < b$, and suppose the nth derivative $f^{(n)}$ exists and is continuous on (a, b). Then for $x \in (a, b)$ we have

$$R_n(x) = \int_0^x \frac{(x-t)^{n-1}}{(n-1)!} f^{(n)}(t)\, dt. \tag{1}$$

Proof
For $n = 1$, equation (1) asserts

$$R_1(x) = f(x) - f(0) = \int_0^x f'(t)\, dt;$$

this holds by Theorem 34.1. For $n \geq 2$, we repeatedly apply integration by parts, i.e., we use mathematical induction. So, assume (1) holds for some n, $n \geq 1$. We evaluate the integral in (1) by Theorem 34.2, using $u(t) = f^{(n)}(t)$, $v'(t) = \frac{(x-t)^{n-1}}{(n-1)!}$, so that $u'(t) = f^{(n+1)}(t)$ and $v(t) = -\frac{(x-t)^n}{n!}$. We obtain

$$R_n(x) = u(x)v(x) - u(0)v(0) - \int_0^x v(t)u'(t)\, dt$$

$$= f^{(n)}(x) \cdot 0 + f^{(n)}(0)\frac{x^n}{n!} + \int_0^x \frac{(x-t)^n}{n!} f^{(n+1)}(t)\, dt. \tag{2}$$

The definition of R_{n+1} in Definition 31.2 shows that

$$R_{n+1}(x) = R_n(x) - \frac{f^{(n)}(0)}{n!} x^n; \tag{3}$$

hence from (2) we see that (1) holds for $n + 1$. ∎

31.6 Corollary.

If f is as in Theorem 31.5, then for each x in (a, b) different from 0 there is some y between 0 and x such that

$$R_n(x) = \frac{(x - y)^{n-1}}{(n - 1)!} f^{(n)}(y) \cdot x. \tag{1}$$

This form of R_n is known as Cauchy's form of the remainder.

Proof

We suppose $x < 0$, the case $x > 0$ being similar. The Intermediate Value Theorem for Integrals 33.9 shows that

$$\int_x^0 \frac{(x - t)^{n-1}}{(n - 1)!} f^{(n)}(t)\, dt = [0 - x]\frac{(x - y)^{n-1}}{(n - 1)!} f^{(n)}(y) \tag{2}$$

for some y in $(x, 0)$. Since the integral in (2) equals $-R_n(x)$ by Theorem 31.5, formula (1) holds. ∎

Recall that the binomial theorem tells us that

$$(a + b)^n = \sum_{k=0}^n \binom{n}{k} a^k b^{n-k}$$

where

$$\binom{n}{k} = \frac{n!}{k!(n - k)!} = \frac{n(n - 1)\cdots(n - k + 1)}{k!} \quad \text{for} \quad 1 \leq k \leq n.$$

Let $a = x$ and $b = 1$; then

$$(1 + x)^n = 1 + \sum_{k=1}^n \frac{n(n - 1)\cdots(n - k + 1)}{k!} x^k.$$

This result holds for some values of x even if the exponent n isn't an integer, provided we allow the series to be an infinite series. We next prove this using Taylor's Theorem 31.5. Our proof follows that in [34].

31.7 Binomial Series Theorem.

If $\alpha \in \mathbb{R}$ and $|x| < 1$, then

$$(1 + x)^\alpha = 1 + \sum_{k=1}^\infty \frac{\alpha(\alpha - 1)\cdots(\alpha - k + 1)}{k!} x^k. \tag{1}$$

Proof

For $k = 1, 2, 3, \ldots$, let $a_k = \frac{\alpha(\alpha-1)\cdots(\alpha-k+1)}{k!}$. If α is a nonnegative integer, then $a_k = 0$ for $k > \alpha$ and (1) holds for all x as noted in our discussion prior to this theorem. Henceforth we assume α is not a nonnegative integer so that $a_k \neq 0$ for all k. Since

$$\lim_{k\to\infty} \left|\frac{a_{k+1}}{a_k}\right| = \lim_{k\to\infty} \left|\frac{\alpha - k}{k+1}\right| = 1,$$

the series in (1) has radius of convergence 1; see Theorem 23.1 and Corollary 12.3. Likewise $\sum k a_k x^{k-1}$ converges for $|x| < 1$; hence

$$\lim_{n\to\infty} n a_n x^{n-1} = 0 \quad \text{for} \quad |x| < 1. \tag{2}$$

Let $f(x) = (1 + x)^\alpha$ for $|x| < 1$. For $n = 1, 2, \ldots$, we have

$$f^{(n)}(x) = \alpha(\alpha - 1) \cdots (\alpha - n + 1)(1 + x)^{\alpha-n} = n! a_n (1 + x)^{\alpha-n}.$$

Thus $f^{(n)}(0) = n! a_n$ for all $n \geq 1$, and the series in (1) is the Taylor series for f. Also, by Theorem 31.5 we have

$$R_n(x) = \int_0^x \frac{(x - t)^{n-1}}{(n - 1)!} n! a_n (1 + t)^{\alpha-n}\, dt$$

$$= \int_0^x n a_n \left[\frac{x - t}{1 + t}\right]^{n-1} (1 + t)^{\alpha-1}\, dt \tag{3}$$

for $|x| < 1$. It is easy to show that

$$\left|\frac{x - t}{1 + t}\right| \leq |x| \quad \text{if} \quad -1 < x \leq t \leq 0 \text{ or } 0 \leq t \leq x < 1.$$

To see this, note that $t = xy$ for some $y \in [0, 1]$, so

$$\left|\frac{x - t}{1 + t}\right| = \left|\frac{x - xy}{1 + xy}\right| = |x| \cdot \left|\frac{1 - y}{1 + xy}\right| \leq |x|$$

since $1 + xy \geq 1 - y$. Thus the integrand in (3) is bounded by

$$n|a_n| \cdot |x|^{n-1}(1 + t)^{\alpha-1};$$

therefore

$$|R_n(x)| \leq n|a_n| \cdot |x|^{n-1} \int_{-|x|}^{|x|} (1 + t)^{\alpha-1}\, dt.$$

Applying (2), we now see that $\lim_{n\to\infty} R_n(x) = 0$ for $|x| < 1$, so that (1) holds. ∎

We next give an example of a function f whose Taylor series exists but does not represent the function. The function f is *infinitely differentiable on* $\mathbb{R}$, i.e., derivatives of all order exist at all points of $\mathbb{R}$. The example may appear artificial, but the existence of such functions [see also Exercise 31.4] is vital to the theory of distributions, an important theory related to recent work in differential equations and Fourier analysis.

Example 3
Let $f(x) = e^{-1/x}$ for $x > 0$ and $f(x) = 0$ for $x \le 0$; see Figure 31.1. Clearly f has derivatives of all orders at all $x \ne 0$. We will prove

$$f^{(n)}(0) = 0 \quad \text{for} \quad n = 0, 1, 2, 3, \dots. \tag{1}$$

Hence the Taylor series for f is identically zero, so f does not agree with its Taylor series in any open interval containing 0. First we show that for each n there is a polynomial p_n of degree $2n$ such that

$$f^{(n)}(x) = e^{-1/x}p_n(1/x) \quad \text{for} \quad x > 0. \tag{2}$$

This is obvious for $n = 0$; simply set $p_0(t) = 1$ for all t. And this is easy for $n = 1$ and $n = 2$; the reader should check that (2) holds with $n = 1$ and $p_1(t) = t^2$ and that (2) holds with $n = 2$ and $p_2(t) = t^4 - 2t^3$. To apply induction, we assume the result is true for n and write

$$p_n(t) = a_0 + a_1 t + a_2 t^2 + \cdots + a_{2n}t^{2n} \quad \text{where} \quad a_{2n} \ne 0.$$

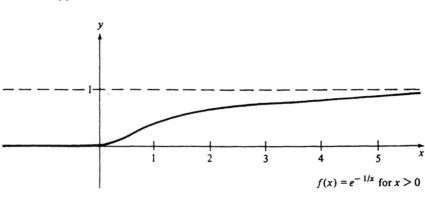

FIGURE 31.1

Then for $x > 0$ we have

$$f^{(n)}(x) = e^{-1/x} \left[\sum_{k=0}^{2n} \frac{a_k}{x^k} \right],$$

and a single differentiation yields

$$f^{(n+1)}(x) = e^{-1/x} \left[0 - \sum_{k=1}^{2n} \frac{k a_k}{x^{k+1}} \right] + \left[\sum_{k=0}^{2n} \frac{a_k}{x^k} \right] e^{-1/x} \cdot \left(\frac{1}{x^2} \right).$$

The assertion (2) is now clear for $n + 1$; in fact, the polynomial p_{n+1} is evidently

$$p_{n+1}(t) = - \sum_{k=1}^{2n} k a_k t^{k+1} + \left[\sum_{k=0}^{2n} a_k t^k \right] \cdot (t^2),$$

which has degree $2n + 2$.

We next prove (1) by induction. Assume that $f^{(n)}(0) = 0$ for some $n \geq 0$. We need to prove

$$\lim_{x \to 0} \frac{f^{(n)}(x) - f^{(n)}(0)}{x - 0} = \lim_{x \to 0} \frac{1}{x} f^{(n)}(x) = 0.$$

Obviously $\lim_{x \to 0^-} \frac{1}{x} f^{(n)}(x) = 0$ since $f^{(n)}(x) = 0$ for all $x < 0$. By Theorem 20.10 it suffices to verify

$$\lim_{x \to 0^+} \frac{1}{x} f^{(n)}(x) = 0.$$

In view of (2), it suffices to show

$$\lim_{x \to 0^+} e^{-1/x} q\left(\frac{1}{x} \right) = 0$$

for any polynomial q. In fact, since $q(1/x)$ is a finite sum of terms of the form $b_k (1/x)^k$, it suffices to show

$$\lim_{x \to 0^+} \left(\frac{1}{x} \right)^k e^{-1/x} = 0 \quad \text{for fixed} \quad k \geq 0.$$

Because of Definition 20.1 we consider a sequence (x_n) of positive numbers such that $\lim x_n = 0$ and show

$$\lim_{n \to \infty} \left(\frac{1}{x_n} \right)^k e^{-1/x_n} = 0.$$

If $y_n = \frac{1}{x_n}$, then $\lim y_n = +\infty$ [by Theorem 9.10] and we need to show $\lim_{n\to\infty} y_n^k e^{-y_n} = 0$ or

$$\lim_{y\to\infty} y^k e^{-y} = 0. \tag{3}$$

To see (3) note that $e^y \geq \frac{y^{k+1}}{(k+1)!}$ for $y > 0$ by Example 1(a) so that

$$y^k e^{-y} \leq y^k (k+1)! \, y^{-k-1} = \frac{(k+1)!}{y} \quad \text{for} \quad y > 0.$$

The limit (3) also can be verified via k applications of L'Hospital's Rule 30.2.

Just as with power series, one can consider Taylor series that are not centered at 0.

31.8 Definition.
Let f be a function defined on some open interval containing $x_0 \in \mathbb{R}$. If f has derivatives of all order at x_0, then the series

$$\sum_{k=0}^{\infty} \frac{f^{(k)}(x_0)}{k!} (x - x_0)^k$$

is called the *Taylor series for f about* x_0.

The theorems in this section are easily transferred to the general Taylor series just defined.

Exercises

31.1. Find the Taylor series for $\cos x$ and indicate why it converges to $\cos x$ for all $x \in \mathbb{R}$.

31.2. Repeat Exercise 31.1 for $\sinh x = \frac{1}{2}(e^x - e^{-x})$ and $\cosh x = \frac{1}{2}(e^x + e^{-x})$.

31.3. In Example 2, why did we apply Theorem 31.3 instead of Corollary 31.4?

31.4. Consider a, b in $\mathbb{R}$ where $a < b$. Show that there exist infinitely differentiable functions f_a, g_b, $h_{a,b}$ and $h_{a,b}^*$ on $\mathbb{R}$ with the following properties. You may assume, without proof, that the sum, product, etc. of infinitely differentiable functions is again infinitely differentiable. The same applies to the quotient provided that the denominator never vanishes.

(a) $f_a(x) = 0$ for $x \le a$ and $f_a(x) > 0$ for $x > a$. Hint: Let $f_a(x) = f(x - a)$ where f is the function in Example 3.

(b) $g_b(x) = 0$ for $x \ge b$ and $g_b(x) > 0$ for $x < b$.

(c) $h_{a,b}(x) > 0$ for $x \in (a, b)$ and $h_{a,b}(x) = 0$ for $x \notin (a, b)$.

(d) $h^*_{a,b}(x) = 0$ for $x \le a$ and $h^*_{a,b}(x) = 1$ for $x \ge b$. Hint: Use $f_a/(f_a + g_b)$.

31.5. Let $g(x) = e^{-1/x^2}$ for $x \ne 0$ and $g(0) = 0$.

(a) Show that $g^{(n)}(0) = 0$ for all $n = 0, 1, 2, 3, \ldots$. Hint: Use Example 3.

(b) Show that the Taylor series for g about 0 agrees with g only at $x = 0$.

31.6. A standard proof of Theorem 31.3 goes as follows. Assume $x > 0$, let M be as in the proof of Theorem 31.3, and let

$$F(t) = f(t) + \sum_{k=1}^{n-1} \frac{(x - t)^k}{k!} f^{(k)}(t) + M \cdot \frac{(x - t)^n}{n!}$$

for $t \in [0, x]$.

(a) Show that F is differentiable on $[0, x]$ and that

$$F'(t) = \frac{(x - t)^{n-1}}{(n - 1)!}[f^{(n)}(t) - M].$$

(b) Show that $F(0) = F(x)$.

(c) Apply Rolle's Theorem 29.2 to F to obtain y in $(0, x)$ such that $f^{(n)}(y) = M$.

6

CHAPTER

Integration

This chapter serves two purposes. it contains a careful development of the Riemann integral, which is the integral studied in standard calculus courses. It also contains an introduction to a generalization of the Riemann integral called the Riemann-Stieltjes integral. The generalization is easy and natural. Moreover, the Riemann-Stieltjes integral is an important tool in probability and statistics, and other areas of mathematics.

§32 The Riemann Integral

The theory of the Riemann integral is no more difficult than several other topics dealt with in this book. The one drawback is that it involves some technical notation and terminology.

32.1 Definition.
Let f be a bounded function on a closed interval $[a, b]$. For $S \subseteq [a, b]$, we adopt the notation

$$M(f, S) = \sup\{f(x) : x \in S\} \quad \text{and} \quad m(f, S) = \inf\{f(x) : x \in S\}.$$

A *partition of* $[a, b]$ is any finite ordered subset P having the form

$$P = \{a = t_0 < t_1 < \cdots < t_n = b\}.$$

The *upper Darboux sum* $U(f, P)$ of f with respect to P is the sum

$$U(f, P) = \sum_{k=1}^{n} M(f, [t_{k-1}, t_k]) \cdot (t_k - t_{k-1})$$

and the *lower Darboux sum* $L(f, P)$ is

$$L(f, P) = \sum_{k=1}^{n} m(f, [t_{k-1}, t_k]) \cdot (t_k - t_{k-1}).$$

Note that

$$U(f, P) \leq \sum_{k=1}^{n} M(f, [a, b]) \cdot (t_k - t_{k-1}) = M(f, [a, b]) \cdot (b - a);$$

likewise $L(f, P) \geq m(f, [a, b]) \cdot (b - a)$, so

$$m(f, [a, b]) \cdot (b - a) \leq L(f, P) \leq U(f, P) \leq M(f, [a, b]) \cdot (b - a). \quad (1)$$

The *upper Darboux integral* $U(f)$ of f over $[a, b]$ is defined by

$$U(f) = \inf\{U(f, P) : P \text{ is a partition of } [a, b]\}$$

and the *lower Darboux integral* is

$$L(f) = \sup\{L(f, P) : P \text{ is a partition of } [a, b]\}.$$

In view of (1), $U(f)$ and $L(f)$ are real numbers.

We will prove in Theorem 32.4 that $L(f) \leq U(f)$. This is not obvious from (1). [Why?] We say that f is *integrable* on $[a, b]$ provided $L(f) = U(f)$. In this case, we write $\int_a^b f$ or $\int_a^b f(x)\, dx$ for this common value:

$$\int_a^b f = \int_a^b f(x)\, dx = L(f) = U(f). \quad (2)$$

Specialists call this integral the *Darboux integral*. Riemann's definition of the integral is a little different [Definition 32.8], but we will show in Theorem 32.9 that the definitions are equivalent. For this reason, we will follow customary usage and call the integral defined above the *Riemann integral*.

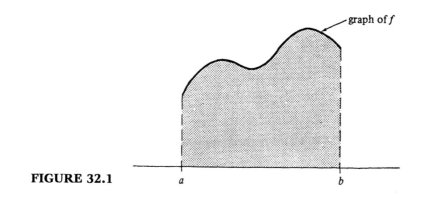

FIGURE 32.1

For nonnegative functions, $\int_a^b f$ is interpreted as the area of the region under the graph of f [see Figure 32.1] for the following reason. Each lower Darboux sum represents the area of a union of rectangles inside the region, and each upper Darboux sum represents the area of a union of rectangles that contains the region. Moreover, $\int_a^b f$ is the unique number that is larger than or equal to all lower Darboux sums and smaller than or equal to all upper Darboux sums. Figure 19.2 on page 137 illustrates the situation for $[a, b] = [0, 1]$ and

$$P = \left\{ 0 < \frac{1}{n} < \frac{2}{n} < \cdots < \frac{n-1}{n} < 1 \right\}.$$

Example 1

The simplest function whose integral is not obvious is $f(x) = x^2$. Consider f on the interval $[0, b]$ where $b > 0$. For a partition

$$P = \{ 0 = t_0 < t_1 < \cdots < t_n = b \},$$

we have

$$U(f, P) = \sum_{k=1}^n \sup\{ x^2 : x \in [t_{k-1}, t_k] \} \cdot (t_k - t_{k-1}) = \sum_{k=1}^n t_k^2 (t_k - t_{k-1}).$$

If we choose $t_k = \frac{kb}{n}$, then we can use Exercise 1.1 to calculate

$$U(f, P) = \sum_{k=1}^n \frac{k^2 b^2}{n^2} \left(\frac{b}{n} \right) = \frac{b^3}{n^3} \sum_{k=1}^n k^2 = \frac{b^3}{n^3} \cdot \frac{n(n+1)(2n+1)}{6}.$$

For large n, this is close to $\frac{b^3}{3}$, so we conclude that $U(f) \leq \frac{b^3}{3}$. For the same partition we find

$$L(f, P) = \sum_{k=1}^{n} \frac{(k-1)^2 b^2}{n^2} \left(\frac{b}{n}\right) = \frac{b^3}{n^3} \cdot \frac{(n-1)(n)(2n-1)}{6},$$

so $L(f) \geq \frac{b^3}{3}$. Therefore $f(x) = x^2$ is integrable on $[0, b]$ and

$$\int_0^b x^2 \, dx = \frac{b^3}{3}.$$

Of course, any calculus student could have calculated this integral using a formula that is based on the Fundamental Theorem of Calculus; see Example 1 in §34.

Example 2
Consider the interval $[0, b]$ and let $f(x) = 1$ for rational x in $[0, b]$, and let $f(x) = 0$ for irrational x in $[0, b]$. For any partition

$$P = \{0 = t_0 < t_1 < \cdots < t_n = b\},$$

we have

$$U(f, P) = \sum_{k=1}^{n} M(f, [t_{k-1}, t_k]) \cdot (t_k - t_{k-1}) = \sum_{k=1}^{n} 1 \cdot (t_k - t_{k-1}) = b$$

and

$$L(f, P) = \sum_{k=1}^{n} 0 \cdot (t_k - t_{k-1}) = 0.$$

It follows that $U(f) = b$ and $L(f) = 0$. The upper and lower Darboux integrals for f do not agree, so f is not integrable!

We next develop some properties of the integral.

32.2 Lemma.
Let f be a bounded function on $[a, b]$. If P and Q are partitions of $[a, b]$ and $P \subseteq Q$, then

$$L(f, P) \leq L(f, Q) \leq U(f, Q) \leq U(f, P). \tag{1}$$

Proof

The middle inequality is obvious. The proofs of the first and third inequalities are similar, so we will prove

$$L(f, P) \leq L(f, Q). \tag{2}$$

An induction argument [Exercise 32.4] shows that we may assume that Q has only one more point, say u, than P. If

$$P = \{a = t_0 < t_1 < \cdots < t_n = b\},$$

then

$$Q = \{a = t_0 < t_1 < \cdots < t_{k-1} < u < t_k < \cdots < t_n = b\}$$

for some $k \in \{1, 2, \ldots, n\}$. The lower Darboux sums for P and Q are the same except for the terms involving t_{k-1} or t_k. In fact, their difference is

$$L(f, Q) - L(f, P) = m(f, [t_{k-1}, u]) \cdot (u - t_{k-1}) + m(f, [u, t_k]) \cdot (t_k - u)$$
$$- m(f, [t_{k-1}, t_k]) \cdot (t_k - t_{k-1}). \tag{3}$$

To establish (2) it suffices to show that this quantity is nonnegative. Using Exercise 4.7(a), we see that

$$m(f, [t_{k-1}, t_k]) \cdot (t_k - t_{k-1})$$
$$= m(f, [t_{k-1}, t_k]) \cdot \{(t_k - u) + (u - t_{k-1})\}$$
$$\leq m(f, [u, t_k]) \cdot (t_k - u) + m(f, [t_{k-1}, u]) \cdot (u - t_{k-1}). \qquad \blacksquare$$

32.3 Lemma.
If f is a bounded function on $[a, b]$, and if P and Q are partitions of $[a, b]$, then $L(f, P) \leq U(f, Q)$.

Proof

The set $P \cup Q$ is also a partition of $[a, b]$. Since $P \subseteq P \cup Q$ and $Q \subseteq P \cup Q$, we can apply Lemma 32.2 to obtain

$$L(f, P) \leq L(f, P \cup Q) \leq U(f, P \cup Q) \leq U(f, Q). \qquad \blacksquare$$

32.4 Theorem.
If f is a bounded function on $[a, b]$, then $L(f) \leq U(f)$.

Proof

Fix a partition P of $[a, b]$. Lemma 32.3 shows that $L(f, P)$ is a lower bound for the set

$$\{U(f, Q) : Q \text{ is a partition of } [a, b]\}.$$

Therefore $L(f, P)$ must be less than or equal to the greatest lower bound [infimum!] of this set. That is,

$$L(f, P) \le U(f). \tag{1}$$

Now (1) shows that $U(f)$ is an upper bound for the set

$$\{L(f, P) : P \text{ is a partition of } [a, b]\},$$

so $U(f) \ge L(f)$. ∎

Note that Theorem 32.4 follows from Lemma 32.3 and Exercise 4.8; see Exercise 32.5. The next theorem gives a "Cauchy criterion" for integrability.

32.5 Theorem.

A bounded function f on $[a, b]$ is integrable if and only if for each $\epsilon > 0$ there exists a partition P of $[a, b]$ such that

$$U(f, P) - L(f, P) < \epsilon. \tag{1}$$

Proof

Suppose first that f is integrable and consider $\epsilon > 0$. There exist partitions P_1 and P_2 of $[a, b]$ satisfying

$$L(f, P_1) > L(f) - \frac{\epsilon}{2} \quad \text{and} \quad U(f, P_2) < U(f) + \frac{\epsilon}{2}.$$

For $P = P_1 \cup P_2$, we apply Lemma 32.2 to obtain

$$U(f, P) - L(f, P) \le U(f, P_2) - L(f, P_1)$$
$$< U(f) + \frac{\epsilon}{2} - \left[L(f) - \frac{\epsilon}{2} \right] = U(f) - L(f) + \epsilon.$$

Since f is integrable, $U(f) = L(f)$, so (1) holds.

Conversely, suppose that for each $\epsilon > 0$ the inequality (1) holds for some partition P. Then we have

$$U(f) \le U(f, P) = U(f, P) - L(f, P) + L(f, P)$$
$$< \epsilon + L(f, P) \le \epsilon + L(f).$$

Since ϵ is arbitrary, we conclude that $U(f) \leq L(f)$. Hence we have $U(f) = L(f)$ by Theorem 32.4, i.e., f is integrable. ∎

The remainder of this section is devoted to establishing the equivalence of Riemann's and Darboux's definitions of integrability. Subsequent sections will depend only on items 32.1–32.5. Therefore the reader who is content with the Darboux integral in Definition 32.1 can safely proceed directly to the next section.

32.6 Definition.
The *mesh* of a partition P is the maximum length of the subintervals comprising P. Thus if

$$P = \{a = t_0 < t_1 < \cdots < t_n = b\},$$

then

$$\mathrm{mesh}(P) = \max\{t_k - t_{k-1} : k = 1, 2, \ldots, n\}.$$

Here is another "Cauchy criterion" for integrability.

32.7 Theorem.
A bounded function f on $[a, b]$ is integrable if and only if for each $\epsilon > 0$ there exists a $\delta > 0$ such that

$$\mathrm{mesh}(P) < \delta \quad implies \quad U(f, P) - L(f, P) < \epsilon \tag{1}$$

for all partitions P of $[a, b]$.

Proof
Theorem 32.5 shows that the ϵ-δ condition in (1) implies integrability.

Conversely, suppose that f is integrable on $[a, b]$. Let $\epsilon > 0$ and select a partition

$$P_0 = \{a = u_0 < u_1 < \cdots < u_m = b\}$$

of $[a, b]$ such that

$$U(f, P_0) - L(f, P_0) < \frac{\epsilon}{2}. \tag{2}$$

Since f is bounded, there exists $B > 0$ such that $|f(x)| \le B$ for all $x \in [a, b]$. Let $\delta = \frac{\epsilon}{8mB}$; m is the number of intervals comprising P_0.

To verify (1), we consider any partition

$$P = \{a = t_0 < t_1 < \cdots < t_n = b\}$$

with $\mathrm{mesh}(P) < \delta$. Let $Q = P \cup P_0$. If Q has one more element than P, then a glance at (3) in the proof of Lemma 32.2 leads us to

$$L(f, Q) - L(f, P) \le B \cdot \mathrm{mesh}(P) - (-B) \cdot \mathrm{mesh}(P) = 2B \cdot \mathrm{mesh}(P).$$

Since Q has at most m elements that are not in P, an induction argument shows that

$$L(f, Q) - L(f, P) \le 2mB \cdot \mathrm{mesh}(P) < 2mB\delta = \frac{\epsilon}{4}.$$

By Lemma 32.2 we have $L(f, P_0) \le L(f, Q)$, so

$$L(f, P_0) - L(f, P) < \frac{\epsilon}{4}.$$

Similarly

$$U(f, P) - U(f, P_0) < \frac{\epsilon}{4},$$

so

$$U(f, P) - L(f, P) < U(f, P_0) - L(f, P_0) + \frac{\epsilon}{2}.$$

Now (2) implies $U(f, P) - L(f, P) < \epsilon$ and we have verified (1). ∎

Now we give Riemann's definition of integrability.

32.8 Definition.
Let f be a bounded function on $[a, b]$, and let $P = \{a = t_0 < t_1 < \cdots < t_n = b\}$ be a partition of $[a, b]$. A *Riemann sum* of f associated with the partition P is a sum of the form

$$\sum_{k=1}^{n} f(x_k)(t_k - t_{k-1})$$

where $x_k \in [t_{k-1}, t_k]$ for $k = 1, 2, \ldots, n$. The choice of x_k's is quite arbitrary, so there are infinitely many Riemann sums associated with a single function and partition.

The function f is *Riemann integrable* on $[a, b]$ if there exists a number r with the following property. For each $\epsilon > 0$ there exists $\delta > 0$ such that

$$|S - r| < \epsilon \qquad (1)$$

for every Riemann sum S of f associated with a partition P having mesh$(P) < \delta$. The number r is the *Riemann integral* of f on $[a, b]$ and will be provisionally written as $\mathcal{R} \int_a^b f$.

32.9 Theorem.
A bounded function f on $[a, b]$ is Riemann integrable if and only if it is [Darboux] integrable, in which case the values of the integrals agree.

Proof
Suppose first that f is [Darboux] integrable on $[a, b]$ in the sense of Definition 32.1. Let $\epsilon > 0$, and let $\delta > 0$ be chosen so that (1) of Theorem 32.7 holds. We show that

$$\left| S - \int_a^b f \right| < \epsilon \qquad (1)$$

for every Riemann sum

$$S = \sum_{k=1}^{n} f(x_k)(t_k - t_{k-1})$$

associated with a partition P having mesh$(P) < \delta$. Clearly we have $L(f, P) \leq S \leq U(f, P)$, so (1) follows from the inequalities

$$U(f, P) < L(f, P) + \epsilon \leq L(f) + \epsilon = \int_a^b f + \epsilon$$

and

$$L(f, P) > U(f, P) - \epsilon \geq U(f) - \epsilon = \int_a^b f - \epsilon.$$

This proves (1); hence f is Riemann integrable and

$$\mathcal{R} \int_a^b f = \int_a^b f.$$

Now suppose that f is Riemann integrable in the sense of Definition 32.8, and consider $\epsilon > 0$. Let $\delta > 0$ and r be as given in

Definition 32.8. Select any partition

$$P = \{a = t_0 < t_1 < \cdots < t_n = b\}$$

with mesh$(P) < \delta$, and for each $k = 1, 2, \ldots, n$, select x_k in $[t_{k-1}, t_k]$ so that

$$f(x_k) < m(f, [t_{k-1}, t_k]) + \epsilon.$$

The Riemann sum S for this choice of x_k's satisfies

$$S \leq L(f, P) + \epsilon(b - a)$$

as well as

$$|S - r| < \epsilon.$$

It follows that

$$L(f) \geq L(f, P) \geq S - \epsilon(b - a) > r - \epsilon - \epsilon(b - a).$$

Since ϵ is arbitrary, we have $L(f) \geq r$. A similar argument shows that $U(f) \leq r$. Since $L(f) \leq U(f)$, we see that $L(f) = U(f) = r$. This shows that f is [Darboux] integrable and that

$$\int_a^b f = r = \mathcal{R} \int_a^b f.$$

∎

Exercises

32.1. Find the upper and lower Darboux integrals for $f(x) = x^3$ on the interval $[0, b]$. *Hint*: Exercise 1.3 and Example 1 in §1 will be useful.

32.2. Let $f(x) = x$ for rational x and $f(x) = 0$ for irrational x.

    (a) Calculate the upper and lower Darboux integrals for f on the interval $[0, b]$.

    (b) Is f integrable on $[0, b]$?

32.3. Repeat Exercise 32.2 for g where $g(x) = x^2$ for rational x and $g(x) = 0$ for irrational x.

32.4. Supply the induction argument needed in the proof of Lemma 32.2.

32.5. Use Exercise 4.8 to prove Theorem 32.4. Specify the sets S and T in this case.

32.6. Let f be a bounded function on $[a, b]$. Suppose there exist sequences (U_n) and (L_n) of upper and lower Darboux sums for f such that $\lim(U_n - L_n) = 0$. Show f is integrable and $\int_a^b f = \lim U_n = \lim L_n$.

32.7. Let f be integrable on $[a, b]$, and suppose that g is a function on $[a, b]$ such that $g(x) = f(x)$ except for finitely many x in $[a, b]$. Show that g is integrable and that $\int_a^b f = \int_a^b g$.

32.8. Show that if f is integrable on $[a, b]$, then f is integrable on every interval $[c, d] \subseteq [a, b]$.

§33 Properties of the Riemann Integral

In this section we establish some basic properties of the Riemann integral and we show that many familiar functions, including piecewise continuous and piecewise monotonic functions, are Riemann integrable.

A function is *monotonic* on an interval if it is either increasing or decreasing on the interval; see Definition 29.6.

33.1 Theorem.
Every monotonic function f on $[a, b]$ is integrable.

Proof
We assume f is increasing on $[a, b]$ and leave the decreasing case to Exercise 33.1. Since $f(a) \leq f(x) \leq f(b)$ for all $x \in [a, b]$, f is clearly bounded on $[a, b]$. In order to apply Theorem 32.5, let $\epsilon > 0$ and select a partition $P = \{a = t_0 < t_1 < \cdots < t_n = b\}$ with mesh less than $\frac{\epsilon}{f(b)-f(a)}$. Then

$$U(f, P) - L(f, P) = \sum_{k=1}^{n} [M(f, [t_{k-1}, t_k]) - m(f, [t_{k-1}, t_k])] \cdot (t_k - t_{k-1})$$

$$= \sum_{k=1}^{n} [f(t_k) - f(t_{k-1})] \cdot (t_k - t_{k-1}).$$

Since mesh$(P) < \epsilon$, we have

$$U(f, P) - L(f, P) < \sum_{k=1}^{n} [f(t_k) - f(t_{k-1})] \cdot \frac{\epsilon}{f(b) - f(a)}$$

$$= [f(b) - f(a)] \cdot \frac{\epsilon}{f(b) - f(a)} = \epsilon.$$

Theorem 32.5 now shows that f is integrable. ∎

33.2 Theorem.
Every continuous function f on $[a, b]$ is integrable.

Proof
Again, in order to apply Theorem 32.5, consider $\epsilon > 0$. Since f is uniformly continuous on $[a, b]$ by Theorem 19.2, there exists $\delta > 0$ such that

$$x, y \in [a, b] \quad \text{and} \quad |x - y| < \delta \quad \text{imply} \quad |f(x) - f(y)| < \frac{\epsilon}{b - a}. \quad (1)$$

Consider any partition $P = \{a = t_0 < t_1 < \cdots < t_n = b\}$ where

$$\max\{t_k - t_{k-1} : k = 1, 2, \ldots, n\} < \delta.$$

Since f assumes its maximum and minimum on each interval $[t_{k-1}, t_k]$ by Theorem 18.1, it follows from (1) that

$$M(f, [t_{k-1}, t_k]) - m(f, [t_{k-1}, t_k]) < \frac{\epsilon}{b - a}$$

for each k. Therefore we have

$$U(f, P) - L(f, P) < \sum_{k=1}^{n} \frac{\epsilon}{b - a}(t_k - t_{k-1}) = \epsilon$$

and Theorem 32.5 shows that f is integrable. ∎

33.3 Theorem.
Let f and g be integrable functions on $[a, b]$, and let c be a real number. Then
   (i) *cf is integrable and $\int_a^b cf = c \int_a^b f$;*
   (ii) *$f + g$ is integrable and $\int_a^b (f + g) = \int_a^b f + \int_a^b g$.*

Exercise 33.8 shows that fg, $\max(f, g)$ and $\min(f, g)$ are also integrable, but there are no formulas giving their integrals in terms of $\int_a^b f$ and $\int_a^b g$.

Proof
The proof of (i) involves three cases: $c > 0$, $c = -1$, and $c < 0$. Of course, (i) is obvious for $c = 0$.

Let $c > 0$ and consider a partition

$$P = \{a = t_0 < t_1 < \cdots < t_n = b\}$$

of $[a, b]$. A simple exercise [Exercise 33.2] shows that

$$M(cf, [t_{k-1}, t_k]) = c \cdot M(f, [t_{k-1}, t_k])$$

for all k, so $U(cf, P) = c \cdot U(f, P)$. Another application of the same exercise shows that $U(cf) = c \cdot U(f)$. Similar arguments show that $L(cf) = c \cdot L(f)$. Since f is integrable, we have $L(cf) = c \cdot L(f) = c \cdot U(f) = U(cf)$. Hence cf is integrable and

$$\int_a^b cf = U(cf) = c \cdot U(f) = c \int_a^b f, \qquad c > 0. \qquad (1)$$

Now we deal with the case $c = -1$. Exercise 5.4 implies that $U(-f, P) = -L(f, P)$ for all partitions P of $[a, b]$. Hence we have

$$\begin{aligned} U(-f) &= \inf\{U(-f, P) : P \text{ is a partition of } [a, b]\} \\ &= \inf\{-L(f, P) : P \text{ is a partition of } [a, b]\} \\ &= -\sup\{L(f, P) : P \text{ is a partition of } [a, b]\} = -L(f). \end{aligned}$$

Replacing f by $-f$, we also obtain $L(-f) = -U(f)$. Since f is integrable, $U(-f) = -L(f) = -U(f) = L(-f)$; hence $-f$ is integrable and

$$\int_a^b (-f) = -\int_a^b f. \qquad (2)$$

The case $c < 0$ is handled by applying (2) and then (1) to $-c$:

$$\int_a^b cf = -\int_a^b (-c)f = -(-c) \int_a^b f = c \int_a^b f.$$

To prove (ii) we will again use Theorem 32.5. Let $\epsilon > 0$. By Theorem 32.5 there exist partitions P_1 and P_2 of $[a, b]$ such that

$$U(f, P_1) - L(f, P_1) < \frac{\epsilon}{2} \quad \text{and} \quad U(g, P_2) - L(g, P_2) < \frac{\epsilon}{2}.$$

Lemma 32.2 shows that if $P = P_1 \cup P_2$, then

$$U(f, P) - L(f, P) < \frac{\epsilon}{2} \quad \text{and} \quad U(g, P) - L(g, P) < \frac{\epsilon}{2}. \tag{3}$$

For any subset S of $[a, b]$, we have

$$\inf\{f(x) + g(x) : x \in S\} \geq \inf\{f(x) : x \in S\} + \inf\{g(x) : x \in S\},$$

i.e., $m(f + g, S) \geq m(f, S) + m(g, S)$. It follows that

$$L(f + g, P) \geq L(f, P) + L(g, P)$$

and similarly we have

$$U(f + g, P) \leq U(f, P) + U(g, P).$$

Therefore from (3) we obtain

$$U(f + g, P) - L(f + g, P) < \epsilon.$$

Theorem 32.5 now shows that $f + g$ is integrable. Since

$$\int_a^b (f + g) = U(f + g) \leq U(f + g, P) \leq U(f, P) + U(g, P)$$

$$< L(f, P) + L(g, P) + \epsilon \leq L(f) + L(g) + \epsilon = \int_a^b f + \int_a^b g + \epsilon$$

and

$$\int_a^b (f + g) = L(f + g) \geq L(f + g, P) \geq L(f, P) + L(g, P)$$

$$> U(f, P) + U(g, P) - \epsilon \geq U(f) + U(g) - \epsilon = \int_a^b f + \int_a^b g - \epsilon,$$

we see that

$$\int_a^b (f + g) = \int_a^b f + \int_a^b g.$$

∎

33.4 Theorem.
If f and g are integrable on $[a, b]$ and if $f(x) \leq g(x)$ for $x \in [a, b]$, then $\int_a^b f \leq \int_a^b g$.

Proof
By Theorem 33.3, $h = g - f$ is integrable on $[a, b]$. Since $h(x) \geq 0$ for all $x \in [a, b]$, it is clear that $L(h, P) \geq 0$ for all partitions P of $[a, b]$, so $\int_a^b h = L(h) \geq 0$. Applying Theorem 33.3 again, we see that

$$\int_a^b g - \int_a^b f = \int_a^b h \geq 0.$$

∎

33.5 Theorem.
If f is integrable on $[a, b]$, then $|f|$ is integrable on $[a, b]$ and

$$\left| \int_a^b f \right| \leq \int_a^b |f|. \tag{1}$$

Proof
This follows easily from Theorem 33.4 provided we know $|f|$ is integrable on $[a, b]$. In fact, $-|f| \leq f \leq |f|$; therefore

$$-\int_a^b |f| \leq \int_a^b f \leq \int_a^b |f|,$$

which implies (1).

We now show that $|f|$ is integrable, a point that was conveniently glossed over in Exercise 25.1. For any subset S of $[a, b]$, we have

$$M(|f|, S) - m(|f|, S) \leq M(f, S) - m(f, S) \tag{2}$$

by Exercise 33.6. From (2) it follows that

$$U(|f|, P) - L(|f|, P) \leq U(f, P) - L(f, P) \tag{3}$$

for all partitions P of $[a, b]$. By Theorem 32.5, for each $\epsilon > 0$ there exists a partition P such that

$$U(f, P) - L(f, P) < \epsilon.$$

In view of (3), the same remark applies to $|f|$, so $|f|$ is integrable by Theorem 32.5. ∎

33.6 Theorem.

Let f be a function defined on $[a, b]$. If $a < c < b$ and f is integrable on $[a, c]$ and on $[c, b]$, then f is integrable on $[a, b]$ and

$$\int_a^b f = \int_a^c f + \int_c^b f. \tag{1}$$

Proof

Since f is bounded on both $[a, c]$ and $[c, b]$, f is bounded on $[a, b]$. In this proof we will decorate upper and lower sums so that it will be clear which intervals we are dealing with. Let $\epsilon > 0$. By Theorem 32.5 there exist partitions P_1 and P_2 of $[a, c]$ and $[c, b]$ such that

$$U_a^c(f, P_1) - L_a^c(f, P_1) < \frac{\epsilon}{2} \quad \text{and} \quad U_c^b(f, P_2) - L_c^b(f, P_2) < \frac{\epsilon}{2}.$$

The set $P = P_1 \cup P_2$ is a partition of $[a, b]$, and it is obvious that

$$U_a^b(f, P) = U_a^c(f, P_1) + U_c^b(f, P_2) \tag{2}$$

with a similar identity for lower sums. It follows that

$$U_a^b(f, P) - L_a^b(f, P) < \epsilon,$$

so f is integrable on $[a, b]$ by Theorem 32.5. Also (1) holds because

$$\int_a^b f \le U_a^b(f, P) = U_a^c(f, P_1) + U_c^b(f, P_2)$$

$$< L_a^c(f, P_1) + L_c^b(f, P_2) + \epsilon \le \int_a^c f + \int_c^b f + \epsilon$$

and similarly $\int_a^b f > \int_a^c f + \int_c^b f - \epsilon$. ∎

Most functions encountered in calculus and analysis are covered by the next definition. However, see Exercises 33.10–33.12.

33.7 Definition.

A function f on $[a, b]$ is *piecewise monotonic* if there is a partition

$$P = \{a = t_0 < t_1 < \cdots < t_n = b\}$$

of $[a, b]$ such that f is monotonic on each interval (t_{k-1}, t_k). The function f is *piecewise continuous* if there is a partition P of $[a, b]$ such that f is uniformly continuous on each interval (t_{k-1}, t_k).

33.8 Theorem.

If f is a piecewise continuous function or a bounded piecewise monotonic function on [a, b], then f is integrable on [a, b].

Proof

Let P be the partition described in Definition 33.7. Consider a fixed interval $[t_{k-1}, t_k]$. If f is piecewise continuous, then its restriction to (t_{k-1}, t_k) can be extended to a continuous function f_k on $[t_{k-1}, t_k]$ by Theorem 19.5. If f is piecewise monotonic, then its restriction to (t_{k-1}, t_k) can be extended to a monotonic function f_k on $[t_{k-1}, t_k]$; for example, if f is increasing on (t_{k-1}, t_k), simply define

$$f_k(t_k) = \sup\{f(x) : x \in (t_{k-1}, t_k)\}$$

and

$$f_k(t_{k-1}) = \inf\{f(x) : x \in (t_{k-1}, t_k)\}.$$

In either case, f_k is integrable on $[t_{k-1}, t_k]$ by Theorem 33.1 or 33.2. Since f agrees with f_k on $[t_{k-1}, t_k]$ except possibly at the endpoints, Exercise 32.7 shows that f is also integrable on $[t_{k-1}, t_k]$. Now Theorem 33.6 and a trivial induction argument show that f is integrable on $[a, b]$. ∎

We close this section with a simple but useful result.

33.9 Intermediate Value Theorem for Integrals.

If f is a continuous function on [a, b], then for at least one x in [a, b] we have

$$f(x) = \frac{1}{b-a} \int_a^b f.$$

Proof

By Theorem 18.1, the function f assumes its maximum value M and its minimum value m on $[a, b]$. Since

$$m \le \frac{1}{b-a} \int_a^b f \le M,$$

the present theorem follows from the Intermediate Value Theorem 18.2 for continuous functions. ∎

Exercises

33.1. Complete the proof of Theorem 33.1 by showing that a decreasing function on $[a, b]$ is integrable.

33.2. This exercise could have appeared just as easily in §4. Let S be a nonempty bounded subset of $\mathbb{R}$. For fixed $c > 0$, let $cS = \{cs : s \in S\}$. Show that $\sup(cS) = c \cdot \sup(S)$ and $\inf(cS) = c \cdot \inf(S)$.

33.3. A function f on $[a, b]$ is called a *step-function* if there exists a partition $P = \{a = u_0 < u_1 < \cdots < u_m = b\}$ of $[a, b]$ such that f is constant on each interval (u_{j-1}, u_j), say $f(x) = c_j$ for x in (u_{j-1}, u_j).

   (a) Show that a step-function f is integrable and evaluate $\int_a^b f$.

   (b) Evaluate the integral $\int_0^4 P(x)\, dx$ for the postage-stamp function P in Exercise 17.10.

33.4. Give an example of a function f on $[0, 1]$ that is *not* integrable for which $|f|$ is integrable. *Hint*: Modify Example 2 in §32.

33.5. Show that $\left| \int_{-2\pi}^{2\pi} x^2 \sin^8(e^x)\, dx \right| \le \frac{16\pi^3}{3}$.

33.6. Prove (2) in the proof of Theorem 33.5. *Hint*: For $x_0, y_0 \in S$, we have $|f(x_0)| - |f(y_0)| \le |f(x_0) - f(y_0)| \le M(f, S) - m(f, S)$.

33.7. Let f be a bounded function on $[a, b]$, so that there exists $B > 0$ such that $|f(x)| \le B$ for all $x \in [a, b]$.

   (a) Show that

$$U(f^2, P) - L(f^2, P) \le 2B[U(f, P) - L(f, P)]$$

   for all partitions P of $[a, b]$. *Hint*: $f(x)^2 - f(y)^2 = [f(x) + f(y)] \cdot [f(x) - f(y)]$.

   (b) Show that if f is integrable on $[a, b]$, then f^2 also is integrable on $[a, b]$.

33.8. Let f and g be integrable functions on $[a, b]$.

   (a) Show that fg is integrable on $[a, b]$. *Hint*: Use $4fg = (f + g)^2 - (f - g)^2$; see Exercise 33.7.

   (b) Show that $\max(f, g)$ and $\min(f, g)$ are integrable on $[a, b]$. *Hint*: Exercise 17.8.

33.9. Let (f_n) be a sequence of integrable functions on $[a, b]$, and suppose that $f_n \to f$ uniformly on $[a, b]$. Prove that f is integrable and that

$$\int_a^b f = \lim_{n\to\infty} \int_a^b f_n.$$

Compare this result with Theorem 25.2.

33.10. Let $f(x) = \sin\frac{1}{x}$ for $x \neq 0$ and $f(0) = 0$. Show that f is integrable on $[-1, 1]$. *Hint*: See the answer to Exercise 33.11(c).

33.11. Let $f(x) = x\,\mathrm{sgn}(\sin\frac{1}{x})$ for $x \neq 0$ and $f(0) = 0$.

(a) Show that f is not piecewise continuous on $[-1, 1]$.

(b) Show that f is not piecewise monotonic on $[-1, 1]$.

(c) Show that f is integrable on $[-1, 1]$.

33.12. Let f be the function described in Exercise 17.14.

(a) Show that f is not piecewise continuous or piecewise monotonic on any interval $[a, b]$.

(b) Show f is integrable on every interval $[a, b]$ and that $\int_a^b f = 0$.

33.13. Suppose f and g are continuous functions on $[a, b]$ such that $\int_a^b f = \int_a^b g$. Prove that there exists x in $[a, b]$ such that $f(x) = g(x)$.

33.14. (a) Suppose f and g are continuous functions on $[a, b]$ and that $g(x) \geq 0$ for all $x \in [a, b]$. Prove that there exists x in $[a, b]$ such that

$$\int_a^b f(t)g(t)\,dt = f(x)\int_a^b g(t)\,dt.$$

(b) Show that Theorem 33.9 is a special case of part (a).

§34 Fundamental Theorem of Calculus

There are two versions of the Fundamental Theorem of Calculus. Each says, roughly speaking, that differentiation and integration are inverse operations. In fact, our first version [Theorem 34.1] says that "the integral of the derivative of a function is given by the function," and our second version [Theorem 34.3] says that "the derivative of the integral of a continuous function is the function." It is somewhat traditional for books to prove our second version first and use it to prove our first version, although some books do avoid this approach. F. Cunningham, Jr. [9] offers some good reasons for avoiding the traditional approach:

(a) Theorem 34.3 implies Theorem 34.1 only for functions g whose derivative g' is continuous; see Exercise 34.1.
(b) Making Theorem 34.1 depend on Theorem 34.3 obscures the fact that the two theorems say different things, have different applications, and may leave the impression that Theorem 34.3 is *the* fundamental theorem.
(c) The need for Theorem 34.1 in calculus is immediate and easily motivated.

In what follows, we say a function h defined on (a, b) is *integrable* on $[a, b]$ if every extension of h to $[a, b]$ is integrable. In view of Exercise 32.7, the value $\int_a^b h$ will not depend on the values of the extensions at a or b.

34.1 Fundamental Theorem of Calculus I.
If g is a continuous function on $[a, b]$ that is differentiable on (a, b), and if g' is integrable on $[a, b]$, then

$$\int_a^b g' = g(b) - g(a). \tag{1}$$

Proof
Let $\epsilon > 0$. By Theorem 32.5, there exists a partition $P = \{a = t_0 < t_1 < \cdots < t_n = b\}$ of $[a, b]$ such that

$$U(g', P) - L(g', P) < \epsilon. \tag{2}$$

We apply the Mean Value Theorem 29.3 to each interval $[t_{k-1}, t_k]$ to obtain $x_k \in (t_{k-1}, t_k)$ for which

$$(t_k - t_{k-1})g'(x_k) = g(t_k) - g(t_{k-1}).$$

Hence we have

$$g(b) - g(a) = \sum_{k=1}^n [g(t_k) - g(t_{k-1})] = \sum_{k=1}^n g'(x_k)(t_k - t_{k-1}).$$

It follows that

$$L(g', P) \le g(b) - g(a) \le U(g', P); \tag{3}$$

see Definition 32.1. Since

$$L(g', P) \le \int_a^b g' \le U(g', P),$$

inequalities (2) and (3) imply that

$$\left| \int_a^b g' - [g(b) - g(a)] \right| < \epsilon.$$

Since ϵ is arbitary, (1) holds. ∎

The integration formulas in calculus all rely in the end on Theorem 34.1.

Example 1
If $g(x) = \frac{x^{n+1}}{n+1}$, then $g'(x) = x^n$, so

$$\int_a^b x^n \, dx = \frac{b^{n+1}}{n+1} - \frac{a^{n+1}}{n+1} = \frac{b^{n+1} - a^{n+1}}{n+1}. \tag{1}$$

In particular,

$$\int_a^b x^2 \, dx = \frac{b^3 - a^3}{3}.$$

Formula (1) is valid for any powers n for which $g(x) = \frac{x^{n+1}}{n+1}$ is defined on $[a, b]$. See Examples 3 and 4 in §28 and Exercises 29.15 and 37.5. For example,

$$\int_a^b \sqrt{x} \, dx = \frac{2}{3}[b^{3/2} - a^{3/2}] \quad \text{for} \quad 0 \le a < b.$$

34.2 Theorem [Integration by Parts].
If u and v are continuous functions on $[a, b]$ that are differentiable on (a, b), and if u' and v' are integrable on $[a, b]$, then

$$\int_a^b u(x)v'(x) \, dx + \int_a^b u'(x)v(x) \, dx = u(b)v(b) - u(a)v(a). \tag{1}$$

Proof

Let $g = uv$; then $g' = uv' + u'v$ by Theorem 28.3. Exercise 33.8 shows that g' is integrable. Now Theorem 34.1 shows that

$$\int_a^b g'(x)\,dx = g(b) - g(a) = u(b)v(b) - u(a)v(a),$$

so (1) holds. ∎

Note that the use of Exercise 33.8 above can be avoided if u' and v' are continuous, which is normally the case.

Example 2

Here is a simple application of integration by parts. To calculate $\int_0^\pi x \cos x\,dx$, we note that the integrand has the form $u(x)v'(x)$ where $u(x) = x$ and $v(x) = \sin x$. Hence

$$\int_0^\pi x \cos x\,dx = u(\pi)v(\pi) - u(0)v(0) - \int_0^\pi 1 \cdot \sin x\,dx = -\int_0^\pi \sin x\,dx = -2.$$

In what follows we use the convention $\int_a^b f = -\int_b^a f$ for $a > b$.

34.3 Fundamental Theorem of Calculus II.

Let f be an integrable function on $[a, b]$. For x in $[a, b]$, let

$$F(x) = \int_a^x f(t)\,dt.$$

Then F is continuous on $[a, b]$. If f is continuous at x_0 in (a, b), then F is differentiable at x_0 and

$$F'(x_0) = f(x_0).$$

Proof

Select $B > 0$ so that $|f(x)| \le B$ for all $x \in [a, b]$. If $x, y \in [a, b]$ and $|x - y| < \frac{\epsilon}{B}$ where $x < y$, say, then

$$|F(y) - F(x)| = \left| \int_x^y f(t)\,dt \right| \le \int_x^y |f(t)|\,dt \le \int_x^y B\,dt = B(y - x) < \epsilon.$$

This shows that F is [uniformly] continuous on $[a, b]$.

Suppose that f is continuous at x_0 in (a, b). Observe that

$$\frac{F(x) - F(x_0)}{x - x_0} = \frac{1}{x - x_0} \int_{x_0}^{x} f(t) \, dt$$

for $x \neq x_0$. The trick is to observe that

$$f(x_0) = \frac{1}{x - x_0} \int_{x_0}^{x} f(x_0) \, dt$$

and therefore

$$\frac{F(x) - f(x_0)}{x - x_0} - f(x_0) = \frac{1}{x - x_0} \int_{x_0}^{x} [f(t) - f(x_0)] \, dt. \tag{1}$$

Let $\epsilon > 0$. Since f is continuous at x_0, there exists $\delta > 0$ such that

$$t \in (a, b) \quad \text{and} \quad |t - x_0| < \delta \quad \text{imply} \quad |f(t) - f(x_0)| < \epsilon;$$

see Theorem 17.2. It follows from (1) that

$$\left| \frac{F(x) - F(x_0)}{x - x_0} - f(x_0) \right| \leq \epsilon$$

for x in (a, b) satisfying $|x - x_0| < \delta$; the cases $x > x_0$ and $x < x_0$ require separate arguments. We have just shown that

$$\lim_{x \to x_0} \frac{F(x) - F(x_0)}{x - x_0} = f(x_0).$$

In other words, $F'(x_0) = f(x_0)$. ∎

A useful technique of integration is known as "substitution." A more accurate description of the process is "change of variable." The technique is the reverse of the chain rule.

34.4 Theorem [Change of Variable].
Let u be a differentiable function on an open interval J such that u' is continuous, and let I be an open interval such that $u(x) \in I$ for all $x \in J$. If f is continuous on I, then $f \circ u$ is continuous on J and

$$\int_{a}^{b} f \circ u(x) u'(x) \, dx = \int_{u(a)}^{u(b)} f(u) \, du \tag{1}$$

for $a, b \in J$.

Note that $u(a)$ need not be less than $u(b)$, even if $a < b$.

Proof
The continuity of $f \circ u$ follows from Theorem 17.5. Fix $c \in I$ and let $F(u) = \int_c^u f(t)\, dt$. Then $F'(u) = f(u)$ for all $u \in I$ by Theorem 34.3. Let $g = F \circ u$. By the Chain Rule 28.4, we have $g'(x) = F'(u(x)) \cdot u'(x) = f(u(x)) \cdot u'(x)$, so by Theorem 34.1

$$\int_a^b f \circ u(x) u'(x)\, dx = \int_a^b g'(x)\, dx = g(b) - g(a) = F(u(b)) - F(u(a))$$

$$= \int_c^{u(b)} f(t)\, dt - \int_c^{u(a)} f(t)\, dt = \int_{u(a)}^{u(b)} f(t)\, dt.$$

This proves (1). ∎

Example 3
Let g be a one-to-one differentiable function on an open interval I. Then $J = g(I)$ is an open interval, and the inverse function g^{-1} is differentiable on J by Theorem 29.9. We show

$$\int_a^b g(x)\, dx + \int_{g(a)}^{g(b)} g^{-1}(u)\, du = bg(b) - ag(a) \tag{1}$$

for $a, b \in I$.

We put $f = g^{-1}$ and $u = g$ in the change of variable formula to obtain

$$\int_a^b g^{-1} \circ g(x) g'(x)\, dx = \int_{g(a)}^{g(b)} g^{-1}(u)\, du.$$

Since $g^{-1} \circ g(x) = x$ for $x \in I$, we obtain

$$\int_{g(a)}^{g(b)} g^{-1}(u)\, du = \int_a^b x g'(x)\, dx.$$

Now integrate by parts with $u(x) = x$ and $v(x) = g(x)$:

$$\int_{g(a)}^{g(b)} g^{-1}(u)\, du = bg(b) - ag(a) - \int_a^b g(x)\, dx.$$

This is formula (1).

Exercises

34.1. Use Theorem 34.3 to prove Theorem 34.1 for the case that g' is continuous. *Hint*: Let $F(x) = \int_a^x g'$; then $F' = g'$. Apply Corollary 29.5.

34.2. Calculate

(a) $\lim_{x \to 0} \frac{1}{x} \int_0^x e^{t^2}\, dt$

(b) $\lim_{h \to 0} \frac{1}{h} \int_3^{3+h} e^{t^2}\, dt$.

34.3. Let f be defined as follows: $f(t) = 0$ for $t < 0$; $f(t) = t$ for $0 \le t \le 1$; $f(t) = 4$ for $t > 1$.

(a) Determine the function $F(x) = \int_0^x f(t)\, dt$.

(b) Sketch F. Where is F continuous?

(c) Where is F differentiable? Calculate F' at the points of differentiability.

34.4. Repeat Exercise 34.3 for f where $f(t) = t$ for $t < 0$; $f(t) = t^2 + 1$ for $0 \le t \le 2$; $f(t) = 0$ for $t > 2$.

34.5. Let f be a continuous function on $\mathbb{R}$ and define

$$F(x) = \int_{x-1}^{x+1} f(t)\, dt \quad \text{for} \quad x \in \mathbb{R}.$$

Show that F is differentiable on $\mathbb{R}$ and compute F'.

34.6. Let f be a continuous function on $\mathbb{R}$ and define

$$G(x) = \int_0^{\sin x} f(t)\, dt \quad \text{for} \quad x \in \mathbb{R}.$$

Show that G is differentiable on $\mathbb{R}$ and compute G'.

34.7. Use change of variables to integrate $\int_0^1 x\sqrt{1 - x^2}\, dx$.

34.8. (a) Use integration by parts to evaluate

$$\int_0^1 x \arctan x\, dx.$$

Hint: Let $u(x) = \arctan x$, so that $u'(x) = \frac{1}{1+x^2}$.

(b) If you used $v(x) = \frac{x^2}{2}$ in part (a), do the computation again with $v(x) = \frac{x^2+1}{2}$. This interesting example is taken from J. L. Borman [6].

34.9. Use Example 3 to show $\int_0^{1/2} \arcsin x\, dx = \frac{\pi}{12} + \frac{\sqrt{3}}{2} - 1$.

34.10. Let g be a strictly increasing continuous function mapping $[0, 1]$ onto $[0, 1]$. Give a geometric argument showing $\int_0^1 g(x)\,dx + \int_0^1 g^{-1}(u)\,du = 1$.

34.11. Suppose that f is a continuous function on $[a, b]$ and that $f(x) \geq 0$ for all $x \in [a, b]$. Show that if $\int_a^b f(x)\,dx = 0$, then $f(x) = 0$ for all x in $[a, b]$.

34.12. Show that if f is a continuous real-valued function on $[a, b]$ satisfying $\int_a^b f(x)g(x)\,dx = 0$ for every continuous function g on $[a, b]$, then $f(x) = 0$ for all x in $[a, b]$.

§35   * Riemann-Stieltjes Integrals

In this long section we introduce a useful generalization of the Riemann integral. In the Riemann integral, all intervals of the same length are given the same weight. For example, in our definition of upper sums

$$U(f, P) = \sum_{k=1}^{n} M(f, [t_{k-1}, t_k]) \cdot (t_k - t_{k-1}), \qquad (*)$$

the factors $(t_k - t_{k-1})$ are the lengths of the intervals involved. In applications such as probability and statistics, it is desirable to modify the definition so as to weight the intervals according to some increasing function F. In other words, the idea is to replace the factors $(t_k - t_{k-1})$ in $(*)$ by $[F(t_k) - F(t_{k-1})]$. The Riemann integral is, then, the special case where $F(t) = t$ for all t.

It is also desirable to allow some *points* to have positive weight. This corresponds to the situations where F has jumps, i.e., where the left-hand and right-hand limits of F differ. In fact, if (c_k) is a sequence of positive numbers for which $\sum c_k < \infty$ and if (u_k) is a sequence in $\mathbb{R}$, then the sums

$$\sum_{k=1}^{\infty} c_k f(u_k)$$

can be viewed as a generalized integral for a suitable F [Exercise 36.14]. In this case, F has a jump at each u_k.

The traditional treatment, in all books that I am aware of, replaces the factors $(t_k - t_{k-1})$ in (*) by $[F(t_k) - F(t_{k-1})]$ and develops the theory from there, though some authors emphasize upper and lower sums while others stress generalized Riemann sums. In this section, we offer a slightly different treatment, so

Warning. Theorems in this section do not necessarily correspond to theorems in other texts.

We deviate from tradition because: (a) Our treatment is more general. Functions that are Riemann-Stieltjes integrable in the traditional sense are integrable in our sense [Theorem 35.20]. (b) In the traditional theory, if f and F have a common discontinuity, then f is not integrable using F. Such unfortunate results disappear in our approach. We will show that piecewise continuous and piecewise monotonic functions are always integrable using F [Theorem 35.17]. We also will observe that if F is a step-function, then *all* bounded functions are integrable; see Example 1. (c) We will give a definition involving Riemann-Stieltjes sums that is equivalent to our definition involving upper and lower sums [Theorem 35.25]. The corresponding standard definitions are not equivalent.

Many of the results in this section are straightforward generalizations of results in §§32 and 33. Accordingly, many proofs will be brief or omitted.

35.1 Notation.
We assume throughout this section that F is an increasing function on a closed interval $[a, b]$. To avoid trivialities we assume $F(a) < F(b)$. All left-hand and right-hand limits exist; see Definition 20.3 and Exercise 35.1. We use the notation

$$F(t^-) = \lim_{x \to t^-} F(x) \quad \text{and} \quad F(t^+) = \lim_{x \to t^+} F(x).$$

For the endpoints we decree

$$F(a^-) = F(a) \quad \text{and} \quad F(b^+) = F(b).$$

Note that $F(t^-) \le F(t^+)$ for all $t \in [a, b]$. If F is continuous at t, then $F(t^-) = F(t) = F(t^+)$. Otherwise $F(t^-) < F(t^+)$ and the difference $F(t^+) - F(t^-)$ is called the *jump* of F at t. The actual value of $F(t)$ at jumps t will play no role in what follows.

In the next definition we employ some of the notation established in Definition 32.1.

35.2 Definition.
For a bounded function f on $[a, b]$ and a partition $P = \{a = t_0 < t_1 < \cdots < t_n = b\}$ of $[a, b]$, we write

$$J_F(f, P) = \sum_{k=0}^{n} f(t_k) \cdot [F(t_k^+) - F(t_k^-)].$$

The *upper Darboux-Stieltjes sum* is

$$U_F(f, P) = J_F(f, P) + \sum_{k=1}^{n} M(f, (t_{k-1}, t_k)) \cdot [F(t_k^-) - F(t_{k-1}^+)]$$

and the *lower Darboux-Stieltjes sum* is

$$L_F(f, P) = J_F(f, P) + \sum_{k=1}^{n} m(f, (t_{k-1}, t_k)) \cdot [F(t_k^-) - F(t_{k-1}^+)].$$

These definitions explicitly take the jump effects of F into account. Note that

$$U_F(f, P) - L_F(f, P)$$
$$= \sum_{k=1}^{n} [M(f, (t_{k-1}, t_k)) - m(f, (t_{k-1}, t_k))][F(t_k^-) - F(t_{k-1}^+)] \qquad (1)$$

and

$$m(f, [a, b]) \cdot [F(b) - F(a)] \leq L_F(f, P) \leq U_F(f, P)$$
$$\leq M(f, [a, b]) \cdot [F(b) - F(a)]. \qquad (2)$$

In checking (2), note that

$$\sum_{k=0}^{n} [F(t_k^+) - F(t_k^-)] + \sum_{k=1}^{n} [F(t_k^-) - F(t_{k-1}^+)] \qquad (3)$$
$$= F(t_n^+) - F(t_0^-) = F(b^+) - F(a^-) = F(b) - F(a).$$

The *upper Darboux-Stieltjes integral* is

$$U_F(f) = \inf\{U_F(f, P) : P \text{ is a partition of } [a, b]\}$$

and the *lower Darboux-Stieltjes integral* is

$$L_F(f) = \sup\{L_F(f, P) : P \text{ is a partition of } [a, b]\}.$$

Theorem 35.5 will show that $L_F(f) \le U_F(f)$. Accordingly, we say f is *Darboux-Stieltjes integrable* on $[a, b]$ with respect to F or, more briefly, *F-integrable* on $[a, b]$, provided $L_F(f) = U_F(f)$; in this case we write

$$\int_a^b f \, dF = \int_a^b f(x) \, dF(x) = L_F(f) = U_F(f).$$

Example 1

For each u in $[a, b]$, let J_u be an increasing step-function with jump 1 at u. For example, we can let

$$J_u(t) = \begin{cases} 0 & \text{for} \quad t < u, \\ 1 & \text{for} \quad t \ge u, \end{cases}$$

for $u > a$, and we can let

$$J_a(t) = \begin{cases} 0 & \text{for} \quad t = a, \\ 1 & \text{for} \quad t > a. \end{cases}$$

Then every bounded function f on $[a, b]$ is J_u-integrable and

$$\int_a^b f \, dJ_u = f(u).$$

More generally, if $u_1, u_2, \ldots, u_m$ are distinct points in $[a, b]$ and if $c_1, c_2, \ldots, c_m$ are positive numbers, then

$$F = \sum_{j=1}^m c_j J_{u_j}$$

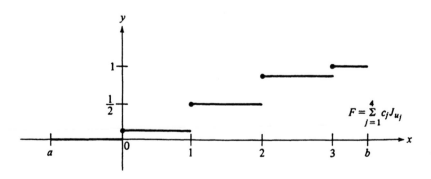

FIGURE 35.1

is an increasing step-function with jumps c_j at u_j. See Figure 35.1 for a special case. Every bounded function f on $[a, b]$ is F-integrable and

$$\int_a^b f \, dF = \sum_{j=1}^m c_j \cdot f(u_j). \tag{1}$$

To check (1), let P be the partition of $[a, b]$ consisting of a, b and all $u_1, u_2, \ldots, u_m$. For this computation we may assume, without loss of generality, that $a = u_1 < u_2 < \cdots < u_m = b$. Then $F(u_j^+) - F(u_j^-) = c_j$ for $j = 1, 2, \ldots, m$ and $F(u_j^-) - F(u_{j-1}^+) = 0$ for $j = 2, 3, \ldots, m$. Therefore

$$U_F(f, P) = L_F(f, P) = J_F(f, P) = \sum_{j=1}^m f(u_j) \cdot c_j$$

for any bounded function f on $[a, b]$. In view of Theorem 35.5, it follows that

$$U_F(f) = L_F(f) = \sum_{j=1}^m f(u_j) \cdot c_j;$$

hence f is F-integrable and (1) holds.

Example 2
We specialize Example 1 to the case $u_1 = 0$, $u_2 = 1$, $u_3 = 2$, $u_4 = 3$; $c_1 = c_4 = \frac{1}{8}$, $c_2 = c_3 = \frac{3}{8}$. Thus we must have $a \leq 0$ and $b \geq 3$; see Figure 35.1. For any bounded function f on $[a, b]$, we have

$$\int_a^b f \, dF = \frac{1}{8} f(0) + \frac{3}{8} f(1) + \frac{3}{8} f(2) + \frac{1}{8} f(3).$$

35.3 Lemma.
Let f be a bounded function on $[a, b]$, and let P and Q be partitions of $[a, b]$ such that $P \subseteq Q$. Then

$$L_F(f, P) \leq L_F(f, Q) \leq U_F(f, Q) \leq U_F(f, P). \tag{1}$$

Proof
We imitate the proof of Lemma 32.2 down to, but not including, formula (3). In the present case, the difference $L_F(f, Q) - L_F(f, P)$

equals

$$f(u) \cdot [F(u^+) - F(u^-)] + m(f, (t_{k-1}, u)) \cdot [F(u^-) - F(t_{k-1}^+)]$$
$$+ m(f, (u, t_k)) \cdot [F(t_k^-) - F(u^+)] \tag{3}$$
$$- m(f, (t_{k-1}, t_k)) \cdot [F(t_k^-) - F(t_{k-1}^+)],$$

and this is nonnegative because

$$m(f, (t_{k-1}, t_k)) \cdot [F(t_k^-) - F(t_{k-1}^+)]$$
$$= m(f, (t_{k-1}, t_k)) \cdot [F(t_k^-) - F(u^+) + F(u^+) - F(u^-)$$
$$+ F(u^-) - F(t_{k-1}^+)]$$
$$\leq m(f, (u, t_k)) \cdot [F(t_k^-) - F(u^+)] + f(u) \cdot [F(u^+) - F(u^-)]$$
$$+ m(f, (t_{k-1}, u)) \cdot [F(u^-) - F(t_{k-1}^+)]. \qquad \blacksquare$$

35.4 Lemma.
If f is a bounded function on $[a, b]$ and if P and Q are partitions of $[a, b]$, then $L_F(f, P) \leq U_F(f, Q)$.

Proof
Imitates the proof of Lemma 32.3. $\qquad \blacksquare$

35.5 Theorem.
For every bounded function f on $[a, b]$, we have $L_F(f) \leq U_F(f)$.

Proof
Imitates the proof of Theorem 32.4. $\qquad \blacksquare$

35.6 Theorem.
A bounded function f on $[a, b]$ is F-integrable if and only if for each $\epsilon > 0$ there exists a partition P such that

$$U_F(f, P) - L_F(f, P) < \epsilon.$$

Proof
Imitates the proof of Theorem 32.5. $\qquad \blacksquare$

We next develop analogues of results in §33; we return later to generalizations of items 32.6–32.9. We begin with the analogue of Theorem 33.2. The analogue of Theorem 33.1 is true, but its proof requires some preparation, so we defer it to Theorem 35.16.

35.7 Theorem.
Every continuous function f on [a, b] is F-integrable.

Proof
To apply Theorem 35.6, let $\epsilon > 0$. Since f is uniformly continuous, there exists $\delta > 0$ such that

$$x, y \in [a, b] \quad \text{and} \quad |x - y| < \delta \quad \text{imply} \quad |f(x) - f(y)| < \frac{\epsilon}{F(b) - F(a)}.$$

Just as in the proof of Theorem 33.2, there is a partition P of $[a, b]$ such that

$$M(f, (t_{k-1}, t_k)) - m(f, (t_{k-1}, t_k)) < \frac{\epsilon}{F(b) - F(a)}$$

for each k. Hence by (1) of Definition 35.2 we have

$$U_F(f, P) - L_F(f, P) \leq \sum_{k=1}^{n} \frac{\epsilon}{F(b) - F(a)} [F(t_k^-) - F(t_{k-1}^+)] \leq \epsilon.$$

Theorem 35.6 now shows that f is F-integrable. ∎

35.8 Theorem.
Let f and g be F-integrable functions on [a, b], and let c be a real number. Then
(i) *cf is F-integrable and $\int_a^b (cf)\, dF = c \int_a^b f\, dF$;*
(ii) *$f + g$ is F-integrable and $\int_a^b (f + g)\, dF = \int_a^b f\, dF + \int_a^b g\, dF$.*

Proof
Imitates the proof of Theorem 33.3, using Theorem 35.6 instead of Theorem 32.5. ∎

35.9 Theorem.
If f and g are F-integrable on [a, b] and if $f(x) \leq g(x)$ for $x \in [a, b]$, then $\int_a^b f\, dF \leq \int_a^b g\, dF$.

Proof
Imitates the proof of Theorem 33.4. ∎

35.10 Theorem.
If f is F-integrable on [a, b], then |f| is F-integrable and

$$\left| \int_a^b f \, dF \right| \leq \int_a^b |f| \, dF.$$

Proof
Imitates the proof of Theorem 33.5 and uses formula (1) of Definition 35.2. ∎

35.11 Theorem.
Let f be a function defined on [a, b]. If a < c < b and f is F-integrable on [a, c] and on [c, b], then f is F-integrable on [a, b] and

$$\int_a^b f \, dF = \int_a^c f \, dF + \int_c^b f \, dF. \tag{1}$$

Proof
Imitates the proof of Theorem 33.6. Note that an upper or lower sum on [a, c] will include the term $f(c)[F(c) - F(c^-)]$ while an upper or lower sum on [c, b] will include the term $f(c)[F(c^+) - F(c)]$. ∎

The next result clearly has no analogue in §32 or §33.

35.12 Theorem.
Let F_1 and F_2 be increasing functions on [a, b]. If f is F_1-integrable and F_2-integrable on [a, b] and if c > 0, then f is cF_1-integrable, f is $(F_1 + F_2)$-integrable,

$$\int_a^b f \, d(cF_1) = c \int_a^b f \, dF_1, \tag{1}$$

and

$$\int_a^b f \, d(F_1 + F_2) = \int_a^b f \, dF_1 + \int_a^b f \, dF_2. \tag{2}$$

Proof
From Theorem 20.4 we see that

$$(F_1 + F_2)(t^+) = \lim_{x \to t^+} [F_1(x) + F_2(x)] = \lim_{x \to t^+} F_1(x) + \lim_{x \to t^+} F_2(x)$$
$$= F_1(t^+) + F_2(t^+)$$

with similar identities for $(F_1+F_2)(t^-)$, $(cF_1)(t^+)$ and $(cF_1)(t^-)$. Hence for any partition P of $[a,b]$, we have

$$U_{F_1+F_2}(f,P) = U_{F_1}(f,P) + U_{F_2}(f,P)$$
$$L_{F_1+F_2}(f,P) = L_{F_1}(f,P) + L_{F_2}(f,P), \tag{3}$$

$U_{cF_1}(f,P) = cU_{F_1}(f,P)$ and $L_{cF_1}(f,P) = cL_{F_1}(f,P)$. It is now clear that f is cF_1-integrable and that (1) holds. To check (2), let $\epsilon > 0$. By Theorem 35.6 and Lemma 35.3, there is a single partition P of $[a,b]$ so that both

$$U_{F_1}(f,P) - L_{F_1}(f,P) < \frac{\epsilon}{2} \quad \text{and} \quad U_{F_2}(f,P) - L_{F_2}(f,P) < \frac{\epsilon}{2}.$$

Hence by (3) we have

$$U_{F_1+F_2}(f,P) - L_{F_1+F_2}(f,P) < \epsilon.$$

This and Theorem 35.6 imply that f is $(F_1 + F_2)$-integrable. The identity (2) follows from

$$\int_a^b f\,d(F_1 + F_2) \le U_{F_1+F_2}(f,P) < L_{F_1+F_2}(f,P) + \epsilon$$

$$= L_{F_1}(f,P) + L_{F_2}(f,P) + \epsilon \le \int_a^b f\,dF_1 + \int_a^b f\,dF_2 + \epsilon$$

and the similar inequality

$$\int_a^b f\,d(F_1 + F_2) > \int_a^b f\,dF_1 + \int_a^b f\,dF_2 - \epsilon. \qquad \blacksquare$$

Example 3
Let (u_n) be a sequence of distinct points in $[a,b]$, and let (c_n) be a sequence of positive numbers such that $\sum c_n < \infty$. Using the notation of Example 1, we define

$$F = \sum_{n=1}^{\infty} c_n J_{u_n}.$$

Then F is an increasing function on $[a,b]$; note that $F(a) = 0$ and $F(b) = \sum_{n=1}^{\infty} c_n < \infty$. Every bounded function f on $[a,b]$ is

F-integrable and

$$\int_a^b f\,dF = \sum_{n=1}^{\infty} c_n f(u_n). \qquad (1)$$

To verify (1), fix f and let $B > 0$ be a bound for $|f|$: $|f(x)| \le B$ for all $x \in [a, b]$. Consider $\epsilon > 0$ and select an integer m so that $\sum_{n=m+1}^{\infty} c_n < \frac{\epsilon}{4B}$. Let

$$F_1 = \sum_{n=1}^{m} c_n J_{u_n} \quad \text{and} \quad F_2 = \sum_{n=m+1}^{\infty} c_n J_{u_n}$$

so that $F = F_1 + F_2$. As noted in Example 1,

$$\int_a^b f\,dF_1 = \sum_{n=1}^{m} c_n f(u_n). \qquad (2)$$

Since

$$F_2(b) - F_2(a) = F_2(b) = \sum_{n=m+1}^{\infty} c_n < \frac{\epsilon}{4B},$$

inequality (2) in Definition 35.2 leads to

$$-\frac{\epsilon}{4} \le L_{F_2}(f, P) \le U_{F_2}(f, P) \le \frac{\epsilon}{4}; \qquad (3)$$

hence

$$U_{F_2}(f, P) - L_{F_2}(f, P) \le \frac{\epsilon}{2}$$

for all partitions P of $[a, b]$. If we select P so that

$$U_{F_1}(f, P) - L_{F_1}(f, P) < \frac{\epsilon}{2},$$

then (3) in the proof of Theorem 35.12 and the identity $F = F_1 + F_2$ imply

$$U_F(f, P) - L_F(f, P) < \epsilon.$$

Theorem 35.6 now shows that f is F-integrable. From (3) we quickly infer that

$$\left| \int_a^b f\,dF_2 \right| \le \frac{\epsilon}{4}. \qquad (4)$$

By Theorem 35.12 and (2) we have

$$\int_a^b f\, dF = \int_a^b f\, dF_1 + \int_a^b f\, dF_2 = \sum_{n=1}^{\infty} c_n f(u_n) - \sum_{n=m+1}^{\infty} c_n f(u_n) + \int_a^b f\, dF_2.$$

Since

$$\left| \sum_{n=m+1}^{\infty} c_n f(u_n) \right| \le B \sum_{n=m+1}^{\infty} c_n < \frac{\epsilon}{4},$$

we use (4) to conclude that

$$\left| \int_a^b f\, dF - \sum_{n=1}^{\infty} c_n f(u_n) \right| < \frac{\epsilon}{2}.$$

Since ϵ is arbitrary, (1) is verified.

The next theorem shows that F-integrals can often be calculated using ordinary Riemann integrals. In fact, most F-integrals encountered in practice are either covered by Example 3 or this theorem.

35.13 Theorem.
Suppose that F is differentiable on $[a, b]$ and that F' is continuous on $[a, b]$. If f is continuous on $[a, b]$, then

$$\int_a^b f\, dF = \int_a^b f(x) F'(x)\, dx. \tag{1}$$

Proof
Note that $f F'$ is Riemann integrable by Theorem 33.2, and f is F-integrable by Theorem 35.7. By Theorems 32.5 and 35.6, there is a partition

$$P = \{a = t_0 < t_1 < \cdots < t_n = b\}$$

such that

$$U(f F', P) - L(f F', P) < \frac{\epsilon}{2} \quad \text{and} \quad U_F(f, P) - L_F(f, P) < \frac{\epsilon}{2}. \tag{2}$$

By the Mean Value Theorem 29.3 applied to F on each interval $[t_{k-1}, t_k]$, there exists x_k in (t_{k-1}, t_k) so that

$$F(t_k) - F(t_{k-1}) = F'(x_k)(t_k - t_{k-1});$$

hence

$$\sum_{k=1}^{n} f(x_k) \cdot [F(t_k) - F(t_{k-1})] = \sum_{k=1}^{n} f(x_k) F'(x_k) \cdot (t_k - t_{k-1}). \quad (3)$$

Since F is continuous, it has no jumps, so by (3)

$$L_F(f, P) \leq U(f F', P) \quad \text{and} \quad L(f F', P) \leq U_F(f, P).$$

Now by (2) we have

$$\int_a^b f \, dF \leq U_F(f, P) < \frac{\epsilon}{2} + L_F(f, P) \leq \frac{\epsilon}{2} + U(f F', P)$$

$$< \frac{\epsilon}{2} + \frac{\epsilon}{2} + L(f F', P) \leq \epsilon + \int_a^b f(x) F'(x) \, dx$$

and similarly $\int_a^b f \, dF > \int_a^b f(x) F'(x) \, dx - \epsilon$. Since $\epsilon > 0$ is arbitrary, (1) holds. ∎

An extension of Theorem 35.13 appears in Exercise 35.10.

Example 4
Let $F(t) = 0$ for $t < 0$, $F(t) = t^2$ for $0 \leq t < 2$, and $F(t) = t + 5$ for $t \geq 2$; see Figure 35.2. We can write $F = F_1 + 3J_2$ where F_1 is continuous. The function F_1 is differentiable except at $t = 2$; the differentiability of F_1 at $t = 0$ is shown in Exercise 28.7. Let f be continuous on $[-3, 3]$, say. Clearly $\int_{-3}^0 f \, dF_1 = 0$. Since F_1 agrees with the differentiable function t^2 on $[0, 2]$, we can apply Theorem 35.13 to obtain

$$\int_0^2 f \, dF_1 = \int_0^2 f(x) \cdot 2x \, dx = 2 \int_0^2 x f(x) \, dx.$$

Similarly we have

$$\int_2^3 f \, dF_1 = \int_2^3 f(x) \cdot 1 \, dx = \int_2^3 f(x) \, dx.$$

Theorem 35.11 now shows that

$$\int_{-3}^3 f \, dF_1 = 2 \int_0^2 x f(x) \, dx + \int_2^3 f(x) \, dx,$$

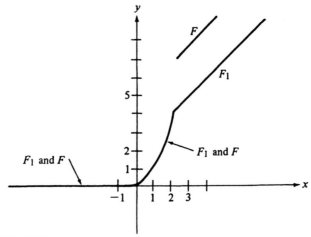

FIGURE 35.2

and then Theorem 35.12 shows that

$$\int_{-3}^{3} f\, dF = \int_{-3}^{3} f\, dF_1 + 3\int_{-3}^{3} f\, dJ_2 = 2\int_{0}^{2} xf(x)\, dx + \int_{2}^{3} f(x)\, dx + 3f(2).$$

As a specific example, if $f(x) = x^3$ then

$$\int_{-3}^{3} f\, dF = 2\int_{0}^{2} x^4\, dx + \int_{2}^{3} x^3\, dx + 3\cdot 8 = \frac{1061}{20} = 53.05.$$

For the results in the remainder of the section we need a more detailed analysis of increasing functions. Some readers may wish to skip the proofs and move on to the next section. We will write $s_n \uparrow s$ to signify that (s_n) is a nondecreasing sequence converging to s, and $s_n \downarrow s$ if (s_n) is a nonincreasing sequence with limit s.

35.14 Lemma.
Let g be an increasing function on $[a, b]$.
    (i) *If $u_n \uparrow u$, then $g(u_n^-) \uparrow g(u^-)$.*
    (ii) *If $u_n \downarrow u$, then $g(u_n^+) \downarrow g(u^+)$.*

Proof
Suppose $u_n \uparrow u$ and let $\epsilon > 0$; here $u \in (a, b]$. There exists $v < u$ such that $v \in [a, b]$ and $g(v) > g(u^-) - \epsilon$. Select N in $\mathbb{N}$ so that $n > N$

implies $u_n > v$. Then

$$n > N \quad \text{implies} \quad g(u_n^-) \geq g(v) > g(u^-) - \epsilon.$$

Since $g(u_n^-) \leq g(u^-)$ for all n, we conclude that $g(u_n^-) \uparrow g(u^-)$. This proves (i), and the proof of (ii) is similar. ∎

35.15 Lemma.
If g is an increasing function on $[a, b]$ and if $\epsilon > 0$, then there exists a partition

$$P = \{a = t_0 < t_1 < \cdots < t_n = b\}$$

such that

$$g(t_k^-) - g(t_{k-1}^+) < \epsilon \quad for \quad k = 1, 2, \ldots, n. \tag{1}$$

Proof
First we show that there exists a partition

$$Q = \{a = s_0 < s_1 < \cdots < s_m = b\}$$

such that

$$g(u^+) - g(u^-) < \epsilon \quad \text{for} \quad u \notin Q. \tag{2}$$

It suffices to show that

$$S = \{s \in (a, b) : g(s^+) - g(s^-) \geq \epsilon\}$$

is finite. Select r in $\mathbb{N}$ so that $r\epsilon > g(b) - g(a)$. If S has more than $r - 1$ elements, we can select

$$a < t_1 < t_2 < \cdots < t_r < b$$

so that $g(t_k^+) - g(t_k^-) \geq \epsilon$ for $k = 1, 2, \ldots, r$. But this implies

$$g(b) - g(a) \geq g(t_r^+) - g(t_1^-) \geq \sum_{k=1}^{r}[g(t_k^+) - g(t_k^-)] \geq r\epsilon > g(b) - g(a).$$

So S is finite and Q can be selected satisfying (2).

Next we show there exists $\delta > 0$ such that

$$u, v \in [s_{j-1}, s_j], \quad u < v, \quad v - u < \delta \quad \text{imply} \quad g(v^-) - g(u^+) < \epsilon. \tag{3}$$

If (3) fails, then for some j there exist sequences (u_n) and (v_n) in $[s_{j-1}, s_j]$ where $u_n < v_n$, $v_n - u_n < \frac{1}{n}$ and $g(v_n^-) - g(u_n^+) \geq \epsilon$. Passing to subsequences, we may suppose that (u_n) and (v_n) are monotonic by Theorem 11.3. Let $u = \lim u_n = \lim v_n$. To obtain a contradiction, we consider four cases.

If $u_n \uparrow u$ and $v_n \uparrow u$, then by Lemma 35.14 we have $g(u_n^-) \to g(u^-)$ and $g(v_n^-) \to g(u^-)$; therefore $[g(v_n^-) - g(u_n^-)] \to 0$. Since

$$g(v_n^-) - g(u_n^-) \geq g(v_n^-) - g(u_n^+) \geq \epsilon$$

for all n, we have a contradiction.

If $u_n \downarrow u$ and $v_n \downarrow u$, then Lemma 35.14 shows that $g(u_n^+) \to g(u^+)$ and $g(v_n^+) \to g(u^+)$, so $[g(v_n^+) - g(u_n^+)] \to 0$. On the other hand, for each n we have

$$g(v_n^+) - g(u_n^+) \geq g(v_n^-) - g(u_n^+) \geq \epsilon,$$

a contradiction.

The case $u_n \downarrow u$ and $v_n \uparrow u$ is impossible since this would imply $u \leq u_n < v_n \leq u$.

Finally, suppose $u_n \uparrow u$ and $v_n \downarrow u$. Then $s_{j-1} < u < s_j$; otherwise (u_n) or (v_n) would be a constant sequence and we could appeal to an earlier case. This time Lemma 35.14 shows that $g(u_n^-) \to g(u^-)$ and $g(v_n^+) \to g(u^+)$, and hence

$$g(u^+) - g(u^-) = \lim_{n \to \infty} [g(v_n^+) - g(u_n^-)] \geq \lim \inf[g(v_n^-) - g(u_n^+)] \geq \epsilon.$$

Since $u \notin Q$, this contradicts (2). We have proved (3).

By adding points to the partition Q we can obtain a partition $P = \{a = t_0 < t_1 < \cdots < t_n = b\}$ such that $P \supseteq Q$ and such that $t_k - t_{k-1} < \delta$ for all $k = 1, 2, \ldots, n$. If k is in $\{1, 2, \ldots, n\}$, both t_{k-1} and t_k belong to some $[s_{j-1}, s_j]$, so by (3) we have $g(t_k^-) - g(t_{k-1}^+) < \epsilon$. ■

35.16 Theorem.
Every monotonic function f on $[a, b]$ is F-integrable.

Proof
We may assume f is increasing. Since $f(a) \leq f(x) \leq f(b)$ for all x in $[a, b]$, f is bounded on $[a, b]$. For $\epsilon > 0$ we apply Lemma 35.15 to obtain

$$P = \{a = t_0 < t_1 < \cdots < t_n = b\}$$

where

$$f(t_k^-) - f(t_{k-1}^+) < \frac{\epsilon}{F(b) - F(a)}$$

for $k = 1, 2, \ldots, n$. Since

$$M(f, (t_{k-1}, t_k)) = f(t_k^-) \quad \text{and} \quad m(f, (t_{k-1}, t_k)) = f(t_{k-1}^+),$$

we have

$$U_F(f, P) - L_F(f, P) = \sum_{k=1}^{n} [f(t_k^-) - f(t_{k-1}^+)] \cdot [F(t_k^-) - F(t_{k-1}^+)]$$

$$< \sum_{k=1}^{n} \frac{\epsilon}{F(b) - F(a)} [F(t_k^-) - F(t_{k-1}^+)] \le \epsilon.$$

Since ϵ is arbitrary, Theorem 35.6 shows that f is F-integrable. ∎

35.17 Theorem.
If f is piecewise continuous or bounded piecewise monotonic on $[a, b]$, then f is F-integrable.

Proof
Just as in the proof of Theorem 33.8, this follows from Theorems 35.7, 35.16 and 35.11, provided we have the following generalization of Exercise 32.7. ∎

35.18 Proposition.
If f is F-integrable on $[a, b]$ and $g(x) = f(x)$ except for finitely many points, then g is F-integrable. Note that we do *not* claim $\int_a^b f \, dF = \int_a^b g \, dF$.

Proof
An induction argument shows that we may assume $g(x) = f(x)$ except for one value $x = u$ in $[a, b]$. For $\epsilon > 0$, Theorem 35.6 shows that

$$U_F(f, P) - L_F(f, P) < \epsilon \tag{1}$$

for some partition $P = \{a = t_0 < t_1 < \cdots < t_n = b\}$. In view of Lemma 35.3, we can add u to P without invalidating (1). Then $u = t_\ell$ for some ℓ in $\{0, 1, 2, \ldots, n\}$. The upper sums for f and g are identical except for the $k = \ell$ term in J_F, so

$$U_F(f, P) - U_F(g, P) = f(u)[F(u^+) - F(u^-)] - g(u)[F(u^+) - F(u^-)].$$

The same remark applies to the lower sums, so

$$L_F(f, P) - L_F(g, P) = U_F(f, P) - U_F(g, P).$$

Therefore

$$U_F(g, P) - L_F(g, P) = U_F(f, P) - L_F(f, P) < \epsilon,$$

so Theorem 35.6 shows that g is F-integrable. ∎

If F_1 and F_2 are increasing functions with continuous derivatives, then Theorem 35.13 allows the formula on integration by parts [Theorem 34.2] to be recast as

$$\int_a^b F_1 \, dF_2 + \int_a^b F_2 \, dF_1 = F_1(b)F_2(b) - F_1(a)F_2(a).$$

There is no hope to prove this in general because if $F(t) = 0$ for $t < 0$ and $F(t) = 1$ for $t \geq 0$, then

$$\int_{-1}^1 F \, dF + \int_{-1}^1 F \, dF = 2 \neq 1 = F(1)F(1) - F(-1)F(-1).$$

The generalization does hold provided the functions in the integrands take the middle values at each of their jumps, as we next prove. The result is a special case of a theorem given by Edwin Hewitt [21].

35.19 Theorem [Integration by Parts].
Suppose that F_1 and F_2 are increasing functions on $[a, b]$ and define

$$F_1^*(t) = \frac{1}{2}[F_1(t^-) + F_1(t^+)] \quad and \quad F_2^*(t) = \frac{1}{2}[F_2(t^-) + F_2(t^+)]$$

for all $t \in [a, b]$. Then

$$\int_a^b F_1^* \, dF_2 + \int_a^b F_2^* \, dF_1 = F_1(b)F_2(b) - F_1(a)F_2(a). \qquad (1)$$

As usual, we decree $F_1(b^+) = F_1(b)$, $F_1(a^-) = F_1(a)$, etc.

Proof

Both integrals in (1) exist in view of Theorem 35.16. For an $\epsilon > 0$, there exists a partition

$$P = \{a = t_0 < t_1 < \cdots < t_n = b\}$$

such that

$$U_{F_1}(F_2^*, P) - L_{F_1}(F_2^*, P) < \epsilon.$$

Some algebraic manipulation [discussed in the next paragraph] shows that

$$U_{F_2}(F_1^*, P) + L_{F_1}(F_2^*, P) = F_1(b)F_2(b) - F_1(a)F_2(a), \qquad (2)$$

so that also

$$U_{F_1}(F_2^*, P) + L_{F_2}(F_1^*, P) = F_1(b)F_2(b) - F_1(a)F_2(a). \qquad (3)$$

It follows from (2) that

$$\int_a^b F_1^* \, dF_2 + \int_a^b F_2^* \, dF_1 \leq U_{F_2}(F_1^*, P) + U_{F_1}(F_2^*, P)$$

$$< U_{F_2}(F_1^*, P) + L_{F_1}(F_2^*, P) + \epsilon$$

$$= F_1(b)F_2(b) - F_1(a)F_2(a) + \epsilon,$$

while (3) leads to

$$\int_a^b F_1^* \, dF_2 + \int_a^b F_2^* \, dF_1 > F_1(b)F_2(b) - F_1(a)F_2(a) - \epsilon.$$

Since ϵ is arbitrary, (1) holds.

To check (2), observe

$$U_{F_2}(F_1^*, P) + L_{F_1}(F_2^*, P) = \sum_{k=0}^n F_1^*(t_k) \cdot [F_2(t_k^+) - F_2(t_k^-)]$$

$$+ \sum_{k=1}^n M(F_1^*, (t_{k-1}, t_k)) \cdot [F_2(t_k^-) - F_2(t_{k-1}^+)]$$

$$+ \sum_{k=0}^n F_2^*(t_k) \cdot [F_1(t_k^+) - F_1(t_k^-)]$$

$$+ \sum_{k=1}^n m(F_2^*, (t_{k-1}, t_k)) \cdot [F_1(t_k^-) - F_1(t_{k-1}^+)]$$

$$= \sum_{k=0}^{n} \frac{1}{2}[F_1(t_k^-) + F_1(t_k^+)] \cdot [F_2(t_k^+) - F_2(t_k^-)]$$

$$+ \sum_{k=1}^{n} F_1(t_k^-) \cdot [F_2(t_k^-) - F(t_{k-1}^+)]$$

$$+ \sum_{k=0}^{n} \frac{1}{2}[F_2(t_k^-) + F_2(t_k^+)] \cdot [F_1(t_k^+) - F_1(t_k^-)]$$

$$+ \sum_{k=1}^{n} F_2(t_{k-1}^+) \cdot [F_1(t_k^-) - F_1(t_{k-1}^+)].$$

The first and third sums add to

$$\sum_{k=0}^{n} [F_1(t_k^+)F_2(t_k^+) - F_1(t_k^-)F_2(t_k^-)], \tag{4}$$

while the second and fourth sums add to

$$\sum_{k=1}^{n} [F_1(t_k^-)F_2(t_k^-) - F_1(t_{k-1}^+)F_2(t_{k-1}^+)]. \tag{5}$$

Since the sums in (4) and (5) add to $F_1(b)F_2(b) - F_1(a)F_2(a)$, equality (2) holds. Of course, this algebra simplifies considerably if F_1 and F_2 are continuous. ∎

We next compare our approach to Riemann-Stieltjes integration to the usual approach. For a bounded function f on $[a, b]$, the usual Darboux-Stieltjes integral is defined via the upper sums

$$\tilde{U}_F(f, P) = \sum_{k=1}^{n} M(f, [t_{k-1}, t_k]) \cdot [F(t_k) - F(t_{k-1})]$$

and the lower sums

$$\tilde{L}_F(f, P) = \sum_{k=1}^{n} m(f, [t_{k-1}, t_k]) \cdot [F(t_k) - F(t_{k-1})].$$

The expressions $\tilde{U}_F(f)$, $\tilde{L}_F(f)$ and $\tilde{\int_a^b} f \, dF$ are defined in analogy to those in Definition 35.2. The usual Riemann-Stieltjes integral is defined via the sums

$$\tilde{S}_F(f, P) = \sum_{k=1}^{n} f(x_k)[F(t_k) - F(t_{k-1})],$$

where $x_k \in [t_{k-1}, t_k]$, and the mesh defined in Definition 32.6; compare Definition 35.24.

The usual Riemann-Stieltjes integrability criterion implies the usual Darboux-Stieltjes integrability criterion; these criteria are not equivalent in general, but they are equivalent if F is continuous. See, for example, [33], §12.2; [34], Chapter 8; or [36], Chapter 6, the most complete treatment being in [34].

35.20 Theorem.
If f is Darboux-Stieltjes integrable on $[a, b]$ with respect to F in the usual sense, then f is F-integrable and the integrals agree.

Proof
For any partition P, $\tilde{L}_F(f, P)$ equals

$$\sum_{k=1}^{n} m(f, [t_{k-1}, t_k]) \cdot [F(t_k) - F(t_k^-) + F(t_k^-) - F(t_{k-1}^+)$$

$$+ F(t_{k-1}^+) - F(t_{k-1})]$$

$$\leq \sum_{k=1}^{n} f(t_k)[F(t_k) - F(t_k^-)]$$

$$+ \sum_{k=1}^{n} m(f, (t_{k-1}, t_k)) \cdot [F(t_k^-) - F(t_{k-1}^+)]$$

$$+ \sum_{k=1}^{n} f(t_{k-1})[F(t_{k-1}^+) - F(t_{k-1})].$$

The first and third sums add to

$$\sum_{k=1}^{n} f(t_k)[F(t_k) - F(t_k^-)] + \sum_{k=0}^{n-1} f(t_k)[F(t_k^+) - F(t_k)]$$

$$= f(t_n)[F(t_n) - F(t_n^-)] + \sum_{k=1}^{n-1} f(t_k)[F(t_k^+) - F(t_k^-)]$$

$$+ f(t_0)[F(t_0^+) - F(t_0)]$$

$$= \sum_{k=0}^{n} f(t_k)[F(t_k^+) - F(t_k^-)] = J_F(f, P).$$

These observations and a glance at the definition of $L_F(f, P)$ now show that $\tilde{L}_F(f, P) \leq L_F(f, P)$. Likewise we have $\tilde{U}_F(f, P) \geq U_F(f, P)$, so

$$U_F(f, P) - L_F(f, P) \leq \tilde{U}_F(f, P) - \tilde{L}_F(f, P). \qquad (1)$$

If $\epsilon > 0$, the usual theory shows that there exists a partition P such that $\tilde{U}_F(f, P) - \tilde{L}_F(f, P) < \epsilon$. By (1) we see that we also have $U_F(f, P) - L_F(f, P) < \epsilon$, so f is F-integrable by Theorem 35.6.

To see equality of the integrals, simply observe that

$$\int_a^{\tilde{b}} f \, dF \leq \tilde{U}_F(f, P) < \tilde{L}_F(f, P) + \epsilon \leq L_F(f, P) + \epsilon \leq \int_a^b f \, dF + \epsilon$$

and similarly

$$\int_a^{\tilde{b}} f \, dF > \int_a^b f \, dF - \epsilon. \qquad \blacksquare$$

We will define Riemann-Stieltjes integrals using a mesh defined in terms of F instead of the usual mesh in Definition 32.6.

35.21 Definition.
The *F-mesh* of a partition P is

$$F\text{-mesh}(P) = \max\{F(t_k^-) - F(t_{k-1}^+) : k = 1, 2, \ldots, n\}.$$

It is convenient to restate Lemma 35.15 for F:

35.22 Lemma.
If $\delta > 0$, there exists a partition P such that F-mesh$(P) < \delta$.

35.23 Theorem.
A bounded function f on $[a, b]$ is F-integrable if and only if for each $\epsilon > 0$ there exists $\delta > 0$ such that

$$F\text{-mesh}(P) < \delta \quad implies \quad U_F(f, P) - L_F(f, P) < \epsilon \qquad (1)$$

for all partitions P of $[a, b]$.

Proof
Suppose that the ϵ-δ condition stated in the theorem holds. If we have $\epsilon > 0$, then (1) applies to some partition P by Lemma 35.22

and hence $U_F(f, P) - L_F(f, P) < \epsilon$. Since this remark applies to all $\epsilon > 0$, Theorem 35.6 implies that f is F-integrable.

The converse is proved just as in Theorem 32.7 with "mesh" replaced by "F-mesh" and references to Lemma 32.2 replaced by references to Lemma 35.3. ∎

35.24 Definition.
Let f be bounded on $[a, b]$, and let

$$P = \{a = t_0 < t_1 < \cdots < t_n = b\}.$$

A *Riemann-Stieltjes sum* of f associated with P and F is a sum of the form

$$J_F(f, P) + \sum_{k=1}^{n} f(x_k)[F(t_k^-) - F(t_{k-1}^+)]$$

where $x_k \in (t_{k-1}, t_k)$ for $k = 1, 2, \ldots, n$.

The function f is *Riemann-Stieltjes integrable on $[a, b]$* if there exists r in $\mathbb{R}$ with the following property. For each $\epsilon > 0$ there exists $\delta > 0$ such that

$$|S - r| < \epsilon \tag{1}$$

for every Riemann-Stieltjes sum S of f associated with a partition P having F-mesh$(P) < \delta$. We call r the *Riemann-Stieltjes integral* of f and temporarily write it as

$$\mathcal{RS} \int_a^b f \, dF.$$

35.25 Theorem.
A bounded function f on $[a, b]$ is F-integrable if and only if it is Riemann-Stieltjes integrable, in which case the integrals are equal.

Proof
The proof that F-integrability implies Riemann-Stieltjes integrability imitates the corresponding proof in Theorem 32.9. The proof of the converse also imitates the corresponding proof, but a little care is needed, so we give it.

Let f be a Riemann-Stieltjes integrable function, and let r be as in Definition 35.24. Consider $\epsilon > 0$, and let $\delta > 0$ be as provided in Definition 35.24. By Lemma 35.22 there exists a partition $P =$ $\{a = t_0 < t_1 < \cdots < t_n = b\}$ with F-mesh$(P) < \delta$. For each $k = 1, 2, \ldots, n$, select x_k in (t_{k-1}, t_k) so that $f(x_k) < m(f, (t_{k-1}, t_k)) + \epsilon$. The Riemann-Stieltjes sum S for this choice of x_k's satisfies

$$S \leq L_F(f, P) + \epsilon[F(b) - F(a)]$$

and also

$$|S - r| < \epsilon;$$

hence $L_F(f) \geq L_F(f, P) > r - \epsilon - \epsilon[F(b) - F(a)]$. It follows that $L_F(f) \geq r$ and similarly $U_F(f) \leq r$. Therefore $L_F(f) = U_F(f) = r$. Thus f is F-integrable and

$$\int_a^b f\,dF = r = \mathcal{RS}\int_a^b f\,dF.$$

■

Exercises

35.1. Let F be an increasing function on $[a, b]$.

    (a) Show that $\lim_{x \to t^-} F(x)$ exists for t in $(a, b]$ and is equal to $\sup\{F(x) : x \in (a, t)\}$.

    (b) Show that $\lim_{x \to t^+} F(x)$ exists for t in $[a, b)$ and is equal to $\inf\{F(x) : x \in (t, b)\}$.

35.2. Calculate $\int_0^3 x^2\,dF(x)$ for the function F in Example 4.

35.3. Let F be the step-function such that $F(t) = n$ for $t \in [n, n+1)$, n an integer. Calculate

    (a) $\int_0^6 x\,dF(x)$,   **(b)** $\int_0^3 x^2\,dF(x)$,

    (c) $\int_{1/4}^{7/4} x^2\,dF(x)$.

35.4. Let $F(t) = \sin t$ for $t \in [-\frac{\pi}{2}, \frac{\pi}{2}]$. Calculate

    (a) $\int_0^{\pi/2} x\,dF(x)$ **(b)** $\int_{-\pi/2}^{\pi/2} x\,dF(x)$.

35.5. Let $f(x) = 1$ for rational x and $f(x) = 0$ for irrational x.

    (a) Show that if F is continuous on $[a, b]$ and $F(a) < F(b)$, then f is not F-integrable on $[a, b]$.

    (b) Observe that f is F-integrable if F is as in Example 1 or 3.

35.6. Let (f_n) be a sequence of F-integrable functions on $[a, b]$, and suppose $f_n \to f$ uniformly on $[a, b]$. Show that f is F-integrable and

$$\int_a^b f \, dF = \lim_{n \to \infty} \int_a^b f_n \, dF.$$

35.7. Let f and g be F-integrable functions on $[a, b]$. Show that

(a) f^2 is F-integrable.

(b) fg is F-integrable.

(c) $\max(f, g)$ and $\min(f, g)$ are F-integrable.

35.8. Let g be continuous on $[a, b]$ where $g(x) \geq 0$ for all $x \in [a, b]$ and define $F(t) = \int_a^t g(x) \, dx$ for $t \in [a, b]$. Show that if f is continuous, then

$$\int_a^b f \, dF = \int_a^b f(x) g(x) \, dx.$$

35.9. Let f be continuous on $[a, b]$.

(a) Show that $\int_a^b f \, dF = f(x)[F(b) - F(a)]$ for some x in $[a, b]$.

(b) Show that Exercise 33.14 is a special case of part (a).

35.10. Suppose F is differentiable on $[a, b]$ and F' is Riemann integrable on $[a, b]$. Prove that a bounded function f on $[a, b]$ is F-integrable if and only if $f F'$ is Riemann integrable, in which case

$$\int_a^b f \, dF = \int_a^b f(x) F'(x) \, dx.$$

Note: The proof is difficult and delicate. A solution is available from the author.

35.11. Here is a "change of variable" formula. Let f be F-integrable on $[a, b]$. Let ϕ be a continuous, strictly increasing function on an interval $[c, d]$ such that $\phi(c) = a$ and $\phi(d) = b$. Define

$$g(u) = f(\phi(u)) \quad \text{and} \quad G(u) = F(\phi(u)) \quad \text{for} \quad u \in [c, d].$$

Show that g is G-integrable and $\int_c^d g \, dG = \int_a^b f \, dF$.

§36 * Improper Integrals

The Riemann integral in §32 has been defined only for functions
that are bounded on a closed interval $[a, b]$. It is convenient to be
able to integrate some functions that are unbounded or are defined
on an unbounded interval.

36.1 Definition.
Consider an interval $[a, b)$ where b is finite or $+\infty$. Suppose that f
is a function on $[a, b)$ that is integrable on each $[a, d]$ for $a < d < b$,
and suppose that the limit

$$\lim_{d \to b^-} \int_a^d f(x)\, dx$$

exists either as a finite number, $+\infty$ or $-\infty$. Then we define

$$\int_a^b f(x)\, dx = \lim_{d \to b^-} \int_a^d f(x)\, dx. \tag{1}$$

If b is finite and f is integrable on $[a, b]$, this definition agrees with
that in Definition 32.1 [Exercise 36.1]. If $b = +\infty$ or if f is not integrable on $[a, b]$, but the limit in (1) exists, then (1) defines an *improper
integral*.

An analogous definition applies if f is defined on $(a, b]$ where a
is finite or $-\infty$ and if f is integrable on each $[c, b]$ for $a < c < b$.
Then we define

$$\int_a^b f(x)\, dx = \lim_{c \to a^+} \int_c^b f(x)\, dx \tag{2}$$

whenever the limit exists.

If f is defined on (a, b) and integrable on all closed subintervals
$[c, d]$, then we fix α in (a, b) and define

$$\int_a^b f(x)\, dx = \int_a^\alpha f(x)\, dx + \int_\alpha^b f(x)\, dx \tag{3}$$

provided the integrals on the right exist and the sum is not of the
form $+\infty + (-\infty)$. Here we agree that $\infty + L = \infty$ if $L \neq -\infty$ and
$(-\infty) + L = -\infty$ if $L \neq \infty$. It is easy [Exercise 36.2] to see that this
definition does not depend on the choice of α.

Whenever the improper integrals defined above exist and are finite, the integrals are said to *converge*. Otherwise they *diverge* to $+\infty$ or to $-\infty$.

Example 1
Consider $f(x) = \frac{1}{x}$ for $x \in (0, \infty)$. For $d > 1$, we have $\int_1^d \frac{1}{x} \, dx = \log d$, so

$$\int_1^\infty \frac{1}{x} \, dx = \lim_{d \to \infty} \log d = +\infty.$$

This improper integral diverges to $+\infty$. For $0 < c < 1$, we have $\int_c^1 \frac{1}{x} \, dx = -\log c$, so

$$\int_0^1 \frac{1}{x} \, dx = \lim_{c \to 0^+} [-\log c] = +\infty.$$

Also we have

$$\int_0^\infty \frac{1}{x} = +\infty.$$

Example 2
Consider $f(x) = x^{-p}$ for $x \in [1, \infty)$ and a fixed positive number $p \neq 1$. For $d > 1$,

$$\int_1^d x^{-p} \, dx = \frac{1}{1-p}[d^{1-p} - 1].$$

It follows that

$$\int_1^\infty x^{-p} \, dx = \frac{1}{1-p}[0 - 1] = \frac{1}{p-1} \quad \text{if } p > 1$$

and

$$\int_1^\infty x^{-p} \, dx = +\infty \quad \text{if } 0 < p < 1.$$

Example 3
We have $\int_0^d \sin x \, dx = 1 - \cos d$ for all d. The value $(1 - \cos d)$ oscillates between 0 and 2, as $d \to \infty$, and therefore the limit

$$\lim_{d \to \infty} \int_0^d \sin x \, dx \quad \textit{does not exist.}$$

Thus the symbol $\int_0^\infty \sin x \, dx$ has no meaning and is not an improper integral. Similarly, $\int_{-\infty}^0 \sin x \, dx$ and $\int_{-\infty}^\infty \sin x \, dx$ have no meaning.
Note that the limit

$$\lim_{a \to \infty} \int_{-a}^a \sin x \, dx$$

clearly exists and equals 0. When such a "symmetric" limit exists even though the improper integral $\int_{-\infty}^\infty$ does not, we have what is called a *Cauchy principal value* of $\int_{-\infty}^\infty$. Thus 0 is the Cauchy principal value of $\int_{-\infty}^\infty \sin x \, dx$, but this is not an improper integral.

It is especially valuable to extend Riemann-Stieltjes integrals to infinite intervals; see the discussion after Theorem 36.4 below. Let F be a bounded increasing function on some interval I. The function F can be extended to all of $\mathbb{R}$ by a simple device: if I is bounded below, define

$$F(t) = \inf\{F(u) : u \in I\} \quad \text{for} \quad t \le \inf I;$$

if I is bounded above, define

$$F(t) = \sup\{F(u) : u \in I\} \quad \text{for} \quad t \ge \sup I.$$

For this reason, we will henceforth assume that F is a bounded increasing function on all of $\mathbb{R}$. We will use the notations

$$F(-\infty) = \lim_{t \to -\infty} F(t) \quad \text{and} \quad F(\infty) = \lim_{t \to \infty} F(t).$$

Improper Riemann-Stieltjes integrals are defined in analogy to improper Riemann integrals.

36.2 Definition.
Suppose that f is F-integrable on each interval $[a, b]$ in $\mathbb{R}$. We make the following definitions whenever the limits exist:

$$\int_0^\infty f \, dF = \lim_{b \to \infty} \int_0^b f \, dF; \qquad \int_{-\infty}^0 f \, dF = \lim_{a \to -\infty} \int_a^0 f \, dF.$$

If both limits exist and their sum does not have the form $\infty + (-\infty)$, we define

$$\int_{-\infty}^\infty f \, dF = \int_{-\infty}^0 f \, dF + \int_0^\infty f \, dF.$$

If this sum is finite, we say f is F-integrable on $\mathbb{R}$. If f is F-integrable on $\mathbb{R}$ for $F(t) = t$ [i.e., the integrals are Riemann integrals], we say f is *integrable* on $\mathbb{R}$.

36.3 Theorem.
If f is F-integrable on each interval $[a, b]$ and if $f(x) \geq 0$ for all $x \in \mathbb{R}$, then f is F-integrable on $\mathbb{R}$ or else $\int_{-\infty}^{\infty} f \, dF = +\infty$.

Proof
We indicate why $\lim_{a \to -\infty} \int_a^0 f \, dF$ exists, and leave the case of $\lim_{b \to \infty} \int_0^b f \, dF$ to the reader. Let $h(a) = \int_a^0 f \, dF$ for $a < 0$, and note that $a' < a < 0$ implies $h(a') \geq h(a)$. This property implies that $\lim_{a \to -\infty} h(a)$ exists and

$$\lim_{a \to -\infty} h(a) = \sup\{h(a) : a \in (-\infty, 0)\}.$$

We omit the simple argument. ∎

36.4 Theorem.
Suppose that $-\infty < F(-\infty) < F(\infty) < \infty$. Let f be a bounded function on $\mathbb{R}$ that is F-integrable on each interval $[a, b]$. Then f is F-integrable on $\mathbb{R}$.

Proof
Select a constant B such that $|f(x)| \leq B$ for all $x \in \mathbb{R}$. Since we have $F(\infty) - F(-\infty) < \infty$, constant functions are F-integrable. Since $0 \leq f + B \leq 2B$, Theorem 36.3 shows that $f + B$ is F-integrable. It follows [Exercise 36.10] that $f = (f + B) + (-B)$ is also F-integrable. ∎

Increasing functions F defined on $\mathbb{R}$ come up naturally in probability and statistics. In these disciplines, F is called a *distribution function* if we also have $F(-\infty) = 0$ and $F(\infty) = 1$. Of course, the function $F(t) = t$ that corresponds to the Riemann integral is not a distribution function. Here is how a distribution function comes up in probability. Consider a random experiment with numerical outcomes; then $F(t)$ can represent the probability that the numerical value will be $\leq t$. As a very simple example, suppose the experiment involves tossing three fair coins and counting the number of heads. The numerical values 0, 1, 2, and 3 will result with probabilities $\frac{1}{8}$,

$\frac{3}{8}, \frac{3}{8},$ and $\frac{1}{8},$ respectively. The corresponding distribution function is defined in Example 2 of §35 and sketched in Figure 35.1.

Frequently a distribution function F has the form

$$F(t) = \int_{-\infty}^{t} g(x)\,dx$$

for an integrable function g satisfying $g(x) \geq 0$ for all $x \in \mathbb{R}$. Then g is called a *density* for F. Note that we must have

$$\int_{-\infty}^{\infty} g(x)\,dx = 1.$$

Also, if g is continuous, then $g(t) = F'(t)$ for all t by Theorem 34.3.

Example 4

It turns out that $\int_{-\infty}^{\infty} e^{-x^2}\,dx = \sqrt{\pi}$ [Exercise 36.7] and hence

$$\int_{-\infty}^{\infty} e^{-x^2/2}\,dx = \sqrt{2\pi}.$$

The most important density in probability is the *normal density*

$$g(x) = \frac{1}{\sqrt{2\pi}} e^{-x^2/2}$$

which gives rise to the *normal distribution*

$$F(t) = \frac{1}{\sqrt{2\pi}} \int_{-\infty}^{t} e^{-x^2/2}\,dx;$$

see Figure 36.1.

Exercises 36.1–36.8 below deal only with Riemann integrals.

Exercises

36.1. Show that if f is integrable on $[a, b]$ as in Definition 32.1, then

$$\lim_{d \to b^-} \int_{a}^{d} f(x)\,dx = \int_{a}^{b} f(x)\,dx.$$

36.2. Show that the definition (3) in Definition 36.1 does not depend on the choice of α.

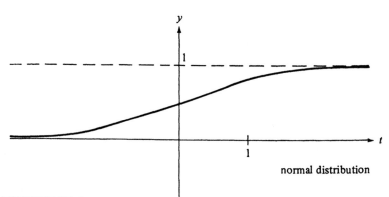

normal distribution

FIGURE 36.1

36.3. (a) Show that

$$\int_0^1 x^{-p}\,dx = \frac{1}{1-p} \quad \text{if } 0 < p < 1 \quad \text{and} \quad \int_0^1 x^{-p}\,dx = +\infty \quad \text{if } p > 1.$$

(b) Show that $\int_0^\infty x^{-p}\,dx = +\infty$ for all $p > 0$.

36.4. Calculate

    (a) $\int_0^1 \log x\,dx$,   **(b)** $\int_2^\infty \frac{\log x}{x}\,dx$,

    (c) $\int_0^\infty \frac{1}{1+x^2}\,dx$.

36.5. Let f be a continuous function on (a, b) such that $f(x) \geq 0$ for all $x \in (a, b)$; a can be $-\infty$, b can be $+\infty$. Show that the improper integral $\int_a^b f(x)\,dx$ exists and equals

$$\sup\left\{\int_c^d f(x)\,dx : [c, d] \subseteq (a, b)\right\}.$$

36.6. Prove the following *comparison tests*. Let f and g be continuous functions on (a, b) such that $0 \leq f(x) \leq g(x)$ for all x in (a, b); a can be $-\infty$, b can be $+\infty$.

    (a) If $\int_a^b g(x)\,dx < \infty$, then $\int_a^b f(x)\,dx < \infty$.

    (b) If $\int_a^b f(x)\,dx = +\infty$, then $\int_a^b g(x)\,dx = +\infty$.

36.7. (a) Use Exercise 36.6 to show $\int_{-\infty}^\infty e^{-x^2}\,dx < \infty$.

    (b) Show that this integral equals $\sqrt{\pi}$. *Hint*: Calculate the double integral $\int_{-\infty}^\infty \int_{-\infty}^\infty e^{-x^2} e^{-y^2}\,dx\,dy$ using polar coordinates.

36.8. Suppose that f is continuous on (a, b) and that $\int_a^b |f(x)| \, dx < \infty$; again a can be $-\infty$, b can be $+\infty$. Show that the integral $\int_a^b f(x) \, dx$ exists and is finite.

36.9. Let F be the normal distribution function in Example 4.

(a) Show that if f is continuous on $\mathbb{R}$ and if the improper integral $\int_{-\infty}^\infty f(x) e^{-x^2/2} \, dx$ exists, then the improper integral $\int_{-\infty}^\infty f \, dF$ exists and

$$\int_{-\infty}^\infty f \, dF = \frac{1}{\sqrt{2\pi}} \int_{-\infty}^\infty f(x) e^{-x^2/2} \, dx.$$

Calculate

(b) $\int_{-\infty}^\infty x^2 \, dF(x)$, **(c)** $\int_{-\infty}^\infty e^{x^2} \, dF(x)$,

(d) $\int_{-\infty}^\infty |x| \, dF(x)$, **(e)** $\int_{-\infty}^\infty x \, dF(x)$.

36.10. Let f and g be F-integrable functions on $\mathbb{R}$. Show that $f + g$ is F-integrable on $\mathbb{R}$ and

$$\int_{-\infty}^\infty (f + g) \, dF = \int_{-\infty}^\infty f \, dF + \int_{-\infty}^\infty g \, dF.$$

36.11. Show that if f and g are F-integrable on $\mathbb{R}$ and if $f(x) \leq g(x)$ for x in $\mathbb{R}$, then $\int_{-\infty}^\infty f \, dF \leq \int_{-\infty}^\infty g \, dF$.

36.12. Generalize Exercise 36.6 to F-integrals on $\mathbb{R}$.

36.13. Generalize Exercise 36.8 to F-integrals on $\mathbb{R}$.

36.14. Let (u_n) be a sequence of distinct points in $\mathbb{R}$, and let (c_n) be a sequence of positive numbers such that $\sum c_n < \infty$.

(a) Show that $F = \sum_{n=1}^\infty c_n J_{u_n}$ is an increasing function on $\mathbb{R}$. *Hint:* See Example 3 in §35.

(b) Show that every bounded function f on $\mathbb{R}$ is F-integrable and

$$\int_{-\infty}^\infty f \, dF = \sum_{n=1}^\infty c_n f(u_n).$$

(c) Show that if (u_n) is an enumeration of the rationals, then F is strictly increasing on $\mathbb{R}$.

(d) When will F be a distribution function?

36.15. (a) Give an example of a sequence (f_n) of integrable functions on $\mathbb{R}$ where $\int_{-\infty}^\infty f_n(x) \, dx = 1$ for all n and yet $f_n \to 0$ uniformly on $\mathbb{R}$.

(b) Suppose that F is a distribution function on $\mathbb{R}$. Show that if (f_n) is a sequence of F-integrable functions on $\mathbb{R}$ and if $f_n \to f$ uniformly on $\mathbb{R}$, then f is F-integrable on $\mathbb{R}$ and

$$\int_{-\infty}^{\infty} f \, dF = \lim_{n \to \infty} \int_{-\infty}^{\infty} f_n \, dF.$$

§37   * A Discussion of Exponents and Logarithms

In this book we have carefully developed the theory but have been casual about using the familiar exponential, logarithmic and trigonometric functions in examples and exercises. Most readers probably found this an acceptable approach, since they are comfortable with these basic functions. In this section, we indicate three ways to develop the exponential and logarithmic functions assuming only the axioms in Chapter 1 and the theoretical results in later chapters. We will provide proofs for the third approach.

Recall that for x in $\mathbb{R}$ and a positive integer n, x^n is the product of x by itself n times. For $x \neq 0$, we have the convention $x^0 = 1$. And for $x \neq 0$ and negative integers $-n$ where $n \in \mathbb{N}$, we define x^{-n} to be the reciprocal of x^n, i.e., $x^{-n} = (x^n)^{-1}$.

37.1 Piecemeal Approach.
This approach starts with Example 2 in §29 and Exercise 29.15 where it is shown that x^r is meaningful whenever $x > 0$ and r is rational, i.e., $r \in \mathbb{Q}$. Moreover,

$$\text{if} \quad h(x) = x^r, \quad \text{then} \quad h'(x) = rx^{r-1}.$$

The algebraic properties $x^r x^s = x^{r+s}$ and $(xy)^r = x^r y^r$ can be verified for $r, s \in \mathbb{Q}$ and positive x and y. For any $t \in \mathbb{R}$ and $x \geq 1$, we *define*

$$x^t = \sup\{x^r : r \in \mathbb{Q} \text{ and } r \leq t\}$$

and

$$\left(\frac{1}{x}\right)^t = (x^t)^{-1}.$$

This defines x^t for $x > 0$. It can be shown that with this definition x^t is finite and the algebraic properties mentioned above still hold. Further, it can be shown that $h(x) = x^t$ is differentiable and $h'(x) = tx^{t-1}$.

Next we can consider a fixed $b > 0$ and the function B defined by $B(x) = b^x$ for $x \in \mathbb{R}$. The function B is differentiable and $B'(x) = c_b B(x)$ for some constant c_b. We elaborate on this last claim. In view of Exercise 28.14, we can write

$$B'(x) = \lim_{h \to 0} \frac{b^{x+h} - b^x}{h} = b^x \cdot \lim_{h \to 0} \frac{b^h - 1}{h}$$

provided these limits exist. Some analysis shows that the last limit does exist, so

$$B'(x) = c_b B(x) \quad \text{where} \quad c_b = \lim_{h \to 0} \frac{1}{h}[b^h - 1].$$

It turns out that $c_b = 1$ for a certain b, known universally as e. Since B is one-to-one if $b \neq 1$, B has an inverse function L which is named $L(y) = \log_b y$. Since B is differentiable, Theorem 29.9 can be applied to show that L is differentiable and

$$L'(y) = \frac{1}{c_b y}.$$

Finally, the familiar properties of $\log_b$ can be established for L.

When all the details are supplied, the above approach is very tedious. It has one, and only one, merit; it is direct without any tricks. One could call it the "brute force approach." The next two approaches begin with some well defined mathematical object [either a power series or an integral] and then work backwards to develop the familiar properties of exponentials and logarithms. In both instances, for motivation we will draw on more advanced facts that we believe but which have *not* been established in this book.

37.2 Exponential Power Series Approach.
This approach is adopted in two of our favorite books: [8], §4.9 and [36], Chapter 8. As noted in Example 1 of §31, we believe

$$e^x = \sum_{k=0}^{\infty} \frac{1}{k!} x^k$$

though we have *not* proved this, since we have not even defined exponentials yet. In this approach, we *define*

$$E(x) = \sum_{k=0}^{\infty} \frac{1}{k!} x^k, \tag{1}$$

and we *define* $e = E(1)$. The series here has radius of convergence $+\infty$ [Example 1, §23], and E is differentiable on $\mathbb{R}$ [Theorem 26.5]. It is easy [Exercise 26.5] to show that $E' = E$. The fundamental property

$$E(x + y) = E(x)E(y) \tag{2}$$

can be established using only the facts observed above. Actually [36] uses a theorem on multiplication of absolutely convergent series, but [8] avoids this. Other properties of E can be quickly established. In particular, E is strictly increasing on $\mathbb{R}$ and has an inverse L. Theorem 29.9 assures us that L is differentiable and $L'(y) = \frac{1}{y}$. For rational r and $x > 0$, x^r was defined in Exercise 29.15. Applying that exercise and the chain rule to $g(x) = L(x^r) - rL(x)$, we find $g'(x) = 0$ for $x > 0$. Since $g(1) = 0$, we conclude that

$$L(x^r) = rL(x) \quad \text{for} \quad r \in \mathbb{Q} \quad \text{and} \quad x > 0. \tag{3}$$

For $b > 0$ and rational r, (3) implies that

$$b^r = E(L(b^r)) = E(rL(b)).$$

Because of this, we *define*

$$b^x = E(xL(b)) \quad \text{for} \quad x \in \mathbb{R}.$$

The familiar properties of exponentials and their inverses [logarithms!] are now easy to prove.

The choice between the approach just outlined and the next approach is really a matter of taste and depends on the appeal of power series. One genuine advantage to the exponential approach is that the series in (1) defining E is equally good for defining $E(z) = e^z$ for complex numbers z.

37.3 Logarithmic Integral Approach.

Let us attempt to solve $f' = f$ where f never vanishes; we expect to obtain $E(x) = e^x$ as one of the solutions. This simple differential equation can be written

$$\frac{f'}{f} = 1. \tag{1}$$

In view of the chain rule, if we could find L satisfying $L'(y) = \frac{1}{y}$, then equation (1) would simplify to

$$(L \circ f)' = 1,$$

so one of the solutions would satisfy

$$L \circ f(y) = y.$$

In other words, one solution f of (1) would be an inverse to L where $L'(y) = \frac{1}{y}$. But by the Fundamental Theorem of Calculus II [Theorem 34.3], we know such a function L exists. Since we also expect $L(1) = 0$, we *define*

$$L(y) = \int_1^y \frac{1}{t}\, dt \quad \text{for} \quad y \in (0, \infty).$$

We use this definition to prove the basic facts about logarithms and exponentials.

37.4 Theorem.
   (i) *The function L is strictly increasing, continuous and differentiable on $(0, \infty)$. We have*

$$L'(y) = \frac{1}{y} \quad \text{for} \quad y \in (0, \infty).$$

   (ii) $L(yz) = L(y) + L(z)$ *for* $y, z \in (0, \infty)$.
   (iii) $L(\frac{y}{z}) = L(y) - L(z)$ *for* $y, z \in (0, \infty)$.
   (iv) $\lim_{y \to \infty} L(y) = +\infty$ *and* $\lim_{y \to 0^+} L(y) = -\infty$.

Proof
It is trivial to show that the function $f(t) = t$ is continuous on $\mathbb{R}$, so its reciprocal $\frac{1}{t}$ is continuous on $(0, \infty)$ by Theorem 17.4. It is easy to see that L is strictly increasing, and the rest of (i) follows immediately from Theorem 34.3.

Assertion (ii) can be proved directly [Exercise 37.1]. Alternatively, fix z and consider $g(y) = L(yz) - L(y) - L(z)$. Since $g(1) = 0$, it suffices to show $g'(y) = 0$ for $y \in (0, \infty)$ [Corollary 29.4]. But since z is fixed, we have

$$g'(y) = \frac{z}{yz} - \frac{1}{y} - 0 = 0.$$

To check (iii), note that $L(\frac{1}{z}) + L(z) = L(\frac{1}{z} \cdot z) = L(1) = 0$, so that $L(\frac{1}{z}) = -L(z)$ and

$$L\left(\frac{y}{z}\right) = L\left(y \cdot \frac{1}{z}\right) = L(y) + L\left(\frac{1}{z}\right) = L(y) - L(z).$$

To see (iv), first observe that $L(2) > 0$ and that $L(2^n) = n \cdot L(2)$ in view of (ii). Thus $\lim_{n \to \infty} L(2^n) = +\infty$. Since L is increasing, it follows that $\lim_{y \to \infty} L(y) = +\infty$. Likewise $L(\frac{1}{2}) < 0$ and $L((\frac{1}{2})^n) = n \cdot L(\frac{1}{2})$, so $\lim_{y \to 0^+} L(y) = -\infty$. ∎

The Intermediate Value Theorem 18.2 shows that L maps $(0, \infty)$ *onto* $\mathbb{R}$. Since L is a strictly increasing function, it has an inverse and the inverse has domain $\mathbb{R}$.

37.5 Definition.
We denote the function inverse to L by E. Thus

$$E(L(y)) = y \quad \text{for} \quad y \in (0, \infty)$$

and

$$L(E(x)) = x \quad \text{for} \quad x \in \mathbb{R}.$$

We also define $e = E(1)$ so that $\int_1^e \frac{1}{t} \, dt = 1$.

37.6 Theorem.
   (i) *The function E is strictly increasing, continuous and differentiable on $\mathbb{R}$. We have*

$$E'(x) = E(x) \quad for \quad x \in \mathbb{R}.$$

   (ii) $E(u + v) = E(u)E(v)$ *for $u, v \in \mathbb{R}$.*
   (iii) $\lim_{x \to \infty} E(x) = +\infty$ *and* $\lim_{x \to -\infty} E(x) = 0$.

Proof

All of (i) follows from Theorem 37.4 in conjunction with Theorem 29.9. In particular,

$$E'(x) = \frac{1}{L'(E(x))} = \frac{1}{\frac{1}{E(x)}} = E(x).$$

If $u, v \in \mathbb{R}$, then $u = L(y)$ and $v = L(z)$ for some $y, z \in (0, \infty)$. Then $u + v = L(yz)$ by (ii) of Theorem 37.4, so

$$E(u + v) = E(L(yz)) = yz = E(L(y))E(L(z)) = E(u)E(v).$$

Assertion (iii) follows from (iv) of Theorem 37.4 [Exercise 37.2]. ∎

Consider $b > 0$ and $r \in \mathbb{Q}$, say $r = \frac{m}{n}$ where $m, n \in \mathbb{Z}$ and $n > 0$. It is customary to write b^r for that positive number a such that $a^n = b^m$. By (ii) of Theorem 37.4, we have $nL(a) = mL(b)$; hence

$$b^r = a = E(L(a)) = E\left(\frac{1}{n} \cdot nL(a)\right) = E\left(\frac{1}{n} \cdot mL(b)\right) = E(rL(b)).$$

This motivates our next definition and also shows that the definition is compatible with the usage of fractional powers in algebra.

37.7 Definition.

For $b > 0$ and $x \in \mathbb{R}$, we define

$$b^x = E(xL(b)).$$

Since $L(e) = 1$, we have $e^x = E(x)$ for all $x \in \mathbb{R}$.

37.8 Theorem.

Fix $b > 0$.

(i) *The function* $B(x) = b^x$ *is continuous and differentiable on* $\mathbb{R}$.

(ii) *If* $b > 1$, *then B is strictly increasing; if* $b < 1$, *then B is strictly decreasing.*

(iii) *If* $b \neq 1$, *then B maps* $\mathbb{R}$ *onto* $(0, \infty)$.

(iv) $b^{u+v} = b^u b^v$ *for* $u, v \in \mathbb{R}$.

Proof

Exercise 37.3. ∎

If $b \neq 1$, the function B has an inverse function.

37.9 Definition.
For $b > 0$ and $b \neq 1$, the inverse of $B(x) = b^x$ is written $\log_b$. The domain of $\log_b$ is $(0, \infty)$ and

$$\log_b y = x \quad \text{if and only if} \quad b^x = y.$$

Note that $\log_e y = L(y)$ for $y > 0$.

37.10 Theorem.
Fix $b > 0$, $b \neq 1$.
  (i) *The function $\log_b$ is continuous and differentiable on $(0, \infty)$.*
  (ii) *If $b > 1$, $\log_b$ is strictly increasing; if $b < 1$, $\log_b$ is strictly decreasing.*
  (iii) $\log_b(yz) = \log_b y + \log_b z$ *for* $y, z \in (0, \infty)$.
  (iv) $\log_b(\frac{y}{z}) = \log_b y - \log_b z$ *for* $y, z \in (0, \infty)$.

Proof
This follows from Theorem 37.4 and the identity $\log_b y = \frac{L(y)}{L(b)}$ [Exercise 37.4]. Note that $L(b)$ is negative if $b < 1$. ∎

The function $E(x) = e^x$ has now been rigorously developed and, as explained in Example 1 of §31, we have

$$E(x) = \sum_{k=0}^{\infty} \frac{1}{k!} x^k.$$

In particular, $e = \sum_{k=0}^{\infty} \frac{1}{k!}$. Also

37.11 Theorem.

$$e = \lim_{h \to 0}(1 + h)^{1/h} = \lim_{n \to \infty}\left(1 + \frac{1}{n}\right)^n.$$

Proof
It suffices to verify the first equality. Since $L'(1) = 1$, we have

$$1 = \lim_{h \to 0} \frac{L(1 + h) - L(1)}{h} = \lim_{h \to 0} \frac{1}{h} L(1 + h) = \lim_{h \to 0} L\left((1 + h)^{1/h}\right).$$

Since E is continuous, we can apply Theorem 20.5 to the function $f(h) = L((1 + h)^{1/h})$ to obtain

$$\lim_{h \to 0}(1 + h)^{1/h} = \lim_{h \to 0} E\left(L\left((1 + h)^{1/h}\right)\right) = E(1) = e.$$

∎

37.12 Trigonometric Functions.

Either approach 37.2 or 37.3 can be modified to rigorously develop the trigonometric functions. They can also be developed using the exponential functions for *complex* values since

$$\sin x = \frac{1}{2i}[e^{ix} - e^{-ix}], \quad \text{etc.};$$

see [36], Chapter 8. The development of the trigonometric functions analogous to approach 37.3 can proceed as follows. Since we believe

$$\arcsin x = \int_0^x (1 - t^2)^{-1/2} \, dt,$$

we can define $A(x)$ as this integral and obtain $\sin x$ from this. Then $\cos x$ and $\tan x$ are easy to obtain. In this development, the number π is defined to be $2 \int_0^1 (1 - t^2)^{-1/2} \, dt$.

In the exercises, use results proved in Theorem 37.4 and subsequent theorems but not the material discussed in 37.1 and 37.2.

Exercises

37.1. Prove directly that

$$\int_1^{yz} \frac{1}{t} \, dt = \int_1^y \frac{1}{t} \, dt + \int_1^z \frac{1}{t} \, dt \quad \text{for} \quad y, z \in (0, \infty).$$

37.2. Prove (iii) of Theorem 37.6.

37.3. Prove Theorem 37.8.

37.4. Consider $b > 0$, $b \neq 1$. Prove $\log_b y = \frac{L(y)}{L(b)}$ for $y \in (0, \infty)$.

37.5. Let p be any real number and define $f(x) = x^p$ for $x > 0$. Show that f is differentiable and $f'(x) = px^{p-1}$; compare Exercise 29.15. *Hint:* $f(x) = E(pL(x))$.

37.6. Show that $x^p y^p = (xy)^p$ for $p \in \mathbb{R}$ and positive x, y.

37.7. **(a)** Show that if $B(x) = b^x$, then $B'(x) = (\log_e b)b^x$.

(b) Find the derivative of $\log_b$.

37.8. For $x > 0$, let $f(x) = x^x$. Show that $f'(x) = [1 + \log_e x] \cdot x^x$.

37.9. **(a)** Show that $\log_e y < y$ for $y > 1$.

(b) Show that

$$\frac{\log_e y}{y} < \frac{2}{\sqrt{y}} \quad \text{for} \quad y > 1. \quad \textit{Hint:} \quad \log_e y = 2\log_e \sqrt{y}.$$

(c) Use part (b) to prove that $\lim_{y \to \infty} \frac{1}{y} \log_e y = 0$. This neat little exercise is based on the paper [20].

Appendix on Set Notation

Consider a set S. The notation $x \in S$ means that x is an element of S; we might also say "x belongs to S" or "x is in S." The notation $x \notin S$ signifies that x is some element but that x does not belong to S. By $T \subseteq S$ we mean that each element of T also belongs to S, i.e., $x \in T$ implies $x \in S$. Thus we have $1 \in \mathbb{N}$, $17 \in \mathbb{N}$, $-3 \notin \mathbb{N}$, $\frac{1}{2} \notin \mathbb{N}$, $\sqrt{2} \notin \mathbb{N}$, $\frac{1}{2} \in \mathbb{Q}$, $\frac{1}{2} \in \mathbb{R}$, $\sqrt{2} \in \mathbb{R}$, $\sqrt{2} \notin \mathbb{Q}$, and $\pi \in \mathbb{R}$. Also we have $\mathbb{N} \subseteq \mathbb{R}$, $\mathbb{Q} \subseteq \mathbb{R}$, and $\mathbb{R} \subseteq \mathbb{R}$.

Small finite sets can be listed using braces $\{\ \}$. For example, $\{2, 3, 5, 7\}$ is the four-element set consisting of the primes less than 10. Sets are often described by properties of their elements via the notation

$$\{\ :\ \}.$$

Before the colon the variable [n or x, for instance] is indicated and after the colon the properties are given. For example,

$$\{n : n \in \mathbb{N} \text{ and } n \text{ is odd}\} \tag{1}$$

represents the set of positive odd integers. The colon is always read "such that," so the set in (1) is read "the set of all n *such that* n is in

309

$\mathbb{N}$ and n is odd." Likewise

$$\{x : x \in \mathbb{R} \text{ and } 1 \leq x < 3\} \tag{2}$$

represents the set of all real numbers that are greater than or equal to 1 and less than 3. In §4 this set is abbreviated $[1, 3)$. Note that $1 \in [1, 3)$ but $3 \notin [1, 3)$. Just to streamline notation, the expressions (1) and (2) may be written as

$$\{n \in \mathbb{N} : n \text{ is odd}\} \quad \text{and} \quad \{x \in \mathbb{R} : 1 \leq x < 3\}.$$

The first set is then read "the set of all n in $\mathbb{N}$ *such that n is odd*."

Another way to list a set is to specify a rule for obtaining its elements using some other set of elements. For example, $\{n^2 : n \in \mathbb{N}\}$ represents the set of all positive integers that are the square of other integers, i.e.,

$$\{n^2 : n \in \mathbb{N}\} = \{m \in \mathbb{N} : m = n^2 \text{ for some } n \in \mathbb{N}\} = \{1, 4, 9, 16, 25, \ldots\}.$$

Similarly $\{\sin \frac{n\pi}{4} : n \in \mathbb{N}\}$ represents the set obtained by evaluating $\sin \frac{n\pi}{4}$ for each positive integer n. Actually this set is finite:

$$\left\{\sin \frac{n\pi}{4} : n \in \mathbb{N}\right\} = \left\{\frac{\sqrt{2}}{2}, 1, 0, -\frac{\sqrt{2}}{2}, -1\right\}.$$

The set in (1) can also be written as $\{2n - 1 : n \in \mathbb{N}\}$. One more example: $\{x^3 : x > 3\}$ is the set of all cubes of all real numbers bigger than 3 and of course equals $\{y \in \mathbb{R} : y > 27\}$, i.e., $(27, \infty)$ in the notation of §5.

For sets S and T, $S \setminus T$ signifies the set $\{x \in S : x \notin T\}$. For a sequence (A_n) of sets, the union $\cup A_n$ and intersection $\cap A_n$ are defined by

$$\bigcup A_n = \{x : x \in A_n \text{ for at least one } n\},$$
$$\bigcap A_n = \{x : x \in A_n \text{ for all } n\}.$$

The *empty set* $\varnothing$ is the set with no elements at all. Thus, for example, $\{n \in \mathbb{N} : 2 < n < 3\} = \varnothing$, $\{r \in \mathbb{Q} : r^2 = 2\} = \varnothing$, $\{x \in \mathbb{R} : x^2 < 0\} = \varnothing$, and $[0, 2] \cap [5, \infty) = \varnothing$.

For functions f and g, the notation $f + g$, fg, $f \circ g$, etc. is explained on page 121.

The end of a proof is indicated by a small black box. This replaces the classical QED.

Selected Hints and

Answers

Notice. These hints and answers should be consulted only after serious attempts have been made to solve the problems. Students who ignore this advice will only cheat themselves.

Many problems can be solved in several ways. Your solution need not agree with that supplied here. Often your solution should be more elaborate.

1.1 *Hint*: The following algebra is needed to verify the induction step:

$$\frac{n(n+1)(2n+1)}{6} + (n+1)^2 = (n+1)\left[\frac{2n^2+n}{6} + n + 1\right] = \cdots$$
$$= \frac{(n+1)(n+2)(2n+3)}{6}.$$

1.3 *Hint*: Suppose the identity holds for n. Then work on the right side of the equation with $n+1$ in place of n. Since $(x+y)^2 = x^2 + 2xy + y^2$,

$$(1 + 2 + \cdots + n + (n+1))^2 = (1 + 2 + \cdots + n)^2$$
$$+ 2(n+1)(1 + 2 + \cdots + n) + (n+1)^2.$$

Use Example 1 to show that the second line has sum $(n+1)^3$; hence

$$(1+2+\cdots+(n+1))^2 = (1+2+\cdots n)^2 + (n+1)^3 = 1^3 + 2^3 + \cdots + (n+1)^3.$$

1.5 *Hint*: $2 - \frac{1}{2^n} + \frac{1}{2^{n+1}} = 2 - \frac{1}{2^{n+1}}$.

1.7 *Hint*: $7^{n+1} - 6(n+1) - 1 = 7(7^n - 6n - 1) + 36n$.

1.9 (a) $n \geq 5$ and also $n = 1$.

(b) Clearly the inequality holds for $n = 5$. Suppose $2^n > n^2$ for some $n \geq 5$. Then $2^{n+1} = 2 \cdot 2^n > 2n^2$, so $2^{n+1} > (n+1)^2$ provided $2n^2 \geq (n+1)^2$ or $n^2 \geq 2n + 1$ for $n \geq 5$._In fact, this holds for $n \geq 3$, which can be verified using calculus or directly: $n^2 \geq 3n = 2n + n > 2n + 1$.

1.11 (a) *Hint:* If $n^2 + 5n + 1$ is even, then so is $(n+1)^2 + 5(n+1) + 1 = n^2 + 5n + 1 + [2n + 6]$.

(b) P_n is false for all n. *Moral:* The basis for induction (I_1) is crucial for mathematical induction.

2.1 *Hint:* Imitate Example 3. You should, of course, verify your assertions concerning nonsolutions. Note that there are sixteen rational candidates for solving $x^2 - 24 = 0$.

2.3 *Hint:* $(2 + \sqrt{2})^{1/2}$ represents a solution of $x^4 - 4x^2 + 2 = 0$.

2.5 *Hint:* $[3 + \sqrt{2}]^{2/3}$ represents a solution of $x^6 - 22x^3 + 49 = 0$.

3.1 (a) A3 and A4 hold for $a \in \mathbb{N}$, but 0 and $-a$ are not in $\mathbb{N}$. Likewise M4 holds for $a \in \mathbb{N}$, but a^{-1} is not in $\mathbb{N}$ unless $a = 1$. These three properties fail for $\mathbb{N}$ since they implicitly require the numbers 0, $-a$ and a^{-1} to be in the system under scrutiny, namely $\mathbb{N}$ in this case.

(b) M4 fails in the sense discussed in (a).

3.3 (iv) Apply (iii), DL, A2, A4, (ii) and A4 again to obtain

$$(-a)(-b) + (-ab) = (-a)(-b) + (-a)b = (-a)[(-b) + b]$$
$$= (-a)[b + (-b)] = (-a) \cdot 0 = 0 = ab + (-ab).$$

Now by (i) we conclude that $(-a)(-b) = ab$.

(v) Suppose $ac = bc$ and $c \neq 0$. By M4 there exists c^{-1} such that $c \cdot c^{-1} = 1$. Now (supply reasons)

$$a = a \cdot 1 = a(c \cdot c^{-1}) = (ac)c^{-1} = (bc)c^{-1} = b(c \cdot c^{-1}) = b \cdot 1 = b.$$

3.5 (a) If $|b| \leq a$, then $-a \leq -|b|$, so $-a \leq -|b| \leq b \leq |b| \leq a$. Now suppose that $-a \leq b \leq a$. If $b \geq 0$, then $|b| = b \leq a$. If $b < 0$, then $|b| = -b \leq a$; the last inequality holds by Theorem 3.2(i) since $-a \leq b$.

(b) By (a), it suffices to prove $-|a - b| \leq |a| - |b| \leq |a - b|$. Each of these inequalities follows from the triangle inequality: $|b| = |(b - a) + a| \leq |b - a| + |a| = |a - b| + |a|$ which implies the first inequality; $|a| = |(a - b) + b| \leq |a - b| + |b|$ which implies the second inequality.

3.7 **(a)** Imitate Exercise 3.5(a).

    (b) By (a), $|a - b| < c$ if and only if $-c < a - b < c$, and this obviously holds [see O4] if and only if $b - c < a < b + c$.

4.1 If the set is bounded above, use any three numbers $\geq$ supremum of the set; see the answers to Exercise 4.3. The sets in (h), (k) and (u) are not bounded above. Note that the set in (i) is simply $[0, 1]$.

4.3 **(a)** 1; **(c)** 7; **(e)** 1; **(g)** 3; **(i)** 1; **(k)** No sup; **(m)** 2; **(o)** 0; **(q)** 16; **(s)** $\frac{1}{2}$; **(u)** No sup; **(w)** $\frac{\sqrt{3}}{2}$. In (s), note that 1 is not prime.

4.5 **Proof** Since $\sup S$ is an upper bound for S, we have $\sup S \geq s$ for all $s \in S$. Also $\sup S \in S$ by assumption. Hence $\sup S$ is the maximum of S, i.e., $\sup S = \max S$.

4.7 **(a)** Suppose $S \subseteq T$. Since $\sup T \geq t$ for all $t \in T$ we obviously have $\sup T \geq s$ for all $s \in S$. So $\sup T$ is an upper bound for the set S. Hence $\sup T$ must be $\geq$ the *least* upper bound for S, i.e., $\sup T \geq \sup S$. A similar argument shows $\inf T \leq \inf S$; give it.

    (b) Since $S \subseteq S \cup T$, $\sup S \leq \sup(S \cup T)$ by (a). Similarly $\sup T \leq \sup(S \cup T)$, so $\max\{\sup S, \sup T\} \leq \sup(S \cup T)$. Since $\sup(S \cup T)$ is the least upper bound for $S \cup T$, we will have equality here provided we show: $\max\{\sup S, \sup T\}$ is an upper bound for the set $S \cup T$. This is easy. If $x \in S$, then $x \leq \sup S \leq \max\{\sup S, \sup T\}$ and if $x \in T$, then $x \leq \sup T \leq \max\{\sup S, \sup T\}$. I.e., $x \leq \max\{\sup S, \sup T\}$ for all $x \in S \cup T$.

4.9 **(1)** If $s \in S$, then $-s \in -S$, so $-s \leq s_0$. Hence we have $s \geq -s_0$ by Theorem 3.2(i).

    (2) Suppose $t \leq s$ for all $s \in S$. Then $-t \geq -s$ for all $s \in S$, i.e., $-t \geq x$ for all $x \in -S$. So $-t$ is an upper bound for the set $-S$. So $-t \geq \sup(-S)$. I.e., $-t \geq s_0$ and hence $t \leq -s_0$.

4.11 **Proof** By 4.7 there is a rational r_1 such that $a < r_1 < b$. By 4.7 again, there is a rational r_2 such that $a < r_2 < r_1$. We continue by induction: If rationals $r_1, \ldots, r_n$ have been selected so that $a < r_n < r_{n-1} < \cdots < r_2 < r_1$, then 4.7 applies to $a < r_n$ to yield a rational r_{n+1} such that $a < r_{n+1} < r_n$. This process yields an infinite set $\{r_1, r_2, \ldots\}$ in $\mathbb{Q} \cap (a, b)$.

Alternative Proof Assume $\mathbb{Q} \cap (a, b)$ is finite. The set is nonempty by 4.7. Let $c = \min(\mathbb{Q} \cap (a, b))$. Then $a < c$, so by 4.7 there is a rational r such that $a < r < c$. Then r belongs to $\mathbb{Q} \cap (a, b)$, so $c \leq r$, a contradiction.

4.13 By Exercise 3.7(b), we have (i) and (ii) equivalent. The equivalence of (ii) and (iii) is obvious from the definition of an open interval.

4.15 Assume $a \le b + \frac{1}{n}$ for all $n \in \mathbb{N}$ but that $a > b$. Then $a - b > 0$, and by the Archimedean property 4.6 we have $n_0(a - b) > 1$ for some $n_0 \in \mathbb{N}$. Then $a > b + \frac{1}{n_0}$ contrary to our assumption.

5.1 (a) $(-\infty, 0)$; (b) $(-\infty, 2]$; (c) $[0, \infty)$; (d) $(-\sqrt{8}, \sqrt{8})$.

5.3 *Hint*: The unbounded sets are in (h), (k), (l), (o), (t) and (u).

5.5 Proof Select $s_0 \in S$. Then $\inf S \le s_0 \le \sup S$ whether these symbols represent $\pm\infty$ or not.

6.1 (a) If $s \le t$, then clearly $s^* \subseteq t^*$. Conversely, assume $s^* \subseteq t^*$ but that $s > t$. Then $t \in s^*$ but $t \notin t^*$, a contradiction.

(b) $s = t$ if and only if both $s \le t$ and $t \le s$ if and only if both $s^* \subseteq t^*$ and $t^* \subseteq s^*$ if and only if $s^* = t^*$.

(c) Consider $r_1 \in s^*$ and $r_2 \in t^*$. Then $r_1 < s$ and $r_2 < t$; so $r_1 + r_2 < s + t$, i.e., $r_1 + r_2 \in (s + t)^*$. Hence $s^* + t^* \subseteq (s + t)^*$. Now consider $r \in (s + t)^*$ so that $r < s + t$. If $r_1 = \frac{1}{2}(s - t + r)$ and $r_2 = \frac{1}{2}(t - s + r)$, then $r_1 < \frac{1}{2}(s - t + s + t) = s$ and $r_2 < \frac{1}{2}(t - s + s + t) = t$. So $r_1 \in s^*$ and $r_2 \in t^*$. Since $r = r_1 + r_2$, we have $r \in s^* + t^*$. Hence $(s + t)^* \subseteq s^* + t^*$.

6.3 (a) If $r \in \alpha$ and $s \in 0^*$, then $r + s < r$, so $r + s \in \alpha$. Hence $\alpha + 0^* \subseteq \alpha$. Conversely, suppose $r \in \alpha$. Since α has no largest element, there is a rational $t \in \alpha$ such that $t > r$. Then $r - t$ is in 0^*, so $r = t + (r - t) \in \alpha + 0^*$. This shows $\alpha \subseteq \alpha + 0^*$.

(b) $-\alpha = \{r \in \mathbb{Q} : s \notin \alpha \text{ for some rational } s < -r\}$.

6.5 (b) No; it corresponds to $(2)^{1/3}$.

(c) This is the Dedekind cut corresponding to $\sqrt{2}$.

7.1 (a) $\frac{1}{4}, \frac{1}{7}, \frac{1}{10}, \frac{1}{13}, \frac{1}{16}$

(c) $\frac{1}{3}, \frac{2}{9}, \frac{1}{9}, \frac{4}{81}, \frac{5}{243}$

7.3 (a) converges to 1; (c) converges to 0; (e) does not converge; (g) does not converge; (i) converges to 0; (k) does not converge; (m) converges to 0 [this sequence is $(0, 0, 0, \ldots)$]; (o) converges to 0; (q) converges to 0 [see Exercise 9.15]; (s) converges to $\frac{4}{3}$.

7.5 (a) Has limit 0 since $s_n = 1/(\sqrt{n^2 + 1} + n)$.

(c) $\sqrt{4n^2 + n} - 2n = n/(\sqrt{4n^2 + n} + 2n)$ and this is close to $\frac{n}{2n + 2n}$ for large n. So limit appears to be $\frac{1}{4}$; it is.

8.1 (a) **Formal Proof** Let $\epsilon > 0$. Let $N = \frac{1}{\epsilon}$. Then $n > N$ implies $\left| \frac{(-1)^n}{n - 0} \right| = \frac{1}{n} < \epsilon$.

(b) *Discussion.* We want $n^{-1/3} < \epsilon$ or $\frac{1}{n} < \epsilon^3$ or $1/\epsilon^3 < n$. So for each $\epsilon > 0$, let $N = 1/\epsilon^3$. You should write out the formal proof.

(c) *Discussion.* We want $|\frac{2n-1}{3n+2} - \frac{2}{3}| < \epsilon$ or $|\frac{-7}{(3n+2)\cdot 3}| < \epsilon$ or $\frac{7}{3(3n+2)} < \epsilon$ or $\frac{7}{3\epsilon} < 3n + 2$ or $\frac{7}{9\epsilon} - \frac{2}{3} < n$. So set N equal to $\frac{7}{9\epsilon} - \frac{2}{3}$.

(d) *Discussion.* We want $(n+6)/(n^2 - 6) < \epsilon$; we assume $n > 2$ so that absolute values can be dropped. As in Example 3 we observe that $n + 6 \le 7n$ and that $n^2 - 6 \ge \frac{1}{2}n^2$ provided $n > 3$. So it suffices to get $7n/(\frac{1}{2}n^2) < \epsilon$ [for $n > 3$] or $\frac{14}{\epsilon} < n$. So try $N = \max\{3, \frac{14}{\epsilon}\}$.

8.3 *Discussion.* We want $\sqrt{s_n} < \epsilon$ or $s_n < \epsilon^2$. But $s_n \to 0$, so we can get $s_n < \epsilon^2$ for large n.

Formal Proof Let $\epsilon > 0$. Since $\epsilon^2 > 0$ and $\lim s_n = 0$, there exists N so that $|s_n - 0| < \epsilon^2$ for $n > N$. Thus $s_n < \epsilon^2$ for $n > N$, so $\sqrt{s_n} < \epsilon$ for $n > N$. I.e., $|\sqrt{s_n} - 0| < \epsilon$ for $n > N$. We conclude that $\lim \sqrt{s_n} = 0$.

8.5 (a) Let $\epsilon > 0$. Our goal is to show that $s - \epsilon < s_n < s + \epsilon$ for large n. Since $\lim a_n = s$, there exists N_1 so that $|a_n - s| < \epsilon$ for $n > N_1$. In particular,

$$n > N_1 \quad \text{implies} \quad s - \epsilon < a_n. \tag{1}$$

Likewise there exists N_2 so that $|b_n - s| < \epsilon$ for $n > N_2$, so

$$n > N_2 \quad \text{implies} \quad b_n < s + \epsilon. \tag{2}$$

Now

$$n > \max\{N_1, N_2\} \quad \text{implies} \quad s - \epsilon < a_n \le s_n \le b_n < s + \epsilon;$$

hence $|s - s_n| < \epsilon$.

(b) It is easy to show that $\lim(-t_n) = 0$ if $\lim t_n = 0$. Now apply (a) to the inequalities $-t_n \le s_n \le t_n$.

8.7 (a) Assume $\lim \cos(\frac{n\pi}{3}) = a$. Then there exists N such that $n > N$ implies $|\cos(\frac{n\pi}{3}) - a| < 1$. Consider $n > N$ and $n + 3$ where n is a multiple of 6; substituting these values in the inequality gives $|1 - a| < 1$ and $|-1 - a| < 1$. By the triangle inequality

$$2 = |(1 - a) - (-1 - a)| \le |1 - a| + |-1 - a| < 1 + 1 = 2,$$

a contradiction.

(b) Assume $\lim(-1)^n n = a$. Then there exists N such that $n > N$ implies $|(-1)^n n - a| < 1$. For an even $n > N$ and for $n + 2$

this tells us that $|n - a| < 1$ and $|n + 2 - a| < 1$. So $2 = |n+2-a-(n-a)| \le |n+2-a|+|n-a| < 2$, a contradiction.

(c) Note that the sequence takes the values $\pm\frac{\sqrt{3}}{2}$ for large n. Assume $\lim \sin(\frac{n\pi}{3}) = a$. Then there must exist N such that

$$n > N \quad \text{implies} \quad |\sin(\frac{n\pi}{3}) - a| < \frac{\sqrt{3}}{2}.$$

Substituting suitable $n > N$, we obtain $|\frac{\sqrt{3}}{2} - a| < \frac{\sqrt{3}}{2}$ and $|\frac{-\sqrt{3}}{2} - a| < \frac{\sqrt{3}}{2}$. By the triangle inequality

$$\sqrt{3} = \left|\frac{\sqrt{3}}{2} - \left(\frac{-\sqrt{3}}{2}\right)\right| \le \left|\frac{\sqrt{3}}{2} - a\right| + \left|a - \left(\frac{-\sqrt{3}}{2}\right)\right|$$

$$< \frac{\sqrt{3}}{2} + \frac{\sqrt{3}}{2} = \sqrt{3},$$

a contradiction.

8.9 (a) *Hint*: There exists N_0 in $\mathbb{N}$ such that $s_n \ge a$ for $n > N_0$. Assume that $s = \lim s_n$ and that $s < a$. Let $\epsilon = a - s$ and select $N \ge N_0$ so that $|s_n - s| < \epsilon$ for $n > N$. Show that $s_n < a$ for $n > N$; a picture might help.

9.1 (a) $\lim(\frac{n+1}{n}) = \lim(1+\frac{1}{n}) = \lim 1 + \lim\frac{1}{n} = 1 + 0 = 1$. The second equality is justified by Theorem 9.3 and the third equality follows from Basic Example 9.7(a).

(b) $\lim(3n + 7)/(6n - 5) = \lim(3 + 7/n)/(6 - 5/n) = (\lim(3 + 7/n))/\lim(6 - 5/n)) = (\lim 3 + 7 \cdot \lim(1/n))/(\lim 6 - 5 \cdot \lim(1/n)) = (3 + 7 \cdot 0)/(6 - 5 \cdot 0) = \frac{1}{2}$. The second equality is justified by Theorem 9.6, the third equality follows from Theorems 9.3 and 9.2, and the fourth equality uses Basic Example 9.7(a).

9.3 First we use Theorem 9.4 twice to obtain $\lim a_n^3 = \lim a_n \cdot \lim a_n^2 = a \cdot \lim a_n^2 = a \cdot \lim a_n \cdot \lim a_n = a \cdot a \cdot a = a^3$. By Theorems 9.3 and 9.2, we have $\lim(a_n^3 + 4a_n) = \lim a_n^3 + 4 \cdot \lim a_n = a^3 + 4a$. Similarly $\lim(b_n^2 + 1) = \lim b_n \cdot \lim b_n + 1 = b^2 + 1$. Since $b^2 + 1 \ne 0$, Theorem 9.6 shows that $\lim s_n = (a^3 + 4a)/(b^2 + 1)$.

9.5 *Hint*: Let $t = \lim t_n$ and show that $t = (t^2 + 2)/2t$. Then show that $t = \sqrt{2}$.

9.7 It has been shown that $s_n < \sqrt{2/(n - 1)}$ for $n \ge 2$, and we need to prove $\lim s_n = 0$.

Discussion. Let $\epsilon > 0$. We want $s_n < \epsilon$, so it suffices to get $\sqrt{2/(n - 1)} < \epsilon$ or $2/(n - 1) < \epsilon^2$ or $2\epsilon^{-2} + 1 < n$.

Formal Proof. Let $\epsilon > 0$ and let $N = 2\epsilon^{-2}+1$. Then $n > N$ implies
$$s_n < \sqrt{2/(n-1)} < \sqrt{2/(2\epsilon^{-2}+1-1)} = \epsilon.$$

9.9 **(a)** Let $M > 0$. Since $\lim s_n = +\infty$ there exists $N \geq N_0$ such that $s_n > M$ for $n > N$. Then clearly $t_n > M$ for $n > N$, since $s_n \leq t_n$ for all n. This shows that $\lim t_n = +\infty$.

     (c) Parts (a) and (b) take care of the infinite limits, so assume (s_n) and (t_n) converge. Since $t_n - s_n \geq 0$ for all $n > N_0$, we have $\lim(t_n - s_n) \geq 0$ by Exercise 8.9(a). Hence $\lim t_n - \lim s_n \geq 0$ by Theorems 9.3 and 9.2.

9.11 **(a)** *Discussion.* Let $M > 0$ and let $m = \inf\{t_n : n \in \mathbb{N}\}$. We want $s_n + t_n > M$ for large n, but it suffices to get $s_n + m > M$ or $s_n > M - m$ for large n. So select N so that $s_n > M - m$ for $n > N$.

     (b) *Hint*: If $\lim t_n > -\infty$, then $\inf\{t_n : n \in \mathbb{N}\} > -\infty$. Use part (a).

9.13 If $|a| < 1$, then $\lim a^n = 0$ by Basic Example 9.7(b). If $a = 1$, then obviously $\lim a^n = 1$.

     Suppose $a > 1$. Then $\frac{1}{a} < 1$, so $\lim(1/a)^n = 0$ as above. Thus $\lim 1/a^n = 0$. Theorem 9.10 [with $s_n = a^n$] now shows that $\lim a^n = +\infty$. [This case can also be handled by applying Exercise 9.12.]

     Suppose $a \leq -1$ and assume $\lim a^n$ exists. For even n, $a^n \geq 1$ and for odd n, $a^n \leq -1$. Clearly $\lim a^n = +\infty$ and $\lim a^n = -\infty$ are impossible. Assume that $\lim a^n = A$ for a real number A. There exists N such that $|a^n - A| < 1$ for $n > N$. For even n this implies $A > 0$ and for odd n this implies $A < 0$, a contradiction.

9.15 Apply Exercise 9.12 with $s_n = a^n/n!$. Then $L = \lim |s_{n+1}/s_n| = \lim \frac{a}{n+1} = 0$, so $\lim s_n = 0$.

9.17 *Discussion.* Let $M > 0$. We want $n^2 > M$ or $n > \sqrt{M}$. So let $N = \sqrt{M}$.

10.1 nondecreasing: (c); nonincreasing: (a), (f); bounded: (a), (b), (d), (f).

10.3 The equality in the hint can be verified by induction; compare Exercise 1.5. Now by (1) in Discussion 10.3 we have

$$s_n = k + \frac{d_1}{10} + \cdots + \frac{d_n}{10^n} \leq k + \frac{9}{10} + \cdots + \frac{9}{10^n} < k+1.$$

10.7 Let $s_0 = \sup S$. Since $s_0 - 1$ is not an upper bound for S, there exists $s_1 \in S$ such that $s_1 > s_0 - 1$. Since $s_0 \notin S$, we have $s_0 - 1 < s_1 < s_0$. Now $\max\{s_0 - \frac{1}{2}, s_1\}$ is not an upper bound for S, so there exists $s_2 \in S$ such that $s_2 > \max\{s_0 - \frac{1}{2}, s_1\}$. Then we have $s_1 < s_2$ and $s_0 - \frac{1}{2} < s_2 < s_0$. We proceed by induction.

Assume that $s_1, s_2, \ldots, s_n$ have been selected in S so that $s_1 <$ $s_2 < \cdots < s_n$ and $s_0 - \frac{1}{n} < s_n < s_0$. Then $\max\{s_0 - \frac{1}{n+1}, s_n\}$ is not an upper bound for S, so there exists $s_{n+1} \in S$ such that $s_{n+1} > \max\{s_0 - \frac{1}{n+1}, s_n\}$. Then $s_1 < s_2 < \cdots < s_{n+1}$ and $s_0 - \frac{1}{n+1} < s_{n+1} < s_0$ and therefore the construction continues. Clearly we have constructed a nondecreasing sequence in S and we also have $\lim s_n = s_0$ since $s_0 - \frac{1}{n} < s_n < s_0$ for all n. [Similar constructions will appear in the next section.]

10.9 **(a)** $s_2 = \frac{1}{2}, s_3 = \frac{1}{6}, s_4 = \frac{1}{48}.$

(b) First we prove

$$0 < s_{n+1} < s_n \le 1 \quad \text{for all} \quad n \ge 1. \tag{1}$$

This is obvious from part (a) for $n = 1, 2, 3$. Assume (1) holds for n. Then $s_{n+1} < 1$, so

$$s_{n+2} = \frac{n+1}{n+2}s_{n+1}^2 = \left(\frac{n+1}{n+2}s_{n+1}\right)s_{n+1} < s_{n+1}$$

since $\left(\frac{n+1}{n+2}\right)s_{n+1} < 1$. Since $s_{n+1} > 0$ we also have $s_{n+2} > 0$. Hence $0 < s_{n+2} < s_{n+1} \le 1$ and (1) holds by induction. Assertion (1) shows that (s_n) is a bounded monotone sequence, so (s_n) converges by Theorem 10.2.

(c) Let $s = \lim s_n$. Using limit theorems we find $s = \lim s_{n+1} = \lim \frac{n}{n+1} \cdot \lim s_n^2 = s^2$. Consequently $s = 1$ or $s = 0$. But $s = 1$ is impossible since $s_n \le \frac{1}{2}$ for $n \ge 2$. So $s = 0$.

10.11 **(a)** Show (t_n) is a bounded monotone sequence.

(b) The answer is not obvious! It turns out that $\lim t_n$ is a Wallis product and has value $\frac{2}{\pi}$ which is about 0.6366. Observe how much easier part (a) is than part (b).

11.1 **(a)** $1, 5, 1, 5, 1, 5, 1, 5$

(b) Let $\sigma(k) = n_k = 2k$. Then (a_{n_k}) is the sequence that takes the single value 5. [There are many other possible choices of σ.]

11.3 **(b)** For (s_n), the set S of subsequential limits is $\{-1, -\frac{1}{2}, \frac{1}{2}, 1\}$. For (t_n), $S = \{0\}$. For (u_n), $S = \{0\}$. For (v_n), $S = \{-1, 1\}$.

(c) $\limsup s_n = 1$, $\liminf s_n = -1$, $\limsup t_n = \liminf t_n = \lim t_n = 0$, $\limsup u_n = \liminf u_n = \lim u_n = 0$, $\limsup v_n = 1$, $\liminf v_n = -1$.

(d) (t_n) and (u_n) converge.

(e) $(s_n), (t_n), (u_n)$ and (v_n) are all bounded.

11.5 **(a)** $[0, 1]$; **(b)** $\limsup q_n = 1$, $\liminf q_n = 0$.

11.7 *Hint*: Use an inductive construction to show that there is a subsequence (r_{n_k}) satisfying $r_{n_k} > k$ for $k \in \mathbb{N}$; compare Example 3.

11.9 **(a)** To show that $[a, b]$ is closed, we need to consider a limit s of a convergent sequence (s_n) from $[a, b]$ and show that s is also in $[a, b]$. But this was done in Exercise 8.9.

  (b) No! $(0, 1)$ is not closed, i.e., $(0, 1)$ does not have the property described in Theorem 11.8. For example, $t_n = \frac{1}{n}$ defines a sequence in $(0, 1)$ such that $t = \lim t_n$ does *not* belong to $(0, 1)$.

12.1 Let $u_N = \inf\{s_n : n > N\}$ and $w_N = \inf\{t_n : n > N\}$. Then (u_N) and (w_N) are nondecreasing sequences and $u_N \leq w_N$ for all $N > N_0$. By Exercise 9.9(c), $\liminf s_n = \lim u_N \leq \lim w_N = \liminf t_n$. The inequality $\limsup s_n \leq \limsup t_n$ can be shown in a similar way or one can apply Exercise 11.8(a).

12.3 **(a)** 0; **(b)** 1; **(c)** 2; **(d)** 3; **(e)** 4; **(f)** 0; **(g)** 2.

12.5 By Exercise 12.4, $\limsup(-s_n - t_n) \leq \limsup(-s_n) + \limsup(-t_n)$, so $-\limsup(-(s_n + t_n)) \geq -\limsup(-s_n) + [-\limsup(-t_n)]$. Now apply Exercise 11.8(a).

12.7 Let (s_{n_j}) be a subsequence of (s_n) such that $\lim_{j \to \infty} s_{n_j} = +\infty$. [We used j here instead of k to avoid confusion with the given $k > 0$.] Then $\lim_{j \to \infty} k s_{n_j} = +\infty$ by Exercise 9.10(a). Since $(k s_{n_j})$ is a subsequence of $(k s_n)$, we conclude that $\limsup(k s_n) = +\infty$.

12.9 **(a)** Since $\liminf t_n > 0$, there exists N_1 such that $m = \inf\{t_n : n > N_1\} > 0$. Now consider $M > 0$. Since $\lim s_n = +\infty$, there exists N_2 such that $s_n > \frac{M}{m}$ for $n > N_2$. Then $n > \max\{N_1, N_2\}$ implies $s_n t_n > (\frac{M}{m}) t_n \geq (\frac{M}{m}) m = M$. Hence $\lim s_n t_n = +\infty$.

12.11 **Partial Proof** Let $M = \liminf |s_{n+1}/s_n|$ and $\beta = \liminf |s_n|^{1/n}$. To show $M \leq \beta$, it suffices to prove that $M_1 \leq \beta$ for all $M_1 < M$. Since

$$\liminf \left| \frac{s_{n+1}}{s_n} \right| = \lim_{N \to \infty} \inf \left\{ \left| \frac{s_{n+1}}{s_n} \right| : n > N \right\} > M_1,$$

there exists N such that

$$\inf \left\{ \left| \frac{s_{n+1}}{s_n} \right| : n > N \right\} > M_1.$$

Now imitate the proof of Theorem 12.2, but note that many of the inequalities will be reversed.

12.13 **Proof** of $\sup A = \liminf s_n$. Consider N in $\mathbb{N}$ and observe that $u_N = \inf\{s_n : n > N\}$ is a number in A, since $\{n \in \mathbb{N} : s_n < u_N\} \subseteq \{1, 2, \ldots, N\}$. So $u_N \leq \sup A$ for all N and consequently $\liminf s_n = \lim u_N \leq \sup A$.

Next consider $a \in A$. Let $N_0 = \max\{n \in \mathbb{N} : s_n < a\} < \infty$. Then $s_n \geq a$ for $n > N_0$. Thus for $N \geq N_0$ we have $u_N = \inf\{s_n : n > N\} \geq a$. It follows that $\liminf s_n = \lim u_N \geq a$. We have just shown that $\liminf s_n$ is an upper bound for the set A. Therefore $\liminf s_n \geq \sup A$.

13.1 **(a)** It is clear that d_1 and d_2 satisfy D1 and D2 of Definition 13.1. If $x, y, z \in \mathbb{R}^k$, then for each $j = 1, 2, \ldots, k$,

$$|x_j - z_j| \leq |x_j - y_j| + |y_j - z_j| \leq d_1(x, y) + d_1(y, z),$$

so $d_1(x, z) \leq d_1(x, y) + d_1(y, z)$. So d_1 satisfies the triangle inequality and a similar argument works for d_2; give it.

(b) For the completeness of d_1 we use Theorem 13.4 and the inequalities

$$d_1(x, y) \leq d(x, y) \leq \sqrt{k}\, d_1(x, y).$$

In fact, if (x_n) is Cauchy for d_1, then the second inequality shows that (x_n) is Cauchy for d. Hence by Theorem 13.4, for some $x \in \mathbb{R}^k$ we have $\lim d(x_n, x) = 0$. By the first inequality, we also have $\lim d_1(x_n, x) = 0$, i.e., (x_n) converges to x in the metric d_1. For d_2, use the completeness of d_1 and the inequalities $d_1(x, y) \leq d_2(x, y) \leq k\, d_1(x, y)$.

13.3 **(b)** No, because $d^*(x, y)$ need not be finite. For example, consider the elements $x = (1, 1, 1, \ldots)$ and $y = (0, 0, 0, \ldots)$.

13.7 **Outline of Proof** Consider an open set $U \subseteq \mathbb{R}$. Let (q_n) be an enumeration of the rationals in U. For each n, let

$$a_n = \inf\{a \in \mathbb{R} : (a, q_n] \subseteq U\}, \quad b_n = \sup\{b \in \mathbb{R} : [q_n, b) \subseteq U\}.$$

Show that $(a_n, b_n) \subseteq U$ for each n and that $U = \bigcup_{n=1}^{\infty}(a_n, b_n)$. Show that

$$(a_n, b_n) \cap (a_m, b_m) \neq \varnothing \quad \text{implies} \quad (a_n, b_n) = (a_m, b_m).$$

Now either there will be only finitely many distinct [and disjoint] intervals or else a subsequence $\{(a_{n_k}, b_{n_k})\}_{k=1}^{\infty}$ of $\{(a_n, b_n)\}$ will consist of disjoint intervals for which $\bigcup_{k=1}^{\infty}(a_{n_k}, b_{n_k}) = U$.

13.9 (a) $\{\frac{1}{n} : n \in \mathbb{N}\} \cup \{0\}$; **(b)** $\mathbb{R}$; **(c)** $[-\sqrt{2}, \sqrt{2}]$.

13.11 Suppose that E is compact, hence closed and bounded by Theorem 13.12. Consider a sequence (x_n) in E. By Theorem 13.5, a subsequence of (x_n) converges to some x in $\mathbb{R}^k$. Since E is closed, x must be in E; see Proposition 13.9(b).

Suppose every sequence in E has a subsequence that converges to a point in E. By Theorem 13.12, it suffices to show E is closed and bounded. If E were unbounded, E would contain a sequence (x_n) where $\lim d(x_n, 0) = +\infty$ and then no subsequence would converge at all. Thus E is bounded. If E were nonclosed, then by Proposition 13.9 there would be a convergent sequence (x_n) in E such that $x = \lim x_n \notin E$. Since every subsequence would also converge to $x \notin E$, we would have a contradiction.

13.13 Assume, for example, that $\sup E \notin E$. The set E is bounded, so by Exercise 10.7, there exists a sequence (s_n) in E where $\lim s_n = \sup E$. Now Proposition 13.9(b) shows that $\sup E \in E$, a contradiction.

13.15 (a) F is bounded because $d(x, 0) \leq 1$ for all $x \in F$ where $0 = (0, 0, 0, \ldots)$. To show F is closed, consider a convergent sequence $(x^{(n)})$ in F. We need to show $x = \lim x^{(n)}$ is in F. For each $j = 1, 2, \ldots$, it is easy to see that $\lim_{n \to \infty} x_j^{(n)} = x_j$. Since each $x_j^{(n)}$ belongs to $[-1, 1]$, x_j belongs to $[-1, 1]$ by Exercise 8.9. It follows that $x \in F$.

(b) For the last assertion of the hint observe that $x^{(n)}, x^{(m)} \in U(x)$ implies $d(x^{(n)}, x^{(m)}) \leq d(x^{(n)}, x) + d(x, x^{(m)}) < 2$ while $d(x^{(n)}, x^{(m)}) = 2$ for $m \neq n$. Now show that no finite subfamily of $\mathcal{U}$ can cover $\{x^{(n)} : n \in \mathbb{N}\}$.

14.1 (a), (b), (c) Converge; use Ratio Test.

(d) Diverges; use Ratio Test or show nth terms don't converge to 0 [see Corollary 14.5].

(e) Compare with $\sum 1/n^2$.

(f) Compare with $\sum \frac{1}{n}$.

14.3 All but (e) converge.

14.5 (a) We assume the series begin with $n = 1$. Let $s_n = \sum_{j=1}^{n} a_j$ and $t_n = \sum_{j=1}^{n} b_j$. We are given $\lim s_n = A$ and $\lim t_n = B$. Hence $\lim(s_n + t_n) = A + B$ by Theorem 9.3. Clearly $s_n + t_n = \sum_{j=1}^{n} (a_j + b_j)$ is the nth partial sum for $\sum(a_n + b_n)$, so $\sum(a_n + b_n) = \lim(s_n + t_n) = A + B$.

(c) The conjecture is not even reasonable for series of two terms: $a_1b_1 + a_2b_2 \neq (a_1 + a_2)(b_1 + b_2)$.

14.7 By Corollary 14.5, there exists N such that $a_n < 1$ for $n > N$. Since $p > 1$, $a_n^p = a_n a_n^{p-1} < a_n$ for $n > N$. Hence $\sum_{n=N+1}^{\infty} a_n^p$ converges by the Comparison Test, so $\sum a_n^p$ also converges.

14.9 *Hint*: Let $N_0 = \max\{n \in N : a_n \neq b_n\} < \infty$. If $n \geq m > N_0$, then $\sum_{k=m}^{n} a_k = \sum_{k=m}^{n} b_k$.

14.11 Assume $a_{n+1}/a_n = r$ for $n \geq 1$. Then $a_2 = ra_1$, $a_3 = r^2a_2$, etc. A simple induction argument shows that $a_n = r^{n-1}a_1$ for $n \geq 1$. Thus $\sum a_n = \sum a_1 r^{n-1}$ is a geometric series.

14.13 **(a)** 2 and $-\frac{2}{5}$.

(b) Note that

$$s_n = \left(1 - \frac{1}{2}\right) + \left(\frac{1}{2} - \frac{1}{3}\right) + \left(\frac{1}{3} - \frac{1}{4}\right) + \cdots + \left(\frac{1}{n} - \frac{1}{n+1}\right) = 1 - \frac{1}{n+1}$$

since the intermediate fractions cancel out. Hence we have $\lim s_n = 1$.

(d) 2.

15.1 **(a)** Converges by Alternating Series Theorem.

(b) Diverges; note that $\lim(n!/2^n) = +\infty$ by Exercise 9.12(b).

15.3 *Hint*: Use integral tests. Note that

$$\lim_{n \to \infty} \int_3^n \frac{1}{x(\log x)^p} \, dx = \lim_{n \to \infty} \int_{\log 3}^{\log n} \frac{1}{u^p} \, du.$$

15.5 There is no smallest $p_0 > 1$, so there is no single series $\sum 1/n^{p_0}$ with which all series $\sum 1/n^p$ [$p > 1$] can be compared.

15.7 **(a)** **Proof** Let $\epsilon > 0$. By the Cauchy criterion, there exists N such that $n \geq m > N$ implies $|\sum_{k=m}^{n} a_k| < \frac{\epsilon}{2}$. In particular,

$$n > N \quad \text{implies} \quad a_{N+1} + \cdots + a_n < \frac{\epsilon}{2}.$$

So $n > N$ implies

$$(n - N)a_n \leq a_{N+1} + \cdots + a_n < \frac{\epsilon}{2}.$$

If $n > 2N$, then $n < 2(n - N)$, so $na_n < 2(n - N)a_n < \epsilon$. This proves $\lim(na_n) = 0$.

16.1 **(a)** In other words, show

$$2 + 7 \cdot 10^{-1} + 4 \cdot 10^{-2} + \sum_{j=3}^{\infty} 9 \cdot 10^{-j} = 2 + 7 \cdot 10^{-1} + 5 \cdot 10^{-2} = \frac{11}{4}.$$

The series is a geometric series; see Example 1 of §14.

   (b) $2.75\bar{0}$

16.3 Let A and B denote the sums of the series. By Exercise 14.5, we have $B-A = \sum(b_n-a_n)$. Since $b_n-a_n \geq 0$ for all n, and $b_n-a_n > 0$ for some n, we clearly have $B - A > 0$.

16.5 (a) $.125\bar{0}$ *and* $.124\bar{9}$; **(c)** $.\bar{6}$; **(e)** $.\overline{54}$

16.7 No.

16.9 (a) $\gamma_n - \gamma_{n+1} = \int_n^{n+1} t^{-1} \, dt - \frac{1}{n+1} > 0$ since $\frac{1}{n+1} < t^{-1}$ for all t in $[n, n+1)$.

   (b) For any n, $\gamma_n \leq \gamma_1 = 1$. Also

$$\gamma_n > \sum_{k=1}^{n} \left(\frac{1}{k} - \int_k^{k+1} t^{-1} \, dt \right) > 0.$$

   (c) Apply Theorem 10.2.

17.1 (a) $\mathrm{dom}(f + g) = \mathrm{dom}(fg) = (-\infty, 4]$, $\mathrm{dom}(f \circ g) = [-2, 2]$, $\mathrm{dom}(g \circ f) = (-\infty, 4]$.

   (b) $f \circ g(0) = 2, g \circ f(0) = 4, f \circ g(1) = \sqrt{3}, g \circ f(1) = 3, f \circ g(2) = 0, g \circ f(2) = 2$.

   (c) No!

   (d) $f \circ g(3)$ is not, but $g \circ f(3)$ is.

17.3 (a) We are given that $f(x) = \cos x$ and $g(x) = x^4$ $[p = 4]$ are continuous. So $g \circ f$ is continuous by Theorem 17.5, i.e., the function $g \circ f(x) = \cos^4 x$ is continuous. Obviously the function identically 1 is continuous [if you do not find this obvious, check it]. Hence $1 + \cos^4 x$ is continuous by Theorem 17.4(i). Finally $\log_e(1 + \cos^4 x)$ is continuous by Theorem 17.5 since this is $h \circ k(x)$ where $k(x) = 1 + \cos^4 x$ and $h(x) = \log_e x$.

   (b) Since we are given $\sin x$ and x^2 are continuous, Theorem 17.5 shows that $\sin^2 x$ is continuous. Similarly, $\cos^6 x$ is continuous. Hence $\sin^2 x + \cos^6 x$ is continuous by Theorem 17.4(i). Since $\sin^2 x + \cos^6 x > 0$ for all x and since x^π is given to be continuous for $x > 0$, we use Theorem 17.5 again to conclude that $[\sin^2 x + \cos^6 x]^\pi$ is continuous.

   (e) We are given that $\sin x$ and $\cos x$ are continuous at each $x \in \mathbb{R}$. So Theorem 17.4(iii) shows that $\frac{\sin x}{\cos x} = \tan x$ is continuous wherever $\cos x \neq 0$, i.e., for $x \neq$ odd multiple of $\frac{\pi}{2}$.

17.5 **(a)** *Remarks.* An ϵ-δ proof can be given based on the identity

$$x^m - y^m = (x - y)(x^{m-1} + x^{m-2}y + \cdots + xy^{m-2} + y^{m-1}).$$

Or the result can be proved by induction on m, as follows. It is easy to prove that $g(x) = x$ is continuous on $\mathbb{R}$. If $f(x) = x^m$ is continuous on $\mathbb{R}$, then so is $(fg)(x) = x^{m+1}$ by Theorem 17.4(ii).

(b) Just use (a) and Theorems 17.4(i) and 17.3.

17.9 **(a)** *Discussion.* Let $\epsilon > 0$. We want $|x^2 - 4| < \epsilon$ for $|x - 2|$ small, i.e., we want $|x - 2| \cdot |x + 2| < \epsilon$ for $|x - 2|$ small. If $|x - 2| < 1$, then $|x + 2| < 5$, so it suffices to get $|x - 2| \cdot 5 < \epsilon$. Set $\delta = \min\{1, \frac{\epsilon}{5}\}$.

(c) For $\epsilon > 0$, let $\delta = \epsilon$ and observe that

$$|x - 0| < \delta \quad \text{implies} \quad \left| x \sin\left(\frac{1}{x}\right) - 0 \right| < \epsilon.$$

17.11 If f is continuous at x_0 and if (x_n) is a monotonic sequence in dom(f) converging to x_0, then we have $\lim f(x_n) = f(x_0)$ by Definition 17.1.

Now assume that

$$\text{if } (x_n) \text{ is monotonic in dom}(f) \text{ and } \lim x_n = x_0, \\ \text{then } \lim f(x_n) = f(x_0), \tag{1}$$

but that f is discontinuous at x_0. Then by Definition 17.1, there exists a sequence (x_n) in dom(f) such that $\lim x_n = x_0$ but $(f(x_n))$ does not converge to $f(x_0)$. Negating Definition 17.1, we see that there exists $\epsilon > 0$ such that

$$\text{for each } N \text{ there is } n > N \text{ satisfying } |f(x_n) - f(x_0)| \geq \epsilon. \tag{2}$$

It is easy to use (2) to obtain a subsequence (x_{n_k}) of (x_n) such that

$$|f(x_{n_k}) - f(x_0)| \geq \epsilon \quad \text{for all} \quad k. \tag{3}$$

Now Theorem 11.3 shows that (x_{n_k}) has a monotonic subsequence $(x_{n_{k_j}})$. By (1) we have $\lim_{j \to \infty} f(x_{n_{k_j}}) = f(x_0)$, but by (3) we have $|f(x_{n_{k_j}}) - f(x_0)| \geq \epsilon$ for all j, a contradiction.

17.13 **(a)** *Hint:* Let $x \in \mathbb{R}$. Select a sequence (x_n) such that $\lim x_n = x$, x_n is rational for even n, and x_n is irrational for odd n. Then $f(x_n)$ is 1 for even n and 0 for odd n, so $(f(x_n))$ cannot converge.

17.15 We abbreviate

(i) f is continuous at x_0,

(ii) $\lim f(x_n) = f(x_0)$ for every sequence (x_n) in $\mathrm{dom}(f) \setminus \{x_0\}$ that converges to x_0.

From Definition 17.1 it is clear that (i) implies (ii). Assume (ii) holds but (i) fails. As in the solution to Exercise 17.11, there is a sequence (x_n) in $\mathrm{dom}(f)$ and an $\epsilon > 0$ such that $\lim x_n = x_0$ and $|f(x_n) - f(x_0)| \geq \epsilon$ for all n. Obviously $x_n \neq x_0$ for all n, i.e., (x_n) is in $\mathrm{dom}(f) \setminus \{x_0\}$. The existence of this sequence contradicts (ii).

18.3 This exercise was deliberately poorly stated, as if f must have a maximum and minimum on $[0, 5)$; see the comments following Theorem 18.1. The minimum of f on $[0, 5)$ is $1 = f(0) = f(3)$, but f has *no maximum* on $[0, 5)$ though $\sup\{f(x) : x \in [0, 5)\} = 21$.

18.5 (a) Let $h = f - g$. Then h is continuous [why?] and $h(b) \leq 0 \leq h(a)$. Now apply Theorem 18.2.

(b) Use the function g defined by $g(x) = x$ for $x \in [0, 1]$.

18.7 *Hint:* Let $f(x) = x2^x$; f is continuous, $f(0) = 0$ and $f(1) = 2$.

18.9 Let $f(x) = a_0 + a_1 x + \cdots + a_n x^n$ where $a_n \neq 0$ and n is odd. We may suppose that $a_n = 1$; otherwise we would work with $(1/a_n)f$. Since f is continuous, Theorem 18.2 shows that it suffices to show that $f(x) < 0$ for some x and $f(x) > 0$ for some other x. This is true because $\lim_{x\to\infty} f(x) = +\infty$ and $\lim_{x\to-\infty} f(x) = -\infty$ [remember $a_n = 1$], but we can avoid these limit notions as follows. Observe that

$$f(x) = x^n \left[1 + \frac{a_0 + a_1 x + \cdots + a_{n-1}x^{n-1}}{x^n}\right]. \tag{1}$$

Let $c = 1 + |a_0| + |a_1| + \cdots + |a_{n-1}|$. If $|x| > c$, then

$$|a_0 + a_1 x + \cdots + a_{n-1}x^{n-1}| \leq (|a_0| + |a_1| + \cdots + |a_{n-1}|)|x|^{n-1} < |x|^n,$$

so the number in brackets in (1) is positive. Now if $x > c$, then $x^n > 0$, so $f(x) > 0$. And if $x < -c$, then $x^n < 0$ [why?], so $f(x) < 0$.

19.1 *Hints:* To decide (a) and (b), use Theorem 19.2. Parts (c), (e), (f) and (g) can be settled using Theorem 19.5. Theorem 19.4 can also be used to decide (e) and (f); compare Example 6. One needs to resort to the definition to handle (d).

19.3 (a) *Discussion.* Let $\epsilon > 0$. We want

$$\left|\frac{x}{x+1} - \frac{y}{y+1}\right| < \epsilon \quad \text{or} \quad \left|\frac{x-y}{(x+1)(y+1)}\right| < \epsilon$$

for $|x - y|$ small, $x, y \in [0, 2]$. Since $x + 1 \geq 1$ and $y + 1 \geq 1$ for $x, y \in [0, 2]$, it suffices to get $|x - y| < \epsilon$. So we let $\delta = \epsilon$.

Formal Proof Let $\epsilon > 0$ and let $\delta = \epsilon$. Then $x, y \in [0, 2]$ and $|x - y| < \delta = \epsilon$ imply

$$|f(x) - f(y)| = \left| \frac{x - y}{(x + 1)(y + 1)} \right| \le |x - y| < \epsilon.$$

(b) *Discussion.* Let $\epsilon > 0$. We want $|g(x) - g(y)| = |\frac{5y-5x}{(2x-1)(2y-1)}| < \epsilon$ for $|x - y|$ small, $x \ge 1$, $y \ge 1$. For $x, y \ge 1$, $2x - 1 \ge 1$ and $2y - 1 \ge 1$, so it suffices to get $|5y - 5x| < \epsilon$. So let $\delta = \frac{\epsilon}{5}$. You should write out the formal proof.

19.5 (a) $\tan x$ is uniformly continuous on $[0, \frac{\pi}{4}]$ by Theorem 19.2.

(b) $\tan x$ is not uniformly continuous on $[0, \frac{\pi}{2})$ by Exercise 19.4(a), since the function is not bounded on that set.

(c) Let $\tilde{h}$ be as in Example 9. Then $(\sin x)\tilde{h}(x)$ is a continuous extension of $(\frac{1}{x}) \sin^2 x$ on $[0, \pi]$. Apply Theorem 19.5.

(e) $\frac{1}{x-3}$ is not uniformly continuous on $(3, 4)$ by Exercise 19.4(a), so it is not uniformly continuous on $(3, \infty)$ either.

(f) *Remark.* It is easy to give an ϵ-δ proof that $\frac{1}{x-3}$ is uniformly continuous on $(4, \infty)$. It is even easier to apply Theorem 19.6.

19.7 (a) We are given that f is uniformly continuous on $[k, \infty)$, and f is uniformly continuous on $[0, k + 1]$ by Theorem 19.2. Let $\epsilon > 0$. There exist δ_1 and δ_2 so that

$$|x - y| < \delta_1, \ x, y \in [k, \infty) \quad \text{imply} \quad |f(x) - f(y)| < \epsilon, (1)$$
$$|x - y| < \delta_2, \ x, y \in [0, k + 1] \quad \text{imply} \quad |f(x) - f(y)| < \epsilon.(2)$$

Let $\delta = \min\{1, \delta_1, \delta_2\}$ and show that

$$|x - y| < \delta, \ x, y \in [0, \infty) \quad \text{imply} \quad |f(x) - f(y)| < \epsilon.$$

19.9 (c) This is tricky, but it turns out that f is uniformly continuous on $\mathbb{R}$. A simple modification of Exercise 19.7(a) shows that it suffices to show that f is uniformly continuous on $[1, \infty)$ and $(-\infty, -1]$. This can be done using Theorem 19.6. Note that we cannot apply Theorem 19.6 on $\mathbb{R}$ because f is not differentiable at $x = 0$; also f' is not bounded near $x = 0$.

19.11 As in the solution to Exercise 19.9(c), it suffices to show that $\tilde{h}$ is uniformly continuous on $[1, \infty)$ and $(-\infty, -1]$. Apply Theorem 19.6.

20.1 $\lim_{x \to \infty} f(x) = \lim_{x \to 0^+} f(x) = 1$; $\lim_{x \to 0^-} f(x) = \lim_{x \to -\infty} f(x) = -1$; $\lim_{x \to 0} f(x)$ does NOT EXIST.

20.3 $\lim_{x\to\infty} f(x) = \lim_{x\to-\infty} f(x) = 0$; $\lim_{x\to 0^+} f(x) = \lim_{x\to 0^-} f(x) = \lim_{x\to 0} f(x) = 1$.

20.5 Let $S = (0, \infty)$. Then $f(x) = 1$ for all $x \in S$. So for any sequence (x_n) in S we have $\lim f(x_n) = 1$. It follows that $\lim_{x\to 0^s} f(x) = \lim_{x\to\infty^s} f(x) = 1$, i.e., $\lim_{x\to 0^+} f(x) = \lim_{x\to\infty} f(x) = 1$. Likewise if $S = (-\infty, 0)$, then $\lim_{x\to 0^s} f(x) = \lim_{x\to-\infty^s} f(x) = -1$, so $\lim_{x\to 0^-} f(x) = \lim_{x\to-\infty} f(x) = -1$. Theorem 20.10 shows that $\lim_{x\to 0} f(x)$ does not exist.

20.7 If (x_n) is a sequence in $(0, \infty)$ and $\lim x_n = +\infty$, then $\lim(1/x_n) = 0$. Since $(\sin x_n)$ is a bounded sequence, we conclude that $\lim(\sin x_n)/x_n = 0$ by Exercise 8.4. Hence $\lim_{x\to\infty} f(x) = 0$. Similarly $\lim_{x\to-\infty} f(x) = 0$. The remaining assertion is $\lim_{x\to 0} \frac{\sin x}{x} = 1$ which is discussed in Example 9 of §19.

20.9 $\lim_{x\to\infty} f(x) = -\infty$; $\lim_{x\to 0^+} f(x) = +\infty$; $\lim_{x\to 0^-} f(x) = -\infty$; $\lim_{x\to-\infty} f(x) = +\infty$; $\lim_{x\to 0} f(x)$ does NOT EXIST.

20.11 **(a)** $2a$; **(c)** $3a^2$.

20.13 First note that if $\lim_{x\to a^s} f(x)$ exists and is finite and if $k \in \mathbb{R}$, then $\lim_{x\to a^s}(kf)(x) = k \cdot \lim_{x\to a^s} f(x)$. This is Theorem 20.4(ii) where f_1 is the constant k and $f_2 = f$.

(a) The remark above and Theorem 20.4 show that

$$\lim_{x\to a}[3f(x) + g(x)^2] = 3\lim_{x\to a} f(x) + [\lim_{x\to a} g(x)]^2 = 3 \cdot 3 + 2^2 = 13.$$

(c) As in (a), $\lim_{x\to a}[3f(x) + 8g(x)] = 25$. There exists an open interval J containing a such that $f(x) > 0$ and $g(x) > 0$ for $x \in J \setminus \{a\}$. Theorem 20.5 applies with $S = J \setminus \{a\}$, $3f + 8g$ in place of f and with $g(x) = \sqrt{x}$ to give $\lim_{x\to a} \sqrt{3f(x) + 8g(x)} = \sqrt{25} = 5$.

20.15 Let (x_n) be a sequence in $(-\infty, 2)$ such that $\lim x_n = -\infty$. We contend that

$$\lim(x_n - 2)^{-3} = 0. \tag{1}$$

We apply Exercises 9.10 and 9.11 and Theorems 9.9 and 9.10 to conclude $\lim(-x_n) = +\infty$, $\lim(2 - x_n) = +\infty$, $\lim(2 - x_n)^3 = +\infty$, $\lim(2 - x_n)^{-3} = 0$, and hence (1) holds.

Now consider a sequence (x_n) in $(2, \infty)$ such that $\lim x_n = 2$. We show

$$\lim(x_n - 2)^{-3} = +\infty. \tag{2}$$

Since $\lim(x_n - 2) = 0$ and each $x_n - 2 > 0$, Theorem 9.10 shows that we have $\lim(x_n - 2)^{-1} = +\infty$ and (2) follows by an application of Theorem 9.9.

20.17 Suppose first that L is finite. We use (1) in Corollary 20.8. Let $\epsilon > 0$. There exist $\delta_1 > 0$ and $\delta_3 > 0$ such that

$$a < x < a + \delta_1 \quad \text{implies} \quad L - \epsilon < f_1(x) < L + \epsilon$$

and

$$a < x < a + \delta_3 \quad \text{implies} \quad L - \epsilon < f_3(x) < L + \epsilon.$$

If $\delta = \min\{\delta_1, \delta_3\}$, then

$$a < x < a + \delta \quad \text{implies} \quad L - \epsilon < f_2(x) < L + \epsilon.$$

So by Corollary 20.8 we have $\lim_{x \to a^+} f_2(x) = L$.

Suppose $L = +\infty$. Let $M > 0$. In view of Discussion 20.9, there exists $\delta > 0$ such that

$$a < x < a + \delta \quad \text{implies} \quad f_1(x) > M.$$

Then clearly

$$a < x < a + \delta \quad \text{implies} \quad f_2(x) > M,$$

and this shows that $\lim_{x \to a^+} f_2(x) = +\infty$. The case $L = -\infty$ is similar.

20.19 Suppose $L_2 = \lim_{x \to a^S} f(x)$ exists with $S = (a, b_2)$. Consider a sequence (x_n) in (a, b_1) with limit a. Then (x_n) is a sequence in (a, b_2) with limit a, so $\lim f(x_n) = L_2$. This shows $\lim_{x \to a^S} f(x) = L_2$ with $S = (a, b_1)$.

Suppose $L_1 = \lim_{x \to a^S} f(x)$ exists with $S = (a, b_1)$, and consider a sequence (x_n) in (a, b_2) with limit a. There exists N so that $n \geq N$ implies $x_n < b_1$. Then $(x_n)_{n=N}^{\infty}$ is a sequence in (a, b_1) with limit a. Hence $\lim f(x_n) = L_1$ whether we begin the sequence at $n = N$ or $n = 1$. This shows $\lim_{x \to a^S} f(x) = L_1$ with $S = (a, b_2)$.

21.1 Let $\epsilon > 0$. For $j = 1, 2, \ldots, k$, there exist $\delta_j > 0$ such that

$$s, t \in \mathbb{R} \quad \text{and} \quad |s - t| < \delta_j \quad \text{imply} \quad |f_j(s) - f_j(t)| < \frac{\epsilon}{\sqrt{k}}.$$

Let $\delta = \min\{\delta_1, \delta_2, \ldots, \delta_k\}$. Then by (1) in the proof of Proposition 21.2,

$$s, t \in \mathbb{R} \quad \text{and} \quad |s - t| < \delta \quad \text{imply} \quad d^*(\gamma(s), \gamma(t)) < \epsilon.$$

21.3 *Hint*: Show that $|d(s, s_0) - d(t, s_0)| \leq d(s, t)$. Hence if $\epsilon > 0$, then

$$s, t \in S \quad \text{and} \quad d(s, t) < \epsilon \quad \text{imply} \quad |f(s) - f(t)| < \epsilon.$$

21.5 (b) By part (a), there is an unbounded continuous real-valued function f on E. Show that $h = \frac{|f|}{1+|f|}$ is continuous, bounded and does not assume its supremum 1 on E.

21.7 (b) γ is continuous at t_0 if for each $t_0 \in [a, b]$ and $\epsilon > 0$ there exists $\delta > 0$ such that

$$t \in [a, b] \quad \text{and} \quad |t - t_0| < \delta \quad \text{imply} \quad d^*(\gamma(t), \gamma(t_0)) < \epsilon.$$

Note: If γ is continuous at each $t_0 \in [a, b]$, then γ is uniformly continuous on $[a, b]$ by Theorem 21.4.

21.9 (a) Use $f(x_1, x_2) = x_1$, say.

(b) This is definitely *not* obvious, but there do exist continuous mappings of $[0, 1]$ onto the unit square. Such functions must be "wild" and are called Peano curves [after the same Peano with the axioms]; see [8], §5.5, or [34], §6.3.

21.11 (a) If a continuous function mapped $[0, 1]$ onto $(0, 1)$, then the image $(0, 1)$ would be compact by Theorem 21.4(i). But $(0, 1)$ is not closed and hence not compact.

22.1 (a) $[0, 1]$ is connected but $[0, 1] \cup [2, 3]$ is not. See Theorem 22.2. Alternatively, apply the Intermediate Value Theorem 18.2.

22.3 Assume that E is connected but that E^- is not. Then there exist disjoint open sets U_1 and U_2 such that $E^- \subseteq U_1 \cup U_2$, $E^- \cap U_1 \neq \emptyset$ and $E^- \cap U_2 \neq \emptyset$. Since $E \subseteq U_1 \cup U_2$, it suffices to show $E \cap U_1 \neq \emptyset$ and $E \cap U_2 \neq \emptyset$. In fact, if $E \cap U_1 = \emptyset$, then $E^- \cap (S \setminus U_1)$ would be a closed set containing E that is smaller than E^-, contrary to the definition of E^-. Likewise $E \cap U_2 \neq \emptyset$.

22.5 (a) Assume disjoint open sets U_1 and U_2 disconnect $E \cup F$. Consider $s_0 \in E \cap F$; s_0 belongs to one of the open sets, say $s_0 \in U_1$. Since $E \subseteq U_1 \cup U_2$, $E \cap U_1 \neq \emptyset$ and E is connected, we must have $E \cap U_2 = \emptyset$. Similarly $F \cap U_2 = \emptyset$. But then $(E \cup F) \cap U_2 = \emptyset$, a contradiction.

(b) No such example exists in $\mathbb{R}$ [why?], but many exist in the plane. For example, consider

$$E = \{(x_1, x_2) : x_1^2 + x_2^2 = 1 \text{ and } x_1 \geq 0\},$$
$$F = \{(x_1, x_2) : x_1^2 + x_2^2 = 1 \text{ and } x_1 \leq 0\}.$$

22.9 *Discussion.* Given $\epsilon > 0$, we need $\delta > 0$ so that

$$s, t \in \mathbb{R} \quad \text{and} \quad |s - t| < \delta \quad \text{imply} \quad d(F(s), F(t)) < \epsilon. \qquad (1)$$

Now

$$\begin{aligned}
d(F(s), F(t)) &= \sup\{|sf(x) + (1 - s)g(x) - tf(x) - (1 - t)g(x)| : x \in S\} \\
&= \sup\{|sf(x) - tf(x) - sg(x) + tg(x)| : x \in S\} \\
&\le |s - t| \cdot \sup\{|f(x)| + |g(x)| : x \in S\}.
\end{aligned}$$

Since f and g are fixed, the last supremum is a constant M. We may assume $M > 0$, in which case $\delta = \frac{\epsilon}{M}$ will make (1) hold.

22.11 **(a)** Let (f_n) be a convergent sequence in $\mathcal{E}$. By Proposition 13.9(b), it suffices to show $f = \lim f_n$ is in $\mathcal{E}$. For each $x \in S$,

$$|f(x)| \le |f(x) - f_n(x)| + |f_n(x)| \le d(f, f_n) + 1.$$

Since $\lim d(f, f_n) = 0$, we have $|f(x)| \le 1$.

(b) It suffices to show that $C(S)$ is path-connected. So use Exercise 22.9.

23.1 Intervals of convergence: **(a)** $(-1, 1)$; **(c)** $[-\frac{1}{2}, \frac{1}{2}]$; **(e)** $\mathbb{R}$; **(g)** $[-\frac{4}{3}, \frac{4}{3})$.

23.3 $(-(2)^{1/3}, (2)^{1/3})$.

23.5 **(a)** Since $|a_n| \ge 1$ for infinitely many n, we have $\sup\{|a_n|^{1/n} : n > N\} \ge 1$ for all N. Thus $\beta = \limsup |a_n|^{1/n} \ge 1$; hence $R = \frac{1}{\beta} \le 1$.

(b) Select c with $0 < c < \limsup |a_n|$. Then $\sup\{|a_n| : n > N\} > c$ for all N. A subsequence (a_{n_k}) of (a_n) has the property that $|a_{n_k}| > c$ for all k. Since $|a_{n_k}|^{1/n_k} > (c)^{1/n_k}$ and $\lim_{k \to \infty} c^{1/n_k} = 1$ [by 9.7(d)], Exercise 12.1 shows that $\limsup |a_{n_k}|^{1/n_k} \ge 1$. It follows that $\beta = \limsup |a_n|^{1/n} \ge 1$ [use Theorem 11.7]. Hence $R = \frac{1}{\beta} \le 1$.

23.9 **(a)** Obviously $\lim f_n(0) = 0$. Consider $0 < x < 1$ and let $s_n = nx^n$. Then $s_{n+1}/s_n = [\frac{n+1}{n}]x$, so $\lim |s_{n+1}/s_n| = x < 1$. Exercise 9.12(a) shows that $0 = \lim s_n = \lim nx^n = \lim f_n(x)$.

24.1 *Discussion.* Let $\epsilon > 0$. We want $|f_n(x) - 0| < \epsilon$ for all x and for large n. It suffices to get $\frac{3}{\sqrt{n}} < \epsilon$ for large n. So consider $n > 9/\epsilon^2 = N$.

24.3 **(a)** $f(x) = 1$ for $0 \le x < 1$; $f(1) = \frac{1}{2}$; $f(x) = 0$ for $x > 1$. See Exercise 9.13.

(b) (f_n) does *not* converge uniformly on $[0, 1]$ by Theorem 24.3.

24.5 **(a)** $f(x) = 0$ for $x \le 1$ and $f(x) = 1$ for $x > 1$. Note that $f_n(x) = 1/[1 + n/x^n]$ and that $\lim_{n \to \infty} n/x^n = 0$ for $x > 1$ by Exercise 9.12 or 9.14.

(b) $f_n \to 0$ uniformly on $[0, 1]$. *Hint*: Show that $|f_n(x)| \le \frac{1}{n}$ for $x \in [0, 1]$.

(c) *Hint*: Use Theorem 24.3.

24.7 **(a)** Yes. $f(x) = x$ for $x < 1$ and $f(1) = 0$.

(b) No, by Theorem 24.3 again.

24.9 **(a)** $f(x) = 0$ for $x \in [0, 1]$. For $x < 1$, $\lim_{n \to \infty} nx^n = 0$ as in Exercise 23.9(a).

(b) Use calculus to show that f_n takes its maximum at $\frac{n}{n+1}$. Thus $\sup\{|f_n(x)| : x \in [0, 1]\} = f_n(\frac{n}{n+1}) = (\frac{n}{n+1})^{n+1}$. As in Example 8, it turns out that $\lim f_n(\frac{n}{n+1}) = 1/e$. So Remark 24.4 shows that (f_n) does not converge uniformly to 0.

(c) $\int_0^1 f_n(x)\, dx = \frac{n}{(n+1)(n+2)} \to 0 = \int_0^1 f(x)\, dx$.

24.15 **(a)** $f(0) = 0$ and $f(x) = 1$ for $x > 0$. **(b)** No. **(c)** Yes.

25.3 **(a)** Since $f_n(x) = (1 + (\cos x)/n)/(2 + (\sin^2 x)/n)$, we have $f_n \to \frac{1}{2}$ pointwise. To obtain uniform convergence, show that

$$\left| f_n(x) - \frac{1}{2} \right| = \left| \frac{2\cos x - \sin^2 x}{2(2n + \sin^2 x)} \right| \le \frac{3}{2(2n)} < \epsilon$$

for all real numbers x and all $n > \frac{3}{4\epsilon}$.

(b) $\int_2^7 \frac{1}{2}\, dx = \frac{5}{2}$, by Theorem 25.2.

25.5 Since $f_n \to f$ uniformly on S, there exists $N \in \mathbb{N}$ such that $n > N$ implies $|f_n(x) - f(x)| < 1$ for all $x \in S$. In particular, $|f_{N+1}(x) - f(x)| < 1$ for $x \in S$. If M bounds $|f_{N+1}|$ on S [i.e., if $|f_{N+1}(x)| \le M$ for $x \in S$], then $M + 1$ bounds $|f|$ on S [why?].

25.7 Let $g_n(x) = n^{-2} \cos nx$. Then we have $|g_n(x)| \le n^{-2}$ for $x \in \mathbb{R}$ and $\sum n^{-2} < \infty$. So $\sum g_n$ converges uniformly on $\mathbb{R}$ by the Weierstrass M-Test 25.7. The limit function is continuous by Theorem 25.5.

25.9 **(a)** The series converges pointwise to $\frac{1}{1-x}$ on $(-1, 1)$ by (2) of Example 1 in §14. The series converges uniformly on $[-a, a]$ by the Weierstrass M-Test since $|x^n| \le a^n$ for $x \in [-a, a]$ and since $\sum a^n < \infty$.

(b) One can show directly that the sequence of partial sums $s_n(x) = \sum_{k=0}^{n} x^k = (1 - x^{n+1})/(1 - x)$ does not converge uniformly on $(-1, 1)$. It is easier to observe that the partial sums s_n are each bounded on $(-1, 1)$, and hence if (s_n) converges uniformly, then the limit function must be bounded by Exercise 25.5. But $\frac{1}{1-x}$ is not bounded on $(-1, 1)$.

25.11 **(b)** *Hint*: Apply the Weierstrass M-Test to $\sum h_n$, where $h_n(x) = (\frac{3}{4})^n g_n(x)$.

25.13 The series $\sum g_k$ and $\sum h_k$ are uniformly Cauchy on S and it suffices to show that $\sum (g_k + h_k)$ is also; see Theorem 25.6. Let $\epsilon > 0$. There exist N_1 and N_2 such that

$$n \geq m > N_1 \quad \text{implies} \quad \left| \sum_{k=m}^{n} g_k(x) \right| < \frac{\epsilon}{2} \quad \text{for} \quad x \in S, \quad (1)$$

$$n \geq m > N_2 \quad \text{implies} \quad \left| \sum_{k=m}^{n} h_k(x) \right| < \frac{\epsilon}{2} \quad \text{for} \quad x \in S. \quad (2)$$

Then

$$n \geq m > \max\{N_1, N_2\} \quad \text{implies} \quad \left| \sum_{k=m}^{n} (g_k + h_k)(x) \right| < \epsilon \quad \text{for} \quad x \in S.$$

25.15 **(a)** Note that $f_n(x) \geq 0$ for all x and n. Assume (f_n) does not converge to 0 uniformly on $[a, b]$. Then there exists $\epsilon > 0$ such that

$$\text{for each } N \text{ there exists } n > N \text{ and } x \in [a, b] \quad (1)$$
$$\text{such that } f_n(x) \geq \epsilon.$$

We claim

$$\text{for each } n \in \mathbb{N} \text{ there is } x_n \in [a, b] \text{ where } f_n(x_n) \geq \epsilon. \quad (2)$$

If not, there is $n_0 \in \mathbb{N}$ such that $f_{n_0}(x) < \epsilon$ for all $x \in [a, b]$. Since $(f_n(x))$ is nonincreasing for each x, we conclude that $f_n(x) < \epsilon$ for all $x \in [a, b]$ and $n \geq n_0$. This clearly contradicts (1). We have now established the hint.

Now by the Bolzano-Weierstrass theorem, the sequence (x_n) given by (2) has a convergent subsequence $(x_{n_k}): x_{n_k} \to x_0$. Since $\lim f_n(x_0) = 0$, there exists m such that $f_m(x_0) < \epsilon$. Since $x_{n_k} \to x_0$ and f_m is continuous at x_0, we have $\lim_{k \to \infty} f_m(x_{n_k}) = f_m(x_0) < \epsilon$. So there exists K such that

$$k > K \quad \text{implies} \quad f_m(x_{n_k}) < \epsilon.$$

If $k > \max\{K, m\}$, then $n_k \geq k > m$, so

$$f_{n_k}(x_{n_k}) \leq f_m(x_{n_k}) < \epsilon.$$

But $f_n(x_n) \geq \epsilon$ for *all* n, so we have a contradiction.

(b) *Hint*: Show that part (a) applies to the sequence g_n where $g_n = f_n - f$.

26.3 (a) Let $f(x) = \sum_{n=1}^{\infty} nx^n = x/(1-x)^2$ for $|x| < 1$. Then by Theorem 26.5

$$\sum_{n=1}^{\infty} n^2 x^{n-1} = f'(x) = \frac{d}{dx}\left[\frac{x}{(1-x)^2}\right] = (1+x)(1-x)^{-3};$$

therefore $\sum_{n=1}^{\infty} n^2 x^n = (x+x^2)(1-x)^{-3}$.

(b) 6 and $\frac{3}{2}$.

26.5 *Hint*: Apply Theorem 26.5.

26.7 No! The power series would be differentiable at each $x \in \mathbb{R}$, but $f(x) = |x|$ is not differentiable at $x = 0$.

27.1 Let ϕ be as in the hint. By Theorem 27.4, there is a sequence (q_n) of polynomials such that $q_n \to f \circ \phi$ uniformly on $[0, 1]$. Note that ϕ is one-to-one and $\phi^{-1}(y) = \frac{y-a}{b-a}$. Let $p_n = q_n \circ \phi^{-1}$. Then each p_n is a polynomial and $p_n \to f$ uniformly on $[a, b]$.

27.3 (a) Assume that a polynomial p satisfies $|p(x) - \sin x| < 1$ for all $x \in \mathbb{R}$. Clearly p cannot be a constant function. But if p is nonconstant, then p is unbounded on $\mathbb{R}$ and the same is true for $p(x) - \sin x$, a contradiction.

(b) Assume that $|e^x - \sum_{k=0}^{n-1} a_k x^k| < 1$ for all $x \in \mathbb{R}$. For $x > 0$ we have

$$e^x - \sum_{k=0}^{n-1} a_k x^k \geq \frac{1}{n!}x^n - \sum_{k=0}^{n-1} |a_k| x^k$$

and for large x the right side will exceed 1.

27.5 (a) $B_n f(x) = x$ for all n. Use (2) in Lemma 27.2.

(b) $B_n f(x) = x^2 + \frac{1}{n}x(1-x)$. Use (4) in Lemma 27.2.

28.1 (a) $\{0\}$; **(b)** $\{0\}$; **(c)** $\{n\pi : n \in \mathbb{Z}\}$; **(d)** $\{0, 1\}$; **(e)** $\{-1, 1\}$; **(f)** $\{2\}$.

28.3 (b) Since $x - a = (x^{1/3} - a^{1/3})(x^{2/3} + a^{1/3}x^{1/3} + a^{2/3})$,

$$f'(a) = \lim_{x \to a}(x^{2/3} + a^{1/3}x^{1/3} + a^{2/3})^{-1} = (3a^{2/3})^{-1} = \frac{1}{3}a^{-2/3}$$

for $a \neq 0$.

(c) f is not differentiable at $x = 0$ since the limit $\lim_{x \to 0} x^{1/3}/x$ does not exist *as a real number*. The limit does exist and equals $+\infty$, which reflects the geometric fact that the graph of f has a vertical tangent at $(0, 0)$.

28.5 **(c)** Let

$$h(x) = \frac{g(f(x)) - g(f(0))}{f(x) - f(0)}.$$

According to Definition 20.3(a), for $\lim_{x \to 0} h(x)$ to be meaningful, h needs to be defined on $J \setminus \{0\}$ for some open interval J containing 0. But the calculation in (b) shows that h is undefined at $(\pi n)^{-1}$ for $n = \pm 1, \pm 2, \ldots$.

28.7 **(d)** f' is continuous on $\mathbb{R}$, but f' is not differentiable at $x = 0$.

28.9 **(b)** $f(x) = x^4 + 13x$ and $g(y) = y^7$. Then

$$h'(x) = g'(f(x)) \cdot f'(x) = 7(x^4 + 13x)^6 \cdot (4x^3 + 13).$$

28.11 With the stated hypotheses, $h \circ g \circ f$ is differentiable at a and $(h \circ g \circ f)'(a) = h'(g \circ f(a)) \cdot g'(f(a)) \cdot f'(a)$. **Proof** By 28.4, $g \circ f$ is differentiable at a and $(g \circ f)'(a) = g'(f(a)) \cdot f'(a)$. Again by 28.4,

$$(h \circ (g \circ f))'(a) = h'((g \circ f)(a)) \cdot (g \circ f)'(a).$$

28.13 There exist positive numbers δ_1 and ϵ so that f is defined on the interval $(a - \delta_1, a + \delta_1)$ and g is defined on $(f(a) - \epsilon, f(a) + \epsilon)$. By Theorem 17.2, there exists $\delta_2 > 0$ so that

$$x \in \operatorname{dom}(f) \quad \text{and} \quad |x - a| < \delta_2 \quad \text{imply} \quad |f(x) - f(a)| < \epsilon.$$

If $|x - a| < \min\{\delta_1, \delta_2\}$, then $x \in \operatorname{dom}(f)$ and $|f(x) - f(a)| < \epsilon$, so $f(x) \in \operatorname{dom}(g)$, i.e., $x \in \operatorname{dom}(g \circ f)$.

29.1 **(a)** $x = \frac{1}{2}$

(c) If $f(x) = |x|$, then $f'(x) = \pm 1$ except at 0. So no x satisfies the equation $f'(x) = \frac{f(2) - f(-1)}{2 - (-1)} = \frac{1}{3}$. Missing hypothesis: f is not differentiable on $(-1, 2)$, since f is not differentiable at $x = 0$.

(e) $x = \sqrt{3}$

29.3 **(a)** Apply Mean Value Theorem to $[0, 2]$.

(b) By the Mean Value Theorem, $f'(y) = 0$ for some $y \in (1, 2)$. In view of this and part (a), Theorem 29.8 shows that f' takes all values between 0 and $\frac{1}{2}$.

29.5 For any $a \in \mathbb{R}$ we have $\left|\frac{f(x) - f(a)}{x - a}\right| \le |x - a|$. It follows easily that $f'(a)$ exists and equals 0 for all $a \in \mathbb{R}$. So f is constant by Corollary 29.4.

29.7 **(a)** Applying 29.4 to f', we find $f'(x) = a$ for some constant a. If $g(x) = f(x) - ax$, then $g'(x) = 0$ for $x \in I$, so by 29.4 there is a constant b such that $g(x) = b$ for $x \in I$.

29.9 *Hint:* Let $f(x) = e^x - ex$ for $x \in \mathbb{R}$. Use f' to show that f is increasing on $[1, \infty)$ and decreasing on $(-\infty, 1]$. Hence f takes its minimum at $x = 1$.

29.13 Let $h(x) = g(x) - f(x)$ and show $h(x) \geq 0$ for $x \geq 0$.

29.15 As in Example 2, let $g(x) = x^{1/n}$. Since $\text{dom}(g) = [0, \infty)$ if n is even and $\text{dom}(g) = \mathbb{R}$ if n is odd, we have $\text{dom}(g) = \text{dom}(h) \cup \{0\}$. Also $h = g^m$. Use the Chain Rule to calculate $h'(x)$.

29.17 Suppose that $f(a) = g(a)$. Then

$$\lim_{x \to a^+} \frac{h(x) - h(a)}{x - a} = g'(a) \quad \text{and} \quad \lim_{x \to a^-} \frac{h(x) - h(a)}{x - a} = f'(a). \quad (1)$$

If also $f'(a) = g'(a)$, then Theorem 20.10 shows that $h'(a)$ exists and, in fact, $h'(a) = f'(a) = g'(a)$.

Now suppose h is differentiable at a. Then h is continuous at a and so $f(a) = \lim_{x \to a^-} f(x) = \lim_{x \to a^-} h(x) = h(a) = g(a)$. Hence (1) holds. But the limits in (1) both equal $h'(a)$, so $f'(a) = g'(a)$.

30.1 **(a)** 2; **(b)** $\frac{1}{2}$; **(c)** 0; **(d)** 1. Sometimes L'Hospital's rule can be avoided. For example, for (d) note that

$$\frac{\sqrt{1+x} - \sqrt{1-x}}{x} = \frac{2}{\sqrt{1+x} + \sqrt{1-x}}.$$

30.3 **(a)** 0; **(b)** 1; **(c)** $+\infty$ **(d)** $-\frac{2}{3}$.

30.5 **(a)** e^2; **(b)** e^2; **(c)** e.

31.1 Differentiate the power series for $\sin x$ term-by-term and cite Theorem 26.5.

31.3 The derivatives do not have a *common* bound on any interval containing 1.

31.5 **(a)** $g(x) = f(x^2)$ for $x \in \mathbb{R}$ where f is as in Example 3. Use induction to prove that there exist polynomials p_{kn}, $1 \leq k \leq n$, so that

$$g^{(n)}(x) = \sum_{k=1}^{n} f^{(k)}(x^2) p_{kn}(x) \quad \text{for} \quad x \in \mathbb{R}, \ n \geq 1.$$

32.1 Use the partition P in Example 1 to show $U(f, P) = b^4 n^2 (n+1)^2 / (4n^4)$ and $L(f, P) = b^4 (n-1)^2 n^2 / (4n^4)$. Conclude that $U(f) = b^4/4$ and $L(f) = b^4/4$.

32.3 **(a)** The upper sums are the same as in Example 1, so $U(g) = b^3/3$. Show that $L(g) = 0$.

(b) No.

32.5 S is all the numbers $L(f, P)$ and T is all $U(f, P)$.

32.7 A simple induction shows that we may assume $g(x) = f(x)$ except at one point $u \in [a, b]$. Let B be a bound for $|f|$ and $|g|$, $B > 0$. If $\epsilon > 0$ there exists a partition P such that $U(f, P) - L(f, P) < \frac{\epsilon}{3}$. We may assume that $t_k - t_{k-1} < \frac{\epsilon}{12B}$ for all k. Since u belongs to at most two intervals $[t_{k-1}, t_k]$, we see that

$$|U(g, P) - U(f, P)| \leq 2 \cdot [B - (-B)] \cdot \max\{t_k - t_{k-1}\} < \frac{\epsilon}{3}.$$

Likewise $|L(g, P) - L(f, P)| < \frac{\epsilon}{3}$, so $U(g, P) - L(g, P) < \epsilon$. Hence g is integrable. The integrals agree since since

$$\int_a^b g \leq U(g, P) < U(f, P) + \frac{\epsilon}{3} < L(f, P) + \frac{2\epsilon}{3} \leq \int_a^b f + \frac{2\epsilon}{3}$$

and similarly $\int_a^b g > \int_a^b f - \frac{2\epsilon}{3}$.

33.1 If f is decreasing on $[a, b]$, then $-f$ is increasing on $[a, b]$, so $-f$ is integrable as proved in Theorem 33.1. Now apply Theorem 33.3 with $c = -1$.

33.3 (b) 138

33.7 (a) For any set $S \subseteq [a, b]$ and $x_0, y_0 \in S$, we have

$$f(x_0)^2 - f(y_0)^2 \leq |f(x_0) + f(y_0)| \cdot |f(x_0) - f(y_0)|$$
$$\leq 2B|f(x_0) - f(y_0)| \leq 2B[M(f, S) - m(f, S)].$$

It follows that $M(f^2, S) - m(f^2, S) \leq 2B[M(f, S) - m(f, S)]$. Use this to show that $U(f^2, P) - L(f^2, P) \leq 2B[U(f, P) - L(f, P)]$.

(b) Use Theorem 32.5 and part (a).

33.9 Select $m \in \mathbb{N}$ so that $|f(x) - f_m(x)| < \frac{\epsilon}{2(b-a)}$ for all $x \in [a, b]$. Then for any partition P

$$-\frac{\epsilon}{2} \leq L(f - f_m, P) \leq U(f - f_m, P) \leq \frac{\epsilon}{2}.$$

Select a partition P_0 so that $U(f_m, P_0) - L(f_m, P_0) < \frac{\epsilon}{2}$. Since $f = (f - f_m) + f_m$, we can use inequalities from the proof of Theorem 33.3 to conclude that $U(f, P_0) - L(f, P_0) < \epsilon$. Now Theorem 32.5 shows that f is integrable. To complete the exercise, proceed as in the proof of Theorem 25.2.

33.11 (a) and **(b)**: Show that f is neither continuous nor monotonic on any interval containing 0.

(c) Let $\epsilon > 0$. Since f is piecewise continuous on $[\frac{\epsilon}{8}, 1]$, there is a partition P_1 of $[\frac{\epsilon}{8}, 1]$ such that $U(f, P_1) - L(f, P_1) < \frac{\epsilon}{4}$. Likewise there is a partition P_2 of $[-1, -\frac{\epsilon}{8}]$ such that $U(f, P_2) -$

$L(f, P_2) < \frac{\epsilon}{4}$. Let $P = P_1 \cup P_2$, a partition of $[-1, 1]$. Since

$$\left\{ M\left(f, \left[-\frac{\epsilon}{8}, \frac{\epsilon}{8}\right]\right) - m\left(f, \left[-\frac{\epsilon}{8}, \frac{\epsilon}{8}\right]\right) \right\} \cdot \left\{ \frac{\epsilon}{8} - \left(-\frac{\epsilon}{8}\right) \right\} \le \frac{\epsilon}{2},$$

we conclude that $U(f, P) - L(f, P) < \epsilon$. Now Theorem 32.5 shows that f is integrable.

33.13 Apply Theorem 33.9 to $f - g$.

34.3 **(a)** $F(x) = 0$ for $x < 0$; $F(x) = x^2/2$ for $0 \le x \le 1$; $F(x) = 4x - \frac{7}{2}$ for $x > 1$.

(c) F is differentiable except possibly at $x = 1$ by Theorem 34.3. To show F is not differentiable at $x = 1$, use Exercise 29.17.

34.5 $F'(x) = f(x + 1) - f(x - 1)$.

34.9 Use $a = 0$, $b = \frac{\pi}{6}$ and $g(x) = \sin x$.

34.11 If f is not identically 0 on $[a, b]$, then $f(x_0) > 0$ for some x_0 in $[a, b]$ which can be taken to be in (a, b). Choose $\delta > 0$ so that $a < x_0 - \delta < x_0 + \delta < b$ and $f(x) > f(x_0)/2$ for $|x - x_0| < \delta$. Let $g(x) = f(x_0)/2$ for $|x - x_0| < \delta$ and $g(x) = 0$ otherwise. Then $f(x) \ge g(x)$ for $x \in [a, b]$, so

$$\int_a^b f(x)\, dx \ge \int_a^b g(x)\, dx = \delta f(x_0) > 0.$$

35.3 **(a)** 21; **(b)** 14; **(c)** 0.

35.5 **(a)** Every upper sum is $F(b) - F(a)$ and every lower sum is 0. Hence $U_F(f) = F(b) - F(a) \ne 0 = L_F(f)$.

35.7 **(a)** Imitate solution to Exercise 33.7.

(b) and **(c)**: Use hints in Exercise 33.8.

35.9 **(a)** Let m and M be the [assumed] minimum and maximum of f on $[a, b]$. Then $\int_a^b m\, dF \le \int_a^b f\, dF \le \int_a^b M\, dF$ or $m \le [F(b) - F(a)]^{-1} \int_a^b f\, dF \le M$. Apply Theorem 18.2.

(b) Consider f and g as in Exercise 33.14, and let F be as in Exercise 35.8. By part (a), for some $x \in [a, b]$ we have

$$\int_a^b f(t)g(t)\, dt = \int_a^b f\, dF = f(x)[F(b) - F(a)] = f(x) \int_a^b g(t)\, dt.$$

35.11 Let $\epsilon > 0$ and select a partition

$$P = \{a = t_0 < t_1 < \cdots < t_n = b\}$$

satisfying $U_F(f, P) - L_F(f, P) < \epsilon$. Let $u_k = \phi^{-1}(t_k)$ and

$$Q = \{c = u_0 < u_1 < \cdots < u_n = d\}.$$

Show that $U_G(g, Q) = U_F(f, P)$ and $L_G(g, Q) = L_F(f, P)$. Then $U_G(g, Q) - L_G(g, Q) < \epsilon$, so g is G-integrable. The equality of the integrals follows easily.

36.1 *Hint*: If B bounds $|f|$, then

$$\left| \int_a^d f(x)\, dx - \int_a^b f(x)\, dx \right| \le B(b - d).$$

36.3 **(b)** Use part (a) and Examples 1 and 2.

36.7 **(a)** It suffices to show $\int_1^\infty e^{-x^2}\, dx < \infty$. But $e^{-x^2} \le e^{-x}$ for $x \ge 1$ and $\int_1^\infty e^{-x}\, dx = \frac{1}{e}$.

(b) The double integral equals $[\int_{-\infty}^\infty e^{-x^2}\, dx]^2$, and it also equals

$$\int_0^\infty \int_0^{2\pi} e^{-r^2} r\, d\theta\, dr = 2\pi \int_0^\infty e^{-r^2} r\, dr = \pi.$$

36.9 **(a)** *Hint*: Use Theorem 35.13.

(b) 1; **(c)** $+\infty$; **(d)** $\sqrt{2/\pi}$; **(e)** 0.

36.13 *Claim*: If f is continuous on $\mathbb{R}$ and $\int_{-\infty}^\infty |f|\, dF < \infty$, then f is F-integrable. **Proof** Since $0 \le f + |f|$, the integral $\int_{-\infty}^\infty [f + |f|]\, dF$ exists, and since $f + |f| \le 2|f|$, this integral is finite, i.e., $f + |f|$ is F-integrable. Since $-|f|$ is F-integrable, Exercise 36.10 shows that the sum of $f + |f|$ and $-|f|$ is F-integrable.

36.15 **(a)** For example, let $f_n(x) = \frac{1}{n}$ for $x \in [0, n]$ and $f_n(x) = 0$ elsewhere.

(b) **Outline of Proof** First, f is F-integrable on each $[a, b]$ by Exercise 35.6. An elaboration of Exercise 25.5 shows that there is a *common* bound B for $|f|$ *and* all $|f_n|$. Consider any $b > 0$ such that $1 - F(b) < \frac{\epsilon}{2B}$. There exists N so that $|\int_0^b f\, dF - \int_0^b f_n\, dF| < \frac{\epsilon}{2}$ for $n > N$. Then

$$n > N \quad \text{implies} \quad \left| \int_0^b f\, dF - \int_0^\infty f_n\, dF \right| < \epsilon. \qquad (1)$$

In particular, $m, n > N$ implies $|\int_0^\infty f_n\, dF - \int_0^\infty f_m\, dF| < 2\epsilon$, so $(\int_0^\infty f_n\, dF)_{n \in \mathbb{N}}$ is a Cauchy sequence with a finite limit L. From (1) it follows that

$$1 - F(b) < \frac{\epsilon}{2B} \quad \text{implies} \quad \left| \int_0^b f\, dF - L \right| \le \epsilon,$$

so $\lim_{b \to \infty} \int_0^b f \, dF = L$. Hence $\int_0^\infty f \, dF$ exists, is finite, and equals $\lim_{n \to \infty} \int_0^\infty f_n \, dF$. A similar argument handles $\int_{-\infty}^0 f \, dF$.

37.1 *Hint*:

$$\int_1^{yz} t^{-1} \, dt - \int_1^y t^{-1} \, dt = \int_y^{yz} t^{-1} \, dt.$$

37.7 **(a)** $B(x) = E(xL(b))$, so by the Chain Rule, we have $B'(x) = E(xL(b)) \cdot L(b) = L(b)b^x = (\log_e b)b^x$.

37.9 **(a)** $\log_e y = L(y) = \int_1^y t^{-1} \, dt \le y - 1 < y$.

References

For the 12th printing, the bibliography below has been substantially updated.

There are many books with goals similar to ours: [5], [8], [10], [12], [14], [16], [17], [31], [32], [37]. Beardon [5] gives a coherent treatment of limits by defining the notion just once, in terms of directed sets. Bressoud [7] is a wonderful book that introduces analysis through its history. Gardiner [14] is a more challenging book, though the subject matter is at the same level; there is a lot about numbers and there is a chapter on geometry. Hijab [24] is another interesting, idiosyncratic and challenging book. Note that many of these books do not provide answers to the exercises, though [24] provides answers to **all** of the exercises.

Rotman [35] is a very nice book that serves as an intermediate course between the standard calculus sequence and the first courses in **both** abstract algebra and real analysis. I always shared Rotman's doubts about inflicting logic on undergraduates, until I saw Wolf's wonderful book [40] that introduces logic as a mathematical tool in an **interesting** way.

More encyclopedic real analysis books that are presented at the same level with great detail and numerous examples are: [13], [27], [29], [39]. I find Mattuck's book [29] very thoughtfully written; he shares many of his insights (acquired over many years) with the reader.

There are several superb texts at a more sophisticated level: [2], [4], [25], [26], [33], [36], [38]. Any of these books can be used to obtain a really thorough understanding of analysis and to prepare for various advanced graduate-level topics in analysis. The possible directions for study after this are too numerous to enumerate here. However, a reader who has no specific needs or goals but who would like an introduction to several important ideas in several branches of mathematics would enjoy and profit from [15].

[1] Apostol, T.M.: *Mathematical Analysis: A Modern Approach to Advanced Calculus*, 2nd edition. Reading: Addison-Wesley 1974.

[2] Beals, R.: *Advanced Mathematical Analysis*. Graduate Texts in Mathematics 12. New York-Heidelberg-Berlin: Springer-Verlag 1973.

[3] Birkhoff, G. and Mac Lane, S.: *A Survey of Modern Algebra*. New York: Macmillan 1953.

[4] Bear, H. S.: *An Introduction to Mathematical Analysis*. San Diego: Academic Press 1997.

[5] Beardon, A. F.: *Limits—a New Approach to Real Analysis*. Undergraduate Texts in Mathematics. New York-Heidelberg-Berlin: Springer-Verlag 1997.

[6] Borman, J.L.: A remark on integration by parts, American Mathematical Monthly **51** (1944), pp. 32-33.

[7] Bressoud, D.: *A Radical Approach to Real Analysis*. Washington, D.C.: The Mathematical Association of America 1994.

[8] Clark, C.: *The Theoretical Side of Calculus*. Belmont: Wadsworth 1972. Reprinted by Krieger, New York 1978.

[9] Cunningham, F., Jr.: The two fundamental theorems of calculus, American Mathematical Monthly **72** (1965), pp. 406-407.

[10] Fitzpatrick, P. M.: *Real Analysis*. Boston: PWS Publishing Company 1995.

[11] Flatto, L.: *Advanced Calculus*. Baltimore: Williams & Wilkins 1976.

[12] Fridy, J. A.: *Introductory Analysis: The Theory of Calculus* San Diego: Harcourt Brace Jovanovich, Publishers 1987.

[13] Fulks, W.: *Advanced Calculus: An Introduction to Analysis*, 3rd edition. New York: John Wiley & Sons 1978.

[14] Gardiner, A.: New York-Heidelberg-Berlin: Springer-Verlag 1982.

[15] Garding, L.: *Encounter with Mathematics*. New York-Heidelberg-Berlin: Springer-Verlag 1977.

[16] Gaskill, H. S. and Narayanaswami, P.P.: *Elements of Real Analysis*. Upper Saddle River: Prentice-Hall 1998.

[17] Gaughan, E. D.: *Introduction to Analysis*, 2nd edition. Belmont: Wadsworth 1975.

[18] Gemignani, M.: *Introduction to Real Analysis*. Philadelphia: W. B. Saunders Co. 1971. Out of print.

[19] Goffman, C.: *Introduction to Real Analysis*. New York: Harper & Row 1966. Out of print.

[20] Greenstein, D.S.: A property of the logarithm, American Mathematical Monthly **72** (1965), page 767.

[21] Hewitt, E.: Integration by parts for Stieltjes integrals, American Mathematical Monthly **67** (1960), pp. 419-423.

[22] Hewitt, E.: The role of compactness in analysis, American Mathematical Monthly **67** (1960), pp. 499-516.

[23] Hewitt, E. and Stromberg, K.: *Real and Abstract Analysis*. Graduate Texts in Mathematics 25. New York-Heidelberg-Berlin: Springer-Verlag 1975.

[24] Hijab, O.: *Introduction to Calculus and Classical Analysis*. Undergraduate Texts in Mathematics. New York-Heidelberg-Berlin: Springer-Verlag 1997.

[25] Hoffman, K.: *Analysis in Euclidean Space*. Englewood Cliffs: Prentice-Hall 1975.

[26] Johnsonbaugh, R. and Pfaffenberger, W. E.: *Foundations of Mathematical Analysis*. New York: Marcel Dekker 1980.

[27] Kosmala, W.: *Advanced Calculus—A Friendly Approach*. Upper Saddle River: Prentice-Hall 1999.

[28] Landau, E.: *Foundations of Analysis*. Chelsea 1951.

[29] Mattuck, A.: *Introduction to Analysis*. Upper Saddle River: Prentice-Hall 1999.

[30] Niven, I.: *Irrational numbers*. Carus Monograph 11. Washington, D.C.: Mathematical Association of America 1956.

[31] Pedrick, G.: *A First Course in Analysis*. Undergraduate Texts in Mathematics. New York-Heidelberg-Berlin: Springer-Verlag 1994.

[32] Pownall, M. W.: *Real Analysis—A First Course with Foundations*. Dubuque: Wm. C. Brown Publishers 1994.

[33] Protter, M. H. and Morrey, C. B.: *A First Course in Real Analysis*. Undergraduate Texts in Mathematics. New York-Heidelberg-Berlin: Springer-Verlag 1977.

[34] Randolph, J. F.: *Basic Real and Abstract Analysis*. New York: Academic Press 1968.

[35] Rotman, J.: *Journey into Mathematics—An Introduction to Proofs.* Upper Saddle River: Prentice-Hall 1998.

[36] Rudin, W.: *Principles of Mathematical Analysis*, 3rd edition. New York: McGraw-Hill 1976.

[37] Schramm, M. J.: *Introduction to Real Analysis.* Upper Saddle River: Prentice-Hall 1996.

[38] Stromberg, K.: *An Introduction to Classical Real Analysis.* Prindle, Weber & Schmidt 1980.

[39] Trench, W. F.: *Advanced Calculus.* New York: Harper & Row 1978.

[40] Wolf, R. S.: *Proof, Logic, and Conjecture: The Mathematician's Toolbox.* New York: W. H. Freeman and Company 1998.

[41] Wolfe, J.: A proof of Taylor's formula, American Mathematical Monthly **60** (1953), page 415.

Symbols Index

e, 35, 303
$\mathbb{N}$ [natural numbers], 1
$\mathbb{Q}$ [rational numbers], 6
$\mathbb{R}$ [real numbers], 14
$\mathbb{R}^k$, 80
$\mathbb{Z}$ [all integers], 6

$B_n f$ [Bernstein polynomial], 202
$C(S)$, 168
d [a metric], 80
$\text{dist}(a, b)$, 16
$\text{dom}(f)$ [domain], 115
F-mesh(P), 288
$J_F(f, P)$, $U_F(f, P)$, $L_F(f, P)$, 270
J_u [jump function at u], 271
$\lim_{x \to a^s} f(x)$, 146
$\lim_{x \to a} f(x)$, $\lim_{x \to a^+} f(x)$,
    $\lim_{x \to \infty} f(x)$, etc., 146
$\lim s_n$, $s_n \to s$, 33, 49
$\lim \sup s_n$, $\lim \inf s_n$, 58, 75
$M(f, S)$, $m(f, S)$, 244
$\max(f, g)$, $\min(f, g)$, 121

$\max S$, $\min S$, 19
$\text{mesh}(P)$, 249
$n!$ [factorial], 6
$\binom{n}{k}$ [binomial coefficients], 6
$R_n(x)$ [remainder], 231
$\text{sgn}(x)$ [signum function], 125
$\sup S$, $\inf S$, 21, 27
$U(f)$, $L(f)$, 244
$U(f, P)$, $L(f, P)$, 244
$U_F(f)$, $L_F(f)$, 270

E°, 83
E^-, 84
f' [derivative of f], 206
$\tilde{f}$ [extension of f], 139
f^{-1} [inverse function], 129
$f + g$, fg, f/g, $f \circ g$, 121
$F(t^-)$, $F(t^+)$, 269
$f_n \to f$ pointwise, 177
$f_n \to f$ uniformly, 178
$f : S \to S^*$, 157
$s_n \to s$, 33

$s_n \uparrow s, s_n \downarrow s$, 280

(s_{n_k}), 64

$\sum a_n$ [summation], 90

$\int_a^b f = \int_a^b f(x)\, dx$, 244, 292

$\int_a^b f\, dF = \int_a^b f(x)\, dF(x)$, 271

$\int_{-\infty}^{\infty} f\, dF$, 295

$[a, b], (a, b), [a, b), (a, b]$, 20

$[a, \infty), (a, \infty), (-\infty, b]$, etc., 27

$+\infty, -\infty$, 27

$\emptyset$ [empty set], 310

Index

Abel's theorem, 196
absolute value, 16
absolutely convergent series, 91
algebraic number, 8
alternating series theorem, 103
Archimedean property, 23
associative laws, 13

basic examples, limits, 46
basis for induction, 3
Bernstein polynomials, 202
binomial series theorem, 236
binomial theorem, 6
Bolzano-Weierstrass theorem, 69
    for $\mathbb{R}^k$, 82
boundary of a set, 84
bounded function, 126
bounded sequence, 43
bounded set, 20
    in $\mathbb{R}^k$, 82
    in a metric space, 89

Cantor set, 85
Cauchy criterion
    for integrals, 248, 249
    for series, 92
    for series of functions, 188
Cauchy form of the remainder of
        a Taylor series, 236
Cauchy principal value, 294
Cauchy sequence, 60
    in a metric space, 81
    uniformly, 185
cell in $\mathbb{R}^k$, 87
chain rule, 209
change of variable, 265, 291
closed interval, 20, 27
closed set, 72
    in a metric space, 84
closure of a set, 84
coefficients of a power series, 171
commutative laws, 13
compact set, 86
comparison test
    for integrals, 297
    for series, 93

complete metric space, 81
completeness axiom, 22
composition of functions, 121
connected set, 165
continuous function, 116, 157
   piecewise, 258
   uniformly, 134, 157
convergence, interval of, 173
convergence, radius of, 172
convergent improper integral, 293
convergent sequence, 33
   in a metric space, 81
convergent series, 90
converges absolutely, 91
converges pointwise, 177
converges uniformly, 178
convex set, 167
cover, 86
curve, 158

Darboux integrals, 244
Darboux sums, 244
Darboux-Stieljtes sums, 270
Darboux-Stieltjes integrable
   function, 271
Darboux-Stieltjes integrals, 270
decimal expansions, 55, 105, 110
decreasing function, 217
Dedekind cuts, 28
definition by induction, 66
deMorgan's laws, 88
denseness of $\mathbb{Q}$, 24
density function, 296
derivative, 205
diameter of a k-cell, 87
differentiable function, 205
Dini's theorem, 191
disconnected set, 165
discontinuous function, 119
distance between real numbers,
   16
distance function, 80

distribution function, 295
distributive law, 13
divergent improper integral, 293
divergent sequence, 33
divergent series, 90
diverges to $+\infty$ or $-\infty$, 49, 90
divides, 9
domain of a function, 115

e, 35, 303
   is irrational, 113
equivalent properties, 26
Euclidean k-space, 80
Euler's constant, 113
exponentials, a definition, 304
extension of a function, 139

factor, 9
factorial, 6
field, 13
   ordered, 13
F-integrable function, 271, 295
fixed point of a function, 128
F-mesh of a partition, 288
formal proof, 37
function, 115
fundamental theorem of calculus,
   262, 264

generalized mean value theorem,
   222
geometric series, 91
greatest lower bound, 21

half-open interval, 20
Heine-Borel theorem, 86
helix, 158

improper integral, 292
  converges, 293
increasing function, 217
indeterminate forms, 222
induction step, 3
induction, mathematical, 2
inductive definition, 66
infimum of a set, 21
infinite series, 90
infinitely differentiable function,
  238
infinity $+\infty$, $-\infty$, 27
integers, 6
  positive, 1
integrable function, 244, 251, 262
  on $\mathbb{R}$, 295
integral tests for series, 102
integration by parts, 263, 284
integration by substitution, 265
interior of a set, 83
intermediate value property, 127
intermediate value theorem, 127
  for derivatives, 217
  for integrals, 259
interval of convergence, 173
intervals, 20, 27
inverse function
  continuity of, 129
  derivative of, 218
irrational numbers, 26

jump of a function, 269

k-cell, 87
k-dimensional Euclidean space,
  80

L'Hospital's rule, 223
Lagrange's form of the remainder,
  232

least upper bound, 21
left-hand limit, 147
Leibniz' rule, 213
lim inf, lim sup, 58, 75
limit of a function, 146
limit of a sequence, 33, 49
limit theorems
  for functions, 150
  for sequences, 43, 50
  for series, 99
limits of basic examples, 46
logarithms, a definition, 305
long division, 105
lower bound of a set, 20
lower Darboux integral, 244
lower Darboux sum, 244
lower Darboux-Stieltjes integral,
  270
lower Darboux-Stieltjes sum, 270

maps, 163
mathematical induction, 2
maximum of a set, 19
mean value theorem, 215
  generalized, 222
mesh of a partition, 249
metric, metric space, 80
minimum of a set, 19
monotone or monotonic
    sequence, 55
monotonic function, 253
  piecewise, 258

natural domain of a function, 115
natural numbers, 1
nondecreasing sequence, 54
nonincreasing function, 102
nonincreasing sequence, 54
normal density, 296
normal distribution, 296

open cover, 86
open interval, 20, 27
open set in a metric space, 83
order properties, 13
ordered field, 13

partial sums, 90
partition of $[a, b]$, 244
parts, integration by, 263, 284
path, 158, 166
path-connected set, 166
Peano Axioms, 2
piecewise continuous function,
     258
piecewise monotonic function,
     258
pointwise convergence, 177
polynomial approximation
     theorem, 202, 203
polynomial function, 124
positive integers, 1
postage-stamp function, 125
power series, 171
prime number, 9
product rule for derivatives,
     208
proof
   formal, 37

quotient rule for derivatives, 208

radius of convergence, 172
ratio test, 94
rational function, 124
rational numbers, 6
   as decimals, 111
   denseness of, 24
rational zeros theorem, 9
real numbers, 14
real-valued function, 115

remainder of a Taylor series, 231
   Cauchy's form, 236
   Lagrange's form, 232
repeating decimals, 110
Riemann integrable function, 251
Riemann integral, 244, 251
Riemann sum, 250
Riemann-Stieltjes integral, 289
Riemann-Stieltjes sum, 289
right-hand limit, 147
Rolle's theorem, 214
root test, 94
roots of numbers, 129

selection function σ, 64
semi-open interval, 20
sequence, 31
series, 90
   of functions, 187
signum function, 125
step-function, 260
Stieltjes integrals, 271, 289
strictly decreasing function, 216
strictly increasing function, 129,
     216
subcover, 86
subsequence, 63
subsequential limit, 70
substitution, integration by, 265
successor, 1
summation notation, 90
supremum of a set, 21

Taylor series, 231, 240
Taylor's theorem, 232, 235
topology, 83
   of pointwise convergence, 170
transitive law, 13
triangle inequality, 18, 80
two-sided limit, 146

unbounded intervals, 27
uniform convergence, 168, 178
uniformly Cauchy sequence, 168, 185
uniformly continuous function, 134, 157
uniformly convergent series of functions, 187
upper bound of a set, 20
upper Darboux integral, 244

upper Darboux sum, 244
upper Darboux-Stieltjes integral, 270
upper Darboux-Stieltjes sum, 270

Weierstrass M-test, 189
Weierstrass's approximation theorem, 202, 203

(continued from page ii)

Frazier: An Introduction to Wavelets Through Linear Algebra

Gamelin: Complex Analysis.

Gordon: Discrete Probability.

Hairer/Wanner: Analysis by Its History. *Readings in Mathematics.*

Halmos: Finite-Dimensional Vector Spaces. Second edition.

Halmos: Naive Set Theory.

Hämmerlin/Hoffmann: Numerical Mathematics. *Readings in Mathematics.*

Harris/Hirst/Mossinghoff: Combinatorics and Graph Theory.

Hartshorne: Geometry: Euclid and Beyond.

Hijab: Introduction to Calculus and Classical Analysis.

Hilton/Holton/Pedersen: Mathematical Reflections: In a Room with Many Mirrors.

Hilton/Holton/Pedersen: Mathematical Vistas: From a Room with Many Windows.

Iooss/Joseph: Elementary Stability and Bifurcation Theory. Second edition.

Irving: Integers, Polynomials, and Rings: A Course in Algebra

Isaac: The Pleasures of Probability. *Readings in Mathematics.*

James: Topological and Uniform Spaces.

Jänich: Linear Algebra.

Jänich: Topology.

Jänich: Vector Analysis.

Kemeny/Snell: Finite Markov Chains.

Kinsey: Topology of Surfaces.

Klambauer: Aspects of Calculus.

Lang: A First Course in Calculus. Fifth edition.

Lang: Calculus of Several Variables. Third edition.

Lang: Introduction to Linear Algebra. Second edition.

Lang: Linear Algebra. Third edition.

Lang: Short Calculus: The Original Edition of "A First Course in Calculus."

Lang: Undergraduate Algebra. Second edition.

Lang: Undergraduate Analysis.

Laubenbacher/Pengelley: Mathematical Expeditions.

Lax/Burstein/Lax: Calculus with Applications and Computing. Volume 1.

LeCuyer: College Mathematics with APL.

Lidl/Pilz: Applied Abstract Algebra. Second edition.

Logan: Applied Partial Differential Equations, Second edition.

Lovász/Pelikán/Vesztergombi: Discrete Mathematics.

Macki-Strauss: Introduction to Optimal Control Theory.

Malitz: Introduction to Mathematical Logic.

Marsden/Weinstein: Calculus I, II, III. Second edition.

Martin: Counting: The Art of Enumerative Combinatorics.

Martin: The Foundations of Geometry and the Non-Euclidean Plane.

Martin: Geometric Constructions.

Martin: Transformation Geometry: An Introduction to Symmetry.

Millman/Parker: Geometry: A Metric Approach with Models. Second edition.

Moschovakis: Notes on Set Theory.

Owen: A First Course in the Mathematical Foundations of Thermodynamics.

Palka: An Introduction to Complex Function Theory.

Pedrick: A First Course in Analysis.

Peressini/Sullivan/Uhl: The Mathematics of Nonlinear Programming.

Undergraduate Texts in Mathematics

Prenowitz/Jantosciak: Join Geometries.
Priestley: Calculus: A Liberal Art.
Second edition.
Protter/Morrey: A First Course in Real
Analysis. Second edition.
Protter/Morrey: Intermediate Calculus.
Second edition.
Pugh: Real Mathematical Analysis.
Roman: An Introduction to Coding and
Information Theory.
Roman: Introduction to the Mathematics
of Finance: From Risk Management to
Options Pricing.
Ross: Differential Equations: An
Introduction with Mathematica®.
Second edition.
Ross: Elementary Analysis: The Theory
of Calculus.
Samuel: Projective Geometry.
Readings in Mathematics.
Saxe: Beginning Functional Analysis
Scharlau/Opolka: From Fermat to
Minkowski.
Schiff: The Laplace Transform: Theory
and Applications.
Sethuraman: Rings, Fields, and Vector
Spaces: An Approach to Geometric
Constructability.
Sigler: Algebra.
Silverman/Tate: Rational Points on
Elliptic Curves.

Simmonds: A Brief on Tensor Analysis.
Second edition.
Singer: Geometry: Plane and Fancy.
Singer/Thorpe: Lecture Notes on
Elementary Topology and
Geometry.
Smith: Linear Algebra. Third edition.
Smith: Primer of Modern Analysis.
Second edition.
Stanton/White: Constructive
Combinatorics.
Stillwell: Elements of Algebra: Geometry,
Numbers, Equations.
Stillwell: Elements of Number Theory.
Stillwell: Mathematics and Its History.
Second edition.
Stillwell: Numbers and Geometry.
Readings in Mathematics.
Strayer: Linear Programming and Its
Applications.
Toth: Glimpses of Algebra and Geometry.
Second Edition.
Readings in Mathematics.
Troutman: Variational Calculus and
Optimal Control. Second edition.
Valenza: Linear Algebra: An Introduction
to Abstract Mathematics.
Whyburn/Duda: Dynamic Topology.
Wilson: Much Ado About Calculus.